FRENCH ADMINISTRATIVE LAW

Costume of the Conseiller d'Etat designed for Napoléon in 1804.
Reproduced by permission of the Conseil d'Etat, Paris. © *Giraudon.*

French Administrative Law

Fourth Edition

L. NEVILLE BROWN OBE, MA, LL M (Cantab.)
Docteur en Droit (Lyon), Solicitor, Professor Emeritus of Comparative Law in the University of Birmingham

JOHN S. BELL MA, D. Phil. (Oxon.)
Professor of Public and Comparative Law in the University of Leeds

with the assistance of
Jean-Michel Galabert, Conseiller d'Etat, Conseil d'Etat, Paris

CLARENDON PRESS · OXFORD
1993

Oxford University Press, Walton Street, Oxford OX2 6DP
Oxford New York Toronto
Delhi Bombay Calcutta Madras Karachi
Kuala Lumpur Singapore Hong Kong Tokyo
Nairobi Dar es Salaam Cape Town
Melbourne Auckland Madrid
and associated companies in
Berlin Ibadan

Oxford is a trade mark of Oxford University Press

Published in the United States
by Oxford University Press Inc., New York

© John Bell and Neville Brown 1993

All rights reserved. No part of this publication may be reproduced, stored in a retrieval system, or transmitted, in any form or by any means, without the prior permission in writing of Oxford University Press. Within the UK, exceptions are allowed in respect of any fair dealing for the purpose of research or private study, or criticism or review, as permitted under the Copyright, Designs and Patents Act, 1988, or in the case of reprographic reproduction in accordance with the terms of the licences issued by the Copyright Licensing Agency. Enquiries concerning reproduction outside these terms and in other countries should be sent to the Rights Department, Oxford University Press, at the address above

This book is sold subject to the condition that it shall not, by way of trade or otherwise, be lent, re-sold, hired out or otherwise circulated without the publisher's prior consent in any form of binding or cover other than that in which it is published and without a similar condition including this condition being imposed on the subsequent purchaser

British Library Cataloguing in Publication Data
Data available

Library of Congress Cataloging in Publication Data
Brown, Lionel Neville.
French administrative law. — 4th ed. / L. Neville Brown,
John S. Bell with the assistance of Jean-Michel Galabert.
Includes bibliographical references.
1. Administrative law—France. I. Bell, John S. II. Galabert, Jean-Michel.
III. Title.
KJV4669.B76   1993   342.44'60—dc20   [344.4026]   92-27365
ISBN 0-19-825290-0

Typeset by Pure Tech Corporation, Pondicherry, India
Printed and bound in Great Britain
on acid-free paper by
Biddles Ltd
Guildford and King's Lynn

Preface

Since the Third Edition of this book appeared in 1983, much has happened to justify a new edition.

The structure of the French administrative courts has been transformed by the reforms of 1987, which introduced an intermediate tier of five Cours Administratives d'Appel to relieve the over-burdened Conseil d'Etat of part of its appellate jurisdiction. In matters of substantive law, there have been significant developments. In particular, the interaction between the case-law of the Conseil d'Etat and the decisions of the Conseil Constitutionnel has prompted new directions in the expanding doctrine of *principes généraux du droit*; and the Conseil d'Etat, like the House of Lords, has had to come to terms with the supremacy of Community law. Again, the subject of the book is acquiring new importance because of the closer legal integration of Europe being brought about by the institutions of the European Community, especially its courts in Luxembourg, and by the Commission and Court of Human Rights in Strasbourg. Since the first edition in 1967 there has been a remarkable growth in British law-schools of courses relating to French law: we hope this account of contemporary *droit administratif* may help encourage such courses.

The authorship of the book has changed: Professor J. F. Garner, who co-authored the three previous editions, felt he could not assume the burden of this edition. His decision was accepted with great regret: the book still owes much to his inspiration. Our publishers too have changed. Continuity, however, has been preserved by the precious assistance again of M. Galabert, President of Sous-section of the Conseil d'Etat, who found time to scrutinize the entire text, chapter by chapter, and provided invaluable comments and corrections.

Help from many others on detailed points is acknowledged, where appropriate, in the text. We are grateful to the administrative and library staff of the Conseil d'Etat for their ready assistance during several periods of research at the Palais-Royal over the past three years of the book's gestation. Our thanks are also due to Christine Taylor, Christine Wigley, Rachel Haist, and Barbara Goodison in the Faculty of Law at Leeds for their efforts in typing and producing the manuscript, and, last but not least, to Richard Hart and Jane Williams of our publishers for patiently guiding our book through the press.

The first co-author gratefully acknowledges the financial help which he has received from the Leverhulme Foundation as an Emeritus Fellow.

The law is stated as at March 1992.

L.N.B.
J.S.B.

May 1992

Contents

TABLE of FRENCH CASES	xiii
ABBREVIATIONS AND CONVENTIONS	xxvi

1. Introduction   1

2. The Constitutional and Administrative Background   8

  1. Introduction   8
  2. The Constitution: The Division of Powers   8
  3. *Le Conseil constitutionnel*   13
  4. *Droit administratif*   22
  5. Modern Administration   24
  6. *Le Médiateur*   29
  7. Local Government   31
  8. The Civil Service   37

3. The Administrative Courts   41

  1. Introduction   41
  2. The Birth of Autonomous Administrative Courts   42
  3. Coping with Overload   47
      Local Administrative Justice   47
      The Reforms of 1953   48
      The Reforms of 1987   49
      Summary of the Present Jurisdiction of the Conseil d'Etat   51
  4. Integrating Administrative Justice   53
  5. Other Administrative Jurisdictions   55

4. The Structure and Membership of the Courts   59

  1. Introduction   59
  2. The Structure of the Conseil d'Etat   61
      The Administrative Role of the Conseil d'Etat   61

The Judicial Role of the Conseil d'Etat: Section du Contentieux	71
Section du Rapport et des Etudes	73
3. The Membership of the Conseil d'Etat	76
4. Recruitment to the Conseil d'Etat	78
5. The Career Structure within the Administrative Courts	80
Conseil d'Etat	80
Cours Administratives d'Appel and Tribunaux Administratifs	83

5. The Procedure of the Courts   86

1. Introduction	86
2. Commencement of Proceedings	87
3. *Instruction*	91
(a) Requests for Information	92
(b) Expert Evidence	94
(c) Site Visits	94
(d) Inquiry into the Facts	95
(e) Further Questions	95
(f) Interlocutory Orders	96
How Inquisitorial is the *instruction*?	96
Formality and Informality	98
4. *Rapport*	99
5. *Séance d'instruction auprès du Conseil d'Etat*	100
6. *Commissaire du gouvernement*	101
7. Judgment	103
Audience publique	104
Délibéré	108
8. The Decision	109
9. Execution	110
10. Appeal or *cassation*	115
11. Special Procedures	116
(a) *Le référé administratif*	116
(b) Stay of Execution	117
12. Adjournment for Preliminary Ruling	119
13. General Observations	120

6. The Jurisdiction of the Courts   122

1. Introduction	122

CONTENTS

2. The Basic Texts ... 123
3. The Search for a Criterion ... 124
4. The Guiding Principle ... 130
5. The Principal Exceptions ... 131
    (a) The Administration of Justice ... 132
    (b) Parliamentary Proceedings ... 133
    (c) International Relations ... 134
    (d) The Doctrine of Flagrant Irregularity ... 135
    (e) Administrative Activities of a Private Character ... 136
    (f) Matters Traditionally Reserved to the Civil Courts ... 139
    (g) Special Statutory Exceptions ... 140
    (h) Illegality as a Defence in the Ordinary Courts ... 142
    (i) Summary ... 144
6. Conflicts Procedure: The Tribunal des Conflits ... 144
    (a) Positive Conflict of Jurisdiction ... 146
    (b) Negative Conflict of Jurisdiction ... 147
    (c) Conflict of Decisions ... 148
    (d) Preliminary Rulings ... 149
7. Conclusion ... 150

7. The Conditions Precedent for Judicial Review ... 152

1. The Nature of the Act under Review ... 152
2. The Rule of the 'Prior Decision' ... 157
3. The *locus standi* of the Plaintiff ... 158
4. The Absence of Parallel Relief ... 160
5. The Time-Limit for Commencing Proceedings ... 161
6. Exclusion of Judicial Review ... 162

8. The Substantive Law: The Principle of Administrative Liability ... 166

1. Introduction ... 166
2. Categories of Litigation Before the Courts ... 168
    Le contentieux de l'annulation ... 168
    Le contentieux de pleine juridiction ... 168
    Le contentieux de l'interprétation ... 169
    Le contentieux de la répression ... 170
3. Criticism of the Traditional Classification ... 171
4. Administrative Legality and Administrative Liability ... 172

5. The Liability of the Administration	173
6. Administrative Torts	174
7. Liability without Fault: Theory of Risk	183
Risks of Assisting in the Public Service	185
Risks Arising from Dangerous Operations	186
Abnormal Burdens Suffered in the Public Interest	188
State Liability Arising out of Legislation	189
Damages in Tort	191
Exclusion of Liability	192
8. Administrative Contracts	192
Formation	193
Terms of the Contract	195
Contractual Equilibrium and Modification	196
Fait du prince	197
Imprévision	198
Force majeure	199
Other Rights of the Contractor	200
Contracts of Employment	200
9. General Observations	201

9. The Substantive Law: The Principle of Administrative Legality — 202

1. Introduction	202
2. The Content of the Principle of Legality	203
3. *Les principes généraux du droit*	205
Categories of General Principles	209
Prerogatives of the Administration	210
Liberties of the Individual	211
Economic and Social Rights	213
Protection of the Environment	214
Equality before the Law	215
Impartiality	217
Audi alteram partem	217
Duty to Give Reasons	218
Proportionality	218
Non-retroactivity	220
Duty to Abrogate Unlawful Regulations	221
The Right to Judicial Review	222
The Inalienability of the Human Body	222

CONTENTS

4. Grounds for Review ... 223
   - *Inexistence* ... 224
   - *Incompétence* ... 226
   - *Vice de forme* ... 227
   - *Violation de la loi* ... 228
   - *Détournement de pouvoir* ... 229
5. Cassation ... 235
6. The Extent of Judicial Review ... 237
   - (a) Where the Administration has no Discretion ... 237
   - (b) Where the Administration has 'Absolute' Discretion ... 238
   - (c) Where the Administration has Limited Discretion ... 240
   - (d) The Doctrine of Manifest Error in Assessment of the Facts ... 245
7. General Observations ... 250

10. The Influence of *droit administratif* Outside France ... 252

1. Introduction ... 252
2. Belgium ... 253
3. The Netherlands ... 254
4. Italy ... 257
5. Germany ... 258
6. Greece ... 261
7. The Court of Justice of the European Communities ... 262
   - (a) Substantive Law ... 263
   - (b) Jurisdiction of the Court ... 264
   - (c) Procedure and Composition of the Court ... 266
8. French Law and Community Law in Conflict: Reaction of the *Conseil d'Etat* ... 267
9. Conclusion ... 270

11. Conclusions ... 271

1. The Character and Success of the French System ... 271
2. Merits of the French System ... 274
3. Defects of the System ... 280
   - A Final Assessment ... 285

CONTENTS

Appendix A. *The Division of Jurisdiction Between the Ordinary Administrative Courts* — 287

Appendix B. *The Membership of the Conseil d'Etat* — 288

Appendix C. *The Structure of the Conseil d'Etat* — 289

Appendix D. *The Principal Specialized Administrative Jurisdictions* — 290

Appendix E. *Statistics of Cases Decided by the Conseil d'Etat* Statuant au Contentieux — 296

Appendix F. *Judicial Statistics: Tribunaux Administratifs and Cours Administratives d'Appel* — 300

Appendix G. *Judicial Statistics: Tribunal des Conflits 1980, 1990, and 1991* — 303

Appendix H. *An* Arrêt *of the Conseil d'Etat (with conclusions)* — 304

Appendix I. *Examples of Typical* Requêtes *before Administrative Courts* — 310

Appendix J. *An* Avis *of the Conseil d'Etat* — 318

Appendix K. *Select Bibliography* — 319

Index — 323

Table of French Cases

In the following Table 'CAA' means Cour Administrative d'Appel, 'CE' means Conseil d'Etat, and 'TC' means Tribunal des Conflits. The name and date of a case are sufficient to enable the case to be found in any standard set of French law-reports, to which reference is given in the Bibliography in Appendix K. The references in heavy type are to the case numbers in M. Long, P. Weil, and G. Braibant, *et al.*, *Les Grands Arrêts de la jurisprudence administrative* (9th edn., Paris, 1990). The references to *Public Law* ('PL') are to the short summaries and commentaries by R. Errera.

A list of decisions of the Constitutional Council appears at the end of this Table.

ACRE DE L'AIGLE, CE 11 July 1960	140
ACTION FRANÇAISE, TC 8 April 1935, **54**	210
ADAM ET AUTRES, CE 12 December 1960 and TC 12 June 1961	150
ADAM ET AUTRES, CE 22 February 1974	249
ADRASSE, CE 13 June 1975	102
ALLAIN, CE 20 February 1989	134
AMOROS, CE 23 January 1970	118
ANGUET, CE 3 February 1911, **26**	177–8
ARAMU, CE 26 October 1945	206
ASSOCIATION DES ANCIENS ELEVES DE L'ECOLE POLYTECHNIQUE, CE 13 July 1948	158–9
ASSOCIATION DE SAUVEGARDE DU QUARTIER NOTRE-DAME, CE 13 February 1976	119
ASSOCIATION DES PROPRIETAIRES DE CHAMPIGNY-SUR-MARNE, CE 13 December 1968	97
ASSOCIATION FAMILIALE DE L'EXTERNAT SAINT-JOSEPH, CE 20 July 1990, [1990] PL 569	211–2
ASSOCIATION GENERALE DES ADMINISTRATEURS CIVILS, CE 16 December 1988	21–2
ASSOCIATION 'LES AMIS DE LA TERRE', CE 8 March 1985	159

TABLE OF FRENCH CASES

ASSOCIATION 'LES AMIS DE SAINT-AUGUSTIN', CE 25 January 1985 — 229
ASSOCIATION 'LES AMIS DU SOCIALISME', CE 26 May 1986 — 159
ASSOCIATION 'LES CIGOGNES', CE 22 January 1988, [1988] PL 284 — 222–3
ASSOCIATION POUR L'OBJECTION DE CONSCIENCE A TOUTE PARTICIPATION A L'AVORTEMENT, CE 21 December 1990, [1991] PL 134 — 160
AVRANCHES ET DESMARETS, TC 5 July 1951, **82** — 142, 170
'BAC D'ELOKA', see SOCIETE COMMERCIALE DE L'OUEST AFRICAIN — 37, 82, 92, 97
BAREL, CE 28 May 1954, **90** — 215, 217, 238–9, 249
BEAUGE, CE 4 July 1924 — 231, 234
BENJAMIN, CE 19 May 1933, **52** — 210, 238
BENZ, CE 11 December 1970 — 243
BERECIARTUA-ECHARRI, CE 1 April 1988, [1989] PL 470, **115** — 133, 212
BERNETTE, CE 5 May 1976, **109** — 247
BESTHELSEMER, CE 18 November 1949 — 178
BERTIN, CE 20 April 1956, **92** — 130, 137–8
BLANCO, TC 8 February 1873, **1** — 5, 23, 125, 150, 174, 175, 176, 183
BLANKAERT, CE 26 September 1986, [1987] PL 120 — 93
BLETON, CE 16 December 1988, [1989] PL 178 — 21–2
BOISDET, CE 24 September 1990, [1990] PL 572 — 269
BONNIOL, CE 16 January 1931 — 200
BOUCHER, CE 6 January 1967 — 234
BOUGUEN, CE 2 April 1943, **63** — 127
BOURGEOIS, CE 27 July 1990, [1990] PL 571 — 183
BRABANT, CE 14 November 1980 — 52
BRIDET ET AL., CE 27 July 1990 — 185
BRUNAUD, CE 20 January 1988 — 97
BRUNO, TC 9 June 1986 — 136, 150
BUNELIER, TC 25 January 1988 — 148
BUNOZ, CE 23 June 1989, [1989] PL 497 — 128
CACHET, CE 3 November 1922 — 221
CADOT, CE 13 December 1889, **5** — 45
CAILLAT, CE 16 December 1966 — 188
CAISSE D'EPARGNE DE COUTANCES, CE 1 February 1980 — 131

TABLE OF FRENCH CASES

CAISSE PRIMAIRE 'AIDE ET PROTECTION', CE 13 May 1938, **59**	127
CAMBUS, CE 20 March 1987	241
CAMES, CE 21 June 1895, **6**	185
CAMINO, CE 14 January 1916, **33**	243
CANAL, CE 19 October 1962, **101**	12, 54, 82, 206–7, 222
CARLIER, CE 18 November 1949	135–6
CASANOVA, CE 29 March 1901, **8**	160, 213
CHALVON-DEMERSY, CE 18 March 1949	153
CHAMBRE SYNDICALE DU COMMERCE EN DETAIL DE NEVERS, CE 30 May 1930, **48**	213
CHAPOU, CE 20 October 1954	154
CHAVENEAU, CE 1 April 1949	211
CIANELLI, TC 13 June 1960	145
CLEMENT, TC 16 November 1964	139
CLOUET, CE 12 February 1990	181–2
COHEN, CE 8 December 1988	182
COHN-BENDIT, CE 22 December 1978, **111**	265–6, 268, 269
COLRAT, CE 4 December 1925	156
COMMISSION DEPARTEMENTALE DU BAS-RHIN, CE 12 April 1935	233–4
COMMUNE D'ARCUEIL, CE 17 April 1964	129
COMMUNE DE BATZ-SUR-MER c. TESSON, CE 29 September 1970	185
COMMUNE DE MENET, CE 15 April 1983	159
COMMUNE DE MONTFERMEIL, CE 9 May 1962	241
COMMUNE DE SAINTE-MARIE (DE LA REUNION), CE 26 July 1991	34, 193
COMMUNE DE SAINT-PRIEST-LA-PLAINE, CE 22 November 1946, **68**	185
COMMUNE DE TARASCON-SUR-ARIEGE, CE 4 February 1991	121
COMMUNE D'ISSY-LES-MOULINEAUX, CE 12 November 1909	227
COMPAGNIE AIR FRANCE c. EPOUX BARBIER, TC 15 January 1968	149
COMPAGNIE ALITALIA, CE 3 February 1989, [1989] PL 650, **116**	205, 216, 221–2, 268–9
COMPAGNIE DES MESSAGERIES MARITIMES, CE 29 January 1909, **21**	200

TABLE OF FRENCH CASES

COMPAGNIE DES SCIERIES AFRICAINES, CE
9 March 1928 — 200

COMPAGNIE DES TRAMWAYS DE CHERBOURG, CE
9 December 1932, **50** — 199–200

COMPAGNIE GENERALE D'ASSURANCES, CE
15 December 1952 — 134

COMPAGNIE GENERALE D'ECLAIRAGE DE BORDEAUX,
CE 30 March 1916, **34** — 198

COMPAGNIE GENERALE D'ENERGIE RADIOELECTRIQUE,
CE 30 March 1966, **103** — 156, 190

COMPAGNIE GENERALE DES EAUX, CE
12 May 1933 — 138, 196

COMPAGNIE GENERALE FRANÇAISE DES
TRAMWAYS, CE 21 March 1910, **25** — 138, 197

COMPAGNIE LUXEMBOURGEOISE DE TELEVISION,
CE 16 April 1986 — 108, 95

COMPAGNIE NOUVELLE DU GAZ DE DEVILLE-
LES-ROUEN, CE 10 January 1902, **9** — 197

COMPAGNIE PYRENEENNE DE TRANSPORTS PAR
TAXIS, CE 10 February 1928 — 231

CONSEIL NATIONAL DE L'ORDRE DES PHARMACIENS,
CE 12 December 1969 — 16, 19

COOK ET FILS, CE 5 May 1899 — 158

COUESPEL DU MESNIL, CE 1 May 1936 — 93

COUITEAS, CE 30 November 1923, **45** — 188

COULON, CE 11 March 1955 — 93

COURAJOUX, CE 23 January 1951 — 224

CREDIT FONCIER DE FRANCE, CE 11 December
1970, **106** — 204, 216

CUVILLIER, TC 2 June 1945 — 122, 151

D'AILLIERES, CE 7 February 1947, **69** — 163, 222

DARAMY, CE 24 June 1949 — 132, 186

DAVID, CE 5 May 1986 — 115

DAVIN, CE 26 January 1966 — 155, 228

DE BENOUVILLE, CE 9 May 1990, [1990] PL 433 — 219–20

DEFAUX, CE 18 November 1949 — 178

DEHAENE, CE 7 July 1950, **78** — 210, 211

DELIARD (ABBÈ), CE 8 February 1908 — 159

DELVILLE, CE 28 July 1951, **83** — 180

DENIZET, CE 13 November 1953 — 241, 245

TABLE OF FRENCH CASES   xvii

DESPUJOL, CE 18 January 1930, **47**	221
DOL ET LAURENT, CE 28 February 1919, **37**	159, 245
DREXEL-DAHLGREN, CE 27 July 1979	249
DREYFUS-SCHMIDT, CE 8 June 1951	164
DUCHATELET, CE 11 January 1838	189
DUJARDIN, CE 16 January 1976	133
DUTRIEUX, CE 10 February 1978	246
EFFIMIEF, TC 28 March 1955, **91**	129
ELECTRICITÉ DE FRANCE c. SPIRE, CE 2 October 1987, [1989] PL 357 & 653	94, 187–8
EPOUX NEEL, CE 23 January 1970	29
EUCAT, TC 9 June 1986	136, 156
FALCO, CE 17 April 1953, **88**	133, 147
FAVERET, CE 11 February 1949	159
FEDERATION DES INDUSTRIES FRANÇAISES D'ARTICLES DE SPORT, CE 22 November 1974	128
FEDERATION FRANÇAISE DE CYCLISME, CE 26 November 1976	128
FEDERATION DES SYNDICATS GENERAUX DE L'EDUCATION NATIONALE, CE 26 June 1989	216, 222
FEUTRY, TC 29 February 1908, **20**	176
FILMS LUTETIA, CE 18 December 1959, **97**	210
FOUERE, CE 3 February 1975	246
FRAMPAR, CE 24 June 1960, **98**	234
GAILLARD, CE 9 October 1970	185
GANDOSSI ET JOLY, CE 26 July 1985, [1986] PL 158	154
GARDE DES SCEAUX c. BANQUE POPULAIRE DE LA REGION ECONOMIQUE DE STRASBOURG, CE 9 April 1987, [1987] PL 465	187
GARRIGOU, CE 16 March 1956	221
GAVILLET, TC 31 March 1950	145
GERVAISE, CE 10 July 1957	102
G.I.E. BROUSE-CARDELL, CE 21 May 1976	131
GIRAUD, CE 27 January 1988, [1988] PL 283	191
GIUDICELLI, CE 3 November 1950	137
GODOT, CE 24 February 1950	119
GOMEL, CE 4 April 1914	243
GOMEZ, CAA Lyons, 21 December 1990, [1991] PL 461	175, 278
GOMBERT, CE 28 March 1947	132

GRASSIN, CE 26 October 1973	249
GRELLIER, CE 22 January 1986, [1987] PL 118	192
GRIZIVATZ, TC 12 January 1987	136, 156
GROUPE D'INFORMATION ET DE SOUTIEN DES TRAVAILLEURS IMMIGRES, CE 8 December 1978, **110**	211
GUIGON, TC 12 June 1966	136, 150, 225
GUILLET, CE 10 October 1990, [1991] PL 300	153
HENIN, CE 5 November 1976	132
HENRY, CE 27 March 1985, [1986] PL 156	187
HEYRIES, CE 28 June 1918, **35**	10
HUBERSCHWILLER, CE 23 December 1988, [1989] PL 496	97
HUGLO, CE 2 July 1982	117
HYVER, CE 10 October 1990, [1991] PL 300	153
IMBACH, CE 14 May 1948	239
INAO c. ROUSSOT, CE 2 March 1979	241
INGREMEAU, CE 19 October 1990	187
JEANNIER, CE 22 March 1957	181
JEUNESSE INDEPENDANTE CHRETIENNE FEMININE, CE 5 March 1948	244
JOURNAL L'AURORE, CE 28 June 1948	221
JUSSEY, CE 16 January 1930	171
K (MME), CE 27 September 1989, [1989] PL 652	182, 191
KALKOWSKI, CE 10 January 1979	132
KIRKWOOD, CE 30 May 1952	134
LABONNE, CE 8 August 1919, **39**	9
LACAMBRE, CE 28 January 1948	217
LACOSTE, CE 20 January 1926	68
LA FLEURETTE, CE 14 January 1938, **58**	189–90
LAMOTTE, CE 17 February 1950, **77**	162–3, 165, 222
LANGLAIS, CE 28 May 1971	246
LARUELLE, CE 28 July 1951, **83**	180
LEBON, CE 9 June 1978	246
LECOMTE, CE 24 June 1949, **74**	132, 186
LEGOFF, CE 27 May 1987, [1987] PL 637	191
LEGROS, CE 21 July 1972	133
LEMONNIER, CE 26 July 1918, **36**	178, 179, 183
LEROUX, CE 6 May 1988	114
L'ETANG, CE 12 July 1969	133

LETENDARD, CE 13 November 1946	237
LETISSERAND, CE 24 November 1961, **99**	191
LITZLER, CE 23 June 1954	179
LOMARDI-SAUVAN, CE 27 May 1987	153
LOREDON, CE 10 December 1986	241
LOT, CE 11 December 1903, **13**	159
MAHLER, CE 12 November 1990, [1991] PL 136	201
MANOUVRIER, CE 23 February 1968	94
MARDIROSSIAN, CE 15 October 1982	126
MARTIN, TC 8 February 1965	148
MARTIN, CE 30 November 1979	159
MARTINOD, CE 26 June 1968	188
MAUGRAS, CE 16 November 1900	234
MAURICE, CE 15 May 1981	225–6
MENNERET, CE 17 May 1985, **112**	114
MERGUI, CE 19 March 1971	188
MIMEUR, CE 18 November 1949, **75**	178
MINISTRE DE JUSTICE c. DELAMARCHE, CE 10 April 1974	246
MINISTRE DE L'AGRICULTURE c. BRUANT, CE 19 April 1961	241
MINISTRE DE L'AGRICULTURE c. CONSORTS GRIMOUARD, CE 20 April 1956, **92**	130
MINISTRE DE LA CULTURE c. MME CUSENIER, CE 12 March 1986	118
MINISTRE DE LA DEFENSE c. RASZEWSKI, CE 18 November 1988	179
MINISTRE DE L'ECONOMIE c. BEAU ET GIRARD, CE 7 February 1968	68
MINISTRE DE L'INTERIEUR c. BENOUARET, CE 8 December 1978	246
MINISTRE DE L'INTERIEUR c. FABRE LUCE, CE 20 December 1967	228
MINISTRE DES AFFAIRES ETRANGERES c. CONSORTS BURGAT, CE 29 October 1976	190
MINISTRE DES AFFAIRES SOCIALES c. SHAUKAT, CE 26 February 1988, [1988] PL 638	244
MINISTRE DU TRAVAIL c. STEPHAN, CE 16 March 1979	247
MOINEAU, CE 2 February 1945, **66**	236
MONTPEURT, CE 31 July 1942, **62**	127

TABLE OF FRENCH CASES

MORGANE, CE 11 January 1991	224
MOSCONI, CE 18 January 1967	13, 228, 233
MOUILHAUD, CE 28 July 1951	182
MURETTE, TC 27 March 1952, **84**	139
'LES TRAVAILLEURS FRANÇAIS', CE 22 December 1924	183
NATIONALE DES ETUDIANTS DE FRANCE, CE 9 May 1951	134
NALIATO, TC 22 January 1955	129
NICOLO, CE 20 October 1989, [1990] PL 134, **117**	224, 268, 269
NOTRE DAME DU KREISKER, CE 19 January 1954, **89**	154
ODDOS, CE 14 May 1982	132
ŒUVRE DE SAINT NICOLAS, CE 7 July 1950	244
OFFICE NATIONAL DES FORETS, TC 25 June 1973; CE 28 November 1975	137
ORDONNEAU, CE 7 July 1989	181
ORDRE DES AVOCATS A LA COUR DE PARIS, CE 15 May 1987	154
PEDARD, CE 2 April 1960	199
PELLETIER, TC 30 July 1873, **2**	124, 177
PESCHAUD c. GROUPEMENT DU FOOTBALL PROFESSIONNEL, CE 7 December 1979 and TC 7 July 1980	127–8, 150
PETALAS, CE 18 November 1955	227
PEYNET, CE 8 June 1973	210, 214
PHILLIPPE, CE 27 June 1930	200
PICHENE, CE 9 May 1990, [1990] PL 433	219
PIERRE c. EDF, CE 16 October 1970	153
PONCIN, CE 22 June 1963 and 17 June 1964	95
PREFET DE DOUBS c. OLMOS QUINTERO, CE 29 June 1990, [1990] PL 432	242
PRINCE NAPOLEON, CE 19 February 1875, **3**	131, 155
RADIODIFFUSION FRANÇAISE, TC 2 February 1950, **76**	146, 150, 155
RATZEL, CE 22 January 1954	133
RAULT, CE 14 March 1934	231, 233
REGNAULT-DESROZIERS, CE 28 March 1919, **38**	186
RETAIL, CE 10 July 1981	30, 153
RODIERE, CE 26 December 1925, **46**	111

TABLE OF FRENCH CASES

ROGNANT, CE 22 June 1987	171
ROLLAND, TC 12 June 1961	149
ROSAN GIRARD, CE 31 May 1957, **95**	224–5
ROSAY, TC 8 May 1933	148
ROTSCHILD c. L'ADMINISTRATION DES POSTES, CE 6 December 1855	174
ROUGEMONT, CE 7 July 1967	246
ROUSSEAU, CE 14 March 1975	133, 236
RUBAN c. SOCIETE DE L'AUTOROUTE ESTEREL-CÔTE D'AZUR, TC 28 June 1965	130
RUBIN DE SERVENS, CE 2 March 1962, **100**	12, 134
RUHLE, CE 2 February 1938	233
S.A. COOPERATIVE D'HABITATION A BON MARCHE DE VICHY-CUSSET-BELLERIVE, CE 24 April 1964	217–8
S.A. DES GRANDS MAGASINS ECONOMIQUES, CE 10 February 1937	215
S.A. LIBRAIRIE FRANÇOIS MASPERO, CE 2 November 1973, **108**	251
S.A. PHILIP MORRIS, CE 28 February 1992, [1992] PL 340	269
S.A. ROTHMANS, CE 28 February 1992, [1992] PL 340	269
SADOUDI, CE 26 October 1973	179
SAFER DE BOURGOGNE, TC 8 December 1969	145
SAIA, CE 29 September 1989, [1989] PL 651	212–3
SAINTE MARIE DE L'ASSOMPTION, CE 20 October 1972	249
SAULZE, CE 6 November 1968	188
SEALINK UK LTD., CE 22 June 1984	189
SECRETAIRE D'ETAT AUX POSTES c. MME DOUBLET, CE 24 April 1981	192
SEPTFONDS, TC 16 June 1923	143, 170
SOCIETE ALIVAR, CE 23 March 1984	134
SOCIETE COMMERCIALE DE L'OUEST AFRICAIN ('LE BAC D'ELOKA'), TC 22 January 1921, **40**	128, 129, 137
SOCIETE D'ASSURANCES MUTUELLE 'LES TRAVAILLEURS FRANÇAIS' CE 22 December 1924	183
SOCIETE DES COMBUSTIBLES ET CARBURANTS NATIONAUX, TC 19 June 1952	138

TABLE OF FRENCH CASES

SOCIETE DES CONCERTS DU CONSERVATOIRE, CE 9 March 1951, **80**	215
SOCIETE DES ETABLISSEMENTS CRUSE, CE 9 May 1980	218
SOCIETE DES PETROLES SHELL-BERRE, CE 19 June 1964	265
SOCIETE ENTREPRISE PEYROT, TC 8 July 1963	137
SOCIETE FRAMPAR, CE 24 June 1960	132
SOCIETE 'HÔTEL DU VIEUX-BEFFROI', TC 17 March 1949, **72**	140
SOCIETE IGNAZIO MESSINA, CE 30 March 1966	156–7
SOCIETE 'ILE-DE-FRANCE MEDIA', CE 13 February 1991	50
SOCIETE IRANEX, CE 6 November 1963	241
SOCIETE 'JOKELSON ET HANDSTAEM', CE 22 June 1984	189
SOCIETE LABORATOIRES DE THERAPEUTIQUE MODERNE, CE 20 May 1988	241
SOCIETE LE BETON, CE 19 October 1956, **93**	130
SOCIETE 'LES EDITIONS DES ARCHERS', CE 17 April 1985	250–1
SOCIETE 'LES FILMS MARCEAU', CE 14 October 1960	94
SOCIETE 'LES TELEPHERIQUES DU MASSIF DU MONT BLANC', CE 9 December 1988, [1989] PL 180	126, 138
SOCIETE 'MAISON GENESTAL', CE 26 January 1968, **105**	239
SOCIETE MICHEL FAURE, CE 20 January 1950	204
SOCIETE NATIONALE DE TELEVISION EN COULEURS 'ANTENNE 2', CE 30 September 1988	159
SOCIETE PHILHARMONIQUE LIBRE DE FUMAY, CE 17 May 1907	231–2
SOCIETE PROPETROL, CE 5 November 1982	199
SOCIETE RIVOLI-SEBASTAPOL, TC 17 March 1949	140
SOCIETE ROBATEL, CE 19 February 1988	135
SOCIETE TONI, CE 27 April 1951	241
SOULIER, CE 5 March 1954	231
SPIRE, CE 17 February 1978	246
STEIN, CE 20 October 1950	138
SUBRAMANIAN, CE 9 October 1981	139
SYNACOMEX, CE 10 July 1970	119, 265

TABLE OF FRENCH CASES

SYNDICAT AGRICOLE DE LALANDE-DE-POMEROL, CE 14 October 1960	241
SYNDICAT CHRETIEN DU CORPS DES OFFICIERS DE POLICE, CE 21 April 1972	215
SYNDICAT CFDT DES MARINS-PECHEURS DE LA RADE DE BREST, CE 25 July 1975	214–5
SYNDICAT DES PROPRIETAIRES DU QUARTIER CROIX-DE-SEGUEY-TIVOLI, CE 21 December 1906, **17**	158
SYNDICAT GENERAL DES FABRICANTS DE SEMOULES DE FRANCE, CE 1 March 1968	268
SYNDICAT GENERAL DES INGENIEURS-CONSEILS, CE 26 June 1959	12, 208
SYNDICAT NATIONAL DE RADIODIFFUSION ET DE TELEVISION, CE 20 January 1975	210
SYNDICAT REGIONAL DES QUOTIDIENS D'ALGERIE, CE 4 April 1952, **85**	227
TABOURET ET LAROCHE, CE 9 July 1943	230
TERRIER, CE 6 February 1903, **12**	23, 126, 138
TEISSIER, CE 13 March 1953, **87**	229
THEROND, CE 4 March 1910, **24**	126, 138
THOMASSON, TC 12 December 1955	149
TISSOT, CE 14 December 1934	244
TOURAINE AIR TRANSPORT, CE 6 November 1985, [1986] PL 345	188, 189
TREBES, CE 4 March 1949	81–2, 217
TROMPIER-GRAVIER, CE 5 May 1944, **65**	218
UNION DES GRANDES PHARMACIES DE FRANCE, CE 16 October 1968	68
UNION DES GROUPEMENTS DE CADRES SUPERIEURS DE LA FONCTION PUBLIQUE, CE 11 July 1984	221
UNION GENERALE DES HALLES CENTRALES, CE 1 June 1962	62
UNION REGIONALE POUR LA DEFENSE DE L'ENVIRONNEMENT, DE LA NATURE, DE LA VIE ET DE LA QUALITE DE LA VIE, CE 26 July 1985	159
UNIVERSITE PARIS-DAUPHINE, CE 27 July 1990	153
VERITER, CE 23 June 1989	284
VICINI, CE 20 January 1965	211, 243
VIDAILLAC, CE 17 April 1953, **88**	133, 147
VILLE DE DIEPPE, CE 8 December 1972	219

xxiv   TABLE OF FRENCH CASES

VILLE DE LIMOGES, CE 12 May 1973	204
VILLE DE MARSEILLE, CE 14 June 1946	182
VILLE DE NANTERRE, CE 20 November 1964	36, 243–4
VILLE DE PARIS, CE 14 February 1936	198
VILLE DE PARIS c. DUVINAGE, CE 17 January 1986, [1986] PL 345	189
VILLE DE PARIS c. MARABOUT, CE 20 October 1972	182
VILLE DE TOULOUSE, CE 23 April 1982	214
VILLE NOUVELLE EST, CE 28 May 1971, **107**	29, 214, 220, 247–9
VINOLAY, CE 26 July 1978	219, 220, 246–7
VO THANH NGHIA, CE 22 December 1978	135
WAHNAPO, CE 27 February 1981	247
ZENARD, CE 24 November 1933	213
ZIMMERMANN, CE 2 April 1971	232
ZOLA, CE 16 March 1988	118

Decisions of the Constitutional Council

Decisions of the Constitutional Council ('CC') are given by date and serial number. There are four series of decisions, two of which are referred to in this book: 'décisions de conformité' under article 61 of the Constitution ('DC'), and decisions on the *délégalisation* ('L'). The numbers in heavy type refer to the number of the decision to be found in L. Favoreu and L. Philip, *Les Grandes Décisions du Conseil constitutionnel* (5th edn., Paris, 1989)

CC decision no. 71–44 DC of 16 July 1991, **19**	16
CC decision no. 73–80 L of 28 November 1973	15
CC decision no. 77–87 DC of 23 November 1977, **27**	212
CC decision no. 79–105 DC of 25 July 1979, **29**	211
CC decision no. 79–110 DC of 24 December 1979, **30**	18
CC decision no. 80–119 L of 2 December 1980	162
CC decision no. 81–132 DC of 16 January 1982, **33**	17
CC decision no. 82–137 DC of 25 February 1982, **34**	34
CC decision no. 82–146 DC of 22 November 1982	20
CC decision no. 84–179 DC of 12 September 1984	22
CC decision no. 86–207 DC of 25–6 June 1986, **41**	17
CC decision no. 86–224 DC of 23 January 1987, **43**	18, 124, 131, 280

CC decision no. 87–161 L of 23 September 1987	171
CC decision no. 88–248 DC of 17 January 1989, 44	208
CC decision no. 89–260 DC of 28 June 1989	162, 192
CC decision no. 89–261 DC of 28 July 1989	124
CC decision no. 92–308 DC of 9 April 1992	14

Abbreviations and Conventions

The abbreviations listed below relate only to French legal publications. (For further details on these publications see Appendix K)

AJDA	*Actualité juridique. Droit administratif*
D.	*Recueil Dalloz, Recueil Dalloz-Sirey*
GA	M. Long, P. Weil, G. Braibant, P. Delvolvé, and B. Genevois, *Les Grands Arrêts de la jurisprudence administrative* (9th edn., Paris, 1991)
GD	L. Favoreu and L. Philip, *Les Grandes Décisions du Conseil constitutionnel* (5th edn., Paris, 1989)
RDP	*Revue de droit public*
RFDA	*Revue française de droit administratif*
S.	*Recueil Sirey*

Conventions

To reduce excessive use of italics, French legal terms appearing frequently in the text have been printed in roman type, e.g. 'Conseil d'Etat'.

We have referred to the 'Constitutional Council' rather than 'le Conseil constitutionnel' in order to avoid confusion in the use of the word 'Conseil' with the 'Conseil d'Etat'.

1

Introduction

Why should a common lawyer study French administrative law? This is a question that the student of English law, to whom this book is primarily addressed, may fairly ask. The answer lies in the value of a comparative approach to a study of one's own law. In order that the solutions given to problems arising in our administrative law may be better understood, it is instructive and valuable to appreciate how those same or similar problems have been or are being resolved by the corresponding institutions of another highly developed legal system. Great benefit is to be derived from a study of other common-law jurisdictions, but it is sometimes even more valuable to go outside the common-law world and make comparisons with a legal system having a quite different history and tradition.

This comparative method is useful in many branches of law, but it is of particular importance in administrative law, because the nature of the leading problems, and in particular the question of how government can be controlled in the interests of both state and citizen, are common to all the developed nations of the West, and present too in many developing countries of the Third World.

The choice of French law as the means of comparison has been made, not simply because the authors share a deep admiration for a highly developed and flexible but logical system, but for a number of less personal and more valid reasons. These reasons fall into two groups, the general (or cultural) and the specific (or technical).

Among the general justifications for cross-Channel comparison, it is evident that France and the United Kingdom are both highly developed industrial countries, faced in the modern world with the same problems of the control of power within the state in the interests of the individual. As Professor H. W. R. Wade wrote at the time of our first edition 'the great problem as we now see it is how far is power to be governed by law'.[1] This problem remains common to both countries. Moreover, the civilization of France, whilst in many details different from that of the United Kingdom, is based on the same essential principles of democracy and the need to observe 'due process' in matters touching the rights of the individual. Also, for two decades now both countries have shared membership of that great catalyst for economic, legal, and (eventually) political integration, the European Communities.

Among the more technical reasons for comparing English and French administrative law we would lay stress on the following. First, *le droit administratif* is that rare phenomenon—an uncodified branch of a civil-law system. For a common lawyer it has the special fascination of appearing familiar, yet at the same time strange. The familiarity comes from its being judge-made; the strangeness resides in the form and content of the judgments which compose this case-law, as well as in the procedural techniques by which such judgments are arrived at. At a time when a new Benthamite wind has been blowing through our own legal system (fanned initially by the Law Commissions Act 1965 and rekindled by the Green and White Papers of 1989) and the codification of much of the common law has been mooted (although enthusiasm for wholesale codification has now abated), it may be salutary to remind British lawyers that a sister country which pioneered codification has been content to allow her administrative judges to shape this vitally important part of French law unconfined by the strait-jacket of any general code,[2] although they are increasingly obliged to take account of a mass of detailed regulations in arriving at their decisions. These decisions, moreover, are made longer by a tendency on the part of the administrative judge, especially when exercising appellate jurisidiction, to express more fully than in the past the reasons which justify his decision, not simply in order to guide the lower

---

[1] (1962) 78 LQR 189.
[2] What publishers call 'le Code administratif' is no more than a compendium of relevant legislation and regulations on administrative topics (e.g. public works).

INTRODUCTION

courts but also to satisfy the demand of the modern litigant for adequate reasons.

In 1960, the late Professor David observed: 'in this field there is no movement in favour of codification'.[3] But new impetus to codification in general is given by the Decree of 12 September 1989, which has set up a new *Commission supérieure de codification*, whose brief includes not only the revising of obsolete codes, but also the extension of codification into new areas.[4] Administrative law, however, seems unlikely to be such an area, at least for the present.

Secondly, *le droit administratif* presents itself to British eyes as a fully developed system of administrative law. In 1963 Lord Reid could declare that in this country 'we do not have a developed system of administrative law—perhaps because until fairly recently we did not need it'.[5] But by 1971 Lord Denning could reply 'it may truly now be said that we have a developed system of administrative law'.[6] Even if we agree with the later assessment, the longer experience of French law in this area has still valuable lessons to teach us. As the late Professor J. D. B. Mitchell expressed it:

> The question, quite bluntly, is whether we want to restore the place of law in government. That restoration demands a susceptible law and a susceptible body which administers that law, a body which at the same time is aware of the real needs of government and of the value of the individual. This is what, behind its technicality, *droit administratif* is about; it is what the Conseil d'Etat tries to be.[7]

Thirdly, the developed system of *droit administratif*, centred upon the Conseil d'Etat, forms the basis of many continental systems, and has influenced such international institutions as the Administrative Tribunals of the United Nations Organization and, more importantly, the Court of Justice of the European Communities—an institution which we share with France and our other European partners.[8] Our subject, therefore, has a

---

[3] *Le droit français* (1960) i. 116
[4] One should distinguish from codification the consolidation of legislative texts upon a particular subject-matter, as in the case of tax law or public health: this process is much used (see Circular of Prime Minister of 15 June 1987 on 'la codification des textes législatifs et réglementaires').
[5] *Ridge* v. *Baldwin* [1963] 2 All ER 66 at p. 76.
[6] *Breen* v. *Amalgamated Engineering Union* [1971] 1 All ER 1153.
[7] Mitchell, 'The Real Argument about Administrative Law' (1968) *Public Administration* 167.
[8] See Ch. 10, p. 262.

wider and practical significance, especially now that the United Kingdom is a long-established member of the European Communities. Indeed, there are clear signs already of certain feedback from English administrative law into the case-law of the Community Court, although the predominant influence on that court remains the administrative law of France, followed closely by that of Federal Germany.

Fourthly, the separation between public law and private law, which is a hallmark of the French system, has proved an attraction (some might say, a fatal attraction) for some common lawyers. Thus in *O'Reilly* v. *Mackman* [1983] AC 237 the late Lord Diplock imported into English law the dichotomy between 'public law' and 'private law' rights, a distinction of which the consequences are still being unravelled in the case-law.[9]

All these reasons have led the authors to follow the advice of the late Professor A. V. Dicey, who, as long ago as 1885, said:

> it is not uninstructive to compare the merits and defects, on the one hand, of our English rule of law, and, on the other, of French droit administratif.[10]

As is well known, Professor Dicey's comparison concluded with a judgment as to the resounding superiority of the English 'Rule of Law', and a correspondingly almost unreserved condemnation of the French system:

> it is difficult, further, for an Englishman to believe that, at any rate where politics are concerned, the administrative courts can from their very nature give that amount of protection to individual freedom which is secured to every English citizen. (ibid. 403)

It has been fashionable for some time to point out Dicey's errors; the late Sir Ivor Jennings did this most effectively many

---

[9] For the different background in Scotland, see A. W. Bradley, 'Applications for Judicial Review: The Scottish Model' [1987] PL 311.

[10] A. V. Dicey, *Law of the Constitution* (10th edn., by E. C. S. Wade, London, 1959) 394. But Professor Hand has recently shown that the Wade edition of Dicey omitted certain important 'notes', in one of which—'English Misconceptions as to *Droit Administratif*'—Dicey qualified his earlier views of the French system: G. Hand, 'A. V. Dicey. Unpublished Materials on the Comparative Study of Constitutions', in G. Hand and J. McBride, *Droit sans frontières: Essays in Honour of L. Neville Brown* (Birmingham, 1991), 77 at p. 90; also discussed by Prof. S. Cassese, *Albert Venn Dicey e il diritto amministrativo* (Milan, 1990), 12.

years ago in relation to Dicey's 'Rule of Law'.[11] In more recent times, however, the late Professor Lawson demonstrated the essential rightness of Dicey's comparison at the time he first made it.[12] For, in extenuation of his strictures on the *droit administratif*, it should be remembered that Dicey was writing only twelve years or so after the decision in BLANCO (TC 8 February 1873), often regarded as the starting-point of the modern jurisdiction of the Conseil d'Etat, and that the full development of such concepts as *détournement de pouvoir* and *les principes généraux du droit* were then some way in the future. Moreover, as M. Errera has pointed out, Dicey would later have been aware of the extensive purge of the Conseil d'Etat in 1879 to make sure (in the words of the Minister of Justice of the day) that it 'be composed of people in total agreement with the Government'[13]: 38 members of the Conseil either were dismissed or resigned.[14]

For simplicity we have adopted Dicey's phrase for the title to this book. It is, of course, inexact. *Droit administratif* is correctly translated into English as 'administrative law', and both expressions include (with much more precision of content in France than in England) the whole of the law relating to the various organs of the administration, and also the law relating to the civil service (*la fonction publique*) which latter in France includes much of what in England would be classified as local-government law. This book, however, is primarily concerned only with *le contentieux administratif* and the jurisdiction of the Conseil d'Etat when it is *statuant au contentieux* and of the regional courts of appeal and first instance; that is to say, we are concerned with litigation between a citizen and some organ of the state in an administrative context. Literally translated, those expressions mean 'administrative litigation' and 'ruling on a contentious matter'. English lawyers speak rather of 'judicial review', but this refers *ex hypothesi* to review or control of the administration by the 'ordinary' courts of law, whereas in France (as we shall see) the Conseil d'Etat is by no means an ordinary court but the head

---

[11] I. Jennings, *The Law and the Constitution* (1st edn., London, 1933).
[12] 'Dicey Revisited' (1959) 7 *Political Studies*, 109, 207.
[13] Cited in R. Errera, 'Dicey and French Administrative Law' [1985] PL 695 at 697.
[14] V. Wright, 'L'épuration du Conseil d'Etat en juillet 1879' (1972) *Revue d'histoire moderne et contemporaine* 621.

of a separate hierarchy of special administrative courts. Moreover, 'judicial review' carries a very different connotation in the United States and certain other parts of the English-speaking world, where it refers to the power of the courts to declare legislation to be unconstitutional—a task reserved in France to the Constitutional Council, albeit in limited form.

The scope, therefore, of this book is limited to a straightforward exposition of those institutions whereby judicial control over the acts of the administration is exercised in modern France, together with some analysis of the more important principles of law that such institutions apply in carrying out this function. Our primary purpose is to expound French law, but some comparative references will be made to English law, mainly in order to stimulate—or provoke—the informed teacher or enquiring student to explore further the comparison. A full-scale comparative treatment is not possible within the compass of a short book, although whenever common lawyers try to describe a civil-law institution or doctrine the approach necessarily becomes a comparative one, simply because by force of training they see their chosen subject differently from the way the civilian sees it.

Our readers must not assume that the pages which follow describe the whole area of *droit administratif*, nor again that they contain the whole of the law relating to such parts as we have selected for examination. Those readers seeking a fuller treatment of our subject or concerned with other aspects of French law and administration may find helpful the works listed in the selective bibliography at the end of this book. Moreover, it has been assumed that a reader of this book already has a basic knowledge of English or Scottish administrative law; this is not a book on those laws, although we hope that readers may be assisted in their comprehension of their own system by our account of the French system.

In our presentation we have taken into account that the original source-materials of *droit administratif* are fairly readily available and that French is the one foreign language studied by most people in the United Kingdom in their youth. For these reasons, in the pages that follow we have not hesitated to employ French legal terms (printed usually, but not invariably, in italics), although accompanied, it is hoped, by an adequate explanation of their meaning.

The Germans would say that in the United Kingdom we have a *Justizstaat*, where conflicts between public authorities and the ordinary citizen are determined by the 'ordinary' courts; France, on the other hand, is a *Rechtsstaat*, where a series of specially constituted administrative courts exercise control over the state.[15] This fundamental difference between the two systems will be examined in the pages that follow. As we shall see, the difference is more than one of institutions; the principles of law applied have also been developed differently in the two countries, although the results in particular cases may be similar.

The secret of the strength of the Conseil d'Etat and the case-law (or *la jurisprudence*) which it administers is to be found in the history of this unique French institution, in the methods adopted for the recruitment of its personnel, and also in its career-structure generally. It will be necessary therefore in the pages that follow to deal fairly fully with historical and organizational matters, before we come to describe the extent of the jurisdiction of the administrative courts or the kind of law they administer. First, however, we must supply the *mise en scène*: the constitutional, administrative, and political background.

---

[15] See B. Chapman, *The Profession of Government* (1959), 183 ff.

2
The Constitutional and Administrative Background

1 INTRODUCTION

The 'duality' of the *droit civil* and the *droit administratif* in France, and more particularly the dual system of courts, cannot be understood without some appreciation of French constitutional history and the present Constitution of the Fifth Republic. This is particularly important because the Conseil d'Etat was the child of the Revolution of 1789 and the period of the Consulate (1798–1802), although *droit administratif* itself was, as we shall see in a subsequent chapter, a later development.

2 THE CONSTITUTION: THE DIVISION OF POWERS

The course of French political history since the Revolution has been charted by repeated shifts of power between the executive and the representative assembly. On the one hand, there has been the authoritarian or Bonapartist tradition (inherited from the Ancien Régime) of autocratic rule based upon a powerful and centralized bureaucracy and acting more or less independently of parliament. On the other hand, there is the parliamentary tradition, whereby the elected assembly imposes its will upon the executive, although still relying upon a strong bureaucracy. This last tradition reached its apotheosis in the Third and Fourth Republics (1875–1940; 1946–58), although (for reasons which cannot be analysed here) neither produced strong and effective government.

The Constitution of the Fifth Republic established in 1958 retains, in theory, the essential features of a parliamentary regime. Although adopting a rigid separation between executive and legislature (a minister cannot be a deputy), it does not set up an American-style presidential system. The Prime Minister remains responsible to parliament, and only parliament has the power to enact statutes (*lois*). The Fifth Republic, however, differs radically from its two predecessors and, at least in its practical operation, occupies a midway position between the Bonapartist and parliamentary traditions.

In the first place, severe restrictions have been imposed upon parliament's power (liberally exercised under the Fourth Republic) to overthrow the government. These legal checks, in conjunction with such political realities as the Algerian crisis and the prolonged parliamentary majority of the Gaullists, have meant that, since 1958, there have been only thirteen governments (of MM Debré, Pompidou, Chaban-Delmas, Couve de Murville, Messmer, Chirac, Barre, Mauroy, Fabius, Chirac, Rocard, Cresson, and Bérégovoy), a remarkable record of stability when compared with the twenty governments in twelve years of the Fourth Republic. Secondly, in proportion as parliament has weakened, so the powers of the executive have been strengthened and distributed, albeit unequally, between President and Prime Minister. The President indeed is elevated to a role which, as events have shown, can become one of crucial importance. In the third place, the Constitution vests in the executive sweeping powers to regulate by decree. This distinction between the *pouvoir réglementaire* of the government and the *pouvoir législatif* of parliament, which is fundamental in French constitutional theory, requires explanation for those accustomed to the absolute sovereignty of parliament in the United Kingdom—subject, of course, to the supremacy of Community law since 1973.

Under the Third and Fourth Republics it was an accepted constitutional principle that in certain circumstances the government enjoyed autonomous powers to make regulations by decree, even when such powers had not been delegated to it by parliament. Thus, in LABONNE (CE 8 August 1919) the President of the Republic had introduced by decree some road traffic regulations; their legality was challenged on the ground that no statute had authorized him to do so, but the Conseil d'Etat accepted that as head of state he was competent to issue regulations required in the interest of public order, even without express delegation from parliament. Support was found for this view in the Constitution of 1875 which stated: 'The President of the Republic supervises and ensures the execution of statutes'.[1] Likewise, in HEYRIES (CE 28 June 1918), the Conseil d'Etat recognized the validity of a governmental decree which suspended, for the duration of the 1914–18 war, a statutory safeguard for civil servants, on the ground that the decree was necessary, in the special circumstances, to ensure the continuity of the public services of the state. And we shall see below how, in their more circumscribed spheres, the prefect and the mayor possess a similar autonomous *pouvoir réglementaire* in matters of police and public order.

On the other hand, it was also established constitutional theory, until 1958, that parliament could legislate on any matter it chose: there was no province reserved for regulation by decree. This principle was apparently swept away by the Constitution of 1958. Article 34 states that 'All statutes (*lois*) shall be passed by parliament', but it proceeds to give a closed list of the matters upon which parliament may legislate. Article 37 then provides, by way of complement, that all matters not listed in Article 34 shall fall exclusively within the regulatory power of the executive. Moreover, under Article 38, even in the domain reserved to parliament, the government may obtain its consent to legislate by *ordonnance* for a limited period.[2]

Article 41 provides that the government may object to the provisions of a bill which lie outside the competence of parliament as set out in Article 34.

This formal restriction on parliament has been diminished significantly by both constitutional practice and decisions of the Constitutional Council. The government has not resorted to Article 41 since 1980. In a typical year, barely 15 of over 1,400

---

[1] The Constitutions of 1946 and 1958 use similar language in vesting in the Prime Minister 'l'exécution des lois'.

[2] Article 34 stipulates, for example, that only a *loi* may establish the *rules* concerning: civil rights and the fundamental guarantees granted to citizens for the exercise of their public rights; nationality, status and legal capacity, matrimonial property, succession and gifts; the definition of felonies (*crimes*) and misdemeanours (*délits*) and the penalties therefor, criminal procedure, the creation of new courts, and the status of the judiciary. *Loi* must also establish the *fundamental principles*, *inter alia*, of the law of property and of civil and commercial obligations. For an explanation and justification of the choice of the matters included in Article 34, see P. M. Gaudemet, 'La loi dans la Constitution de 1958' [1961] PL 386. On the distinction between *loi* and *règlement*, see B. Nicholas, 'Loi, Règlement and Judicial Review in the Fifth Republic' [1970] PL 251.

decrees enacted by the government are based on Article 37.[3] Unless the government takes a formal objection to the Constitutional Council, that body will refuse to strike down a provision of a *loi* merely because it infringes the competence of the executive to make regulations by decree. The requirement that a *loi* set out the principles, or even the rules, in a wide range of areas has been interpreted so widely that a leading commentator has written that autonomous decrees do not exist, and that all decrees are either delegated legislation or merely implement the provisions of a *loi*.

As a result, two issues are of current interest. The first is the process of declassification (*délégalisation*), whereby the Constitutional Council rules on whether provisions of a *loi* enacted after 1958[4] fall outside the legislative domain of Article 34 and so may be amended or repealed by decree. The procedure under Article 37(2) not only places before the Council the *loi* which is being called in question but also enables the Council to clarify the dividing-line between *loi* and *règlement*. Under this Article, therefore, the Council may also be expressing a view, by implication, upon the constitutionality of the proposed *règlement* to amend or abrogate the *loi* in question, and in particular whether it trespasses upon the domain reserved for *la loi* by Article 34. Hence, the comment of Favoreu and Philip that *délégalisation* is also 'a technique for confining and controlling the power to legislate by *règlement*'.[5]

Secondly, in recent years, the government has frequently sought the delegation of power under Article 38 to make decrees in matters reserved to parliament under Article 34. This was the case especially when there was a change in the party in government in 1981 and 1986. For example, in 1986, the government obtained powers to deal with competition, to end price and rent controls, and to privatize a list of 65 companies within the

---

[3] Favoreu, ' "Les règlements autonomes n'existent pas" ', *RFDA* 1987, 871. Governmental lawmaking can take a variety of forms—*décrets, règlements, règlements d'administration publique, ordonnances, arrêtés*. This text uses 'decree' as a shorthand way of referring to all of these.

[4] Pre-1958 *lois* are not subject to the procedure of *délégalisation*: Article 37 (2) provides for their amendment by decree after consulting the Conseil d'Etat (*après avis du Conseil d'Etat*).

[5] L. Favoreu and L. Philip, *Le Conseil Constitutionnel* (2nd edn., Paris, 1980), 101.

ensuing five years. As long as the objectives of the powers conferred are clearly stated, and no infringement of constitutional values is authorized, the Constitutional Council has refused to declare invalid such wide-ranging delegations of power under a *loi d'habilitation*. The government has to present for ratification the decrees which it has made under the *loi d'habilitation* within the time-limit set out in the enabling text, but parliament need not actually vote approving the *projet de ratification*.

Decrees made under a specific delegation, or under Articles 37 or 38, are classified as administrative decisions and, accordingly, are always subject to the jurisdiction of the administrative courts, which will ensure that they neither trespass upon the domain of parliament, nor infringe either constitutional principles or 'general principles of law' (see Chapter 9, and especially CE 26 June 1959, SYNDICAT GENERAL DES INGENIEURS-CONSEILS). The Conseil d'Etat may similarly impose compliance with these general principles even upon the measures which the President of the Republic is empowered to take in time of national emergency pursuant to Article 16 of the Constitution (see CE 2 March 1962, RUBIN DE SERVENS), as well as upon decrees passed to implement a statute enacted after the holding of a referendum (CE 19 October 1962, CANAL). This control *ex post facto* by the Conseil d'Etat in its judicial capacity (*statuant au contentieux*) is available notwithstanding the fact that the measure in question may have been submitted in draft for the opinion of an administrative section of the Conseil d'Etat—and approved (on this see p. 62).

Once a statute (as distinct from a decree) has been enacted and promulgated, the courts (whether civil or administrative) have no power to question its constitutionality. Thus, while resembling the United States in having a written constitution, France differs from that country and follows the British pattern in having no judicial review in the ordinary courts of the constitutionality of statutes. On the other hand, certain principles are stated or referred to in the Constitution of the Fifth Republic (see especially the Preamble, with its reference to the Declaration of the Rights of Man and to the Preamble of the Constitution of 1946), and these are bound to have some effect on statutory interpretation. In its decisions the Constitutional Council interprets not only the Constitution, but also the statutes which it is reviewing in the light of the Constitution. These interpretations of both Constitution and statutes have an increasing influence on the decisions of the civil, criminal, and administrative courts.

It may now be appreciated that the English distinction between legislation (that is, Acts of Parliament) and subordinate or delegated legislation cannot be rigidly applied in France. In a sense, the *décrets* which the government issues under Article 37 of the Constitution are just as much legislation as the *lois* which parliament enacts under Article 34. Moreover, as we shall see, prefects and mayors may issue *arrêtés* under their inherent *pouvoir réglementaire* which are in no sense *delegated* legislation.[6] Again, under constitutional principles, ministers and other public authorities are recognized as possessing authority to complete by decree the framework of legislation, whether or not this is explicitly stated in the legislation in question.

3 LE CONSEIL CONSTITUTIONNEL

Whilst preserving the supremacy of parliament within the powers entrusted to it by the Constitution, the draftsmen of 1958 wanted to ensure that parliament did not become overpowerful, a factor which was seen as a major weakness under the Third and Fourth Republics. Therefore, a new institution was created by the Constitution of 1958.[7]

The Constitutional Council was given four main functions:

1. To adjudicate upon the validity of presidential and parliamentary elections and upon the conduct of referenda (Articles 58, 59, and 60).

2. To express an opinion, prior to their promulgation, on the legality under the Constitution of all *lois organiques* approved by parliament (Article 61(1)).

3. If an ordinary *loi* or international treaty is challenged as contrary to the Constitution, to decide upon its constitutionality, again prior to its promulgation or ratification (Articles 61(2) and 54).

---

[6] See, for example, CE 18 January 1967, MOSCONI, reproduced in App. H.
[7] A brief and authoritative account in French is Favoreu and Philip, *Le Conseil Constitutionnel*; also B. Genevois, *La Jurisprudence du Conseil constitutionnel* (Paris, 1988). See also J. Bell, *French Constitutional Law* (Oxford, 1992), ch. 1.

4. To ensure that both parliament and government keep within the domains reserved for their respective legislative activities under Articles 34 and 37 of the Constitution.

Outside these headings the Council has no general or inherent jurisdiction: as the French say, it has only *une compétence d'attribution* (i.e. a jurisdiction expressly assigned to it). So far as concerns the Council's functions in controlling the constitutionality of legislation (heads 2 and 3 above), a distinction is made, as is apparent, between *lois organiques* and *lois ordinaires*.

Lois organiques are laws of particular importance which affect the powers and interrelationship of such constitutional authorities as the President of the Republic, parliament, the Constitutional Council itself, and the judiciary. Organic laws as so defined *must*, prior to their promulgation, be submitted in every instance to the Council for a declaration that they conform to the Constitution (Article 61(1)).

Ordinary laws are not subject to automatic scrutiny by the Council. However, an ordinary law may (but only prior to promulgation) be referred to the Council by the following four high officers of state: the President of the Republic, the Prime Minister, the President of the Senate, and the President of the National Assembly (Article 61(2)). In addition, since 1974, the Council may be seised by a collective submission of 60 deputies or 60 senators, a most important innovation and the method most often used in practice.

The Council may also be called upon to adjudicate as to the compatibility with the Constitution of international conventions and treaties to which France is becoming a signatory (Article 54). Since 1992, the Council may not only be seized of the matter by one of the four high officers of state, but also by sixty deputies (or senators). An example of this role of the Council arose in 1992 when the President of the Republic called upon the Constitutional Council to decide upon the constitutionality of the Treaty of Maastricht subscribed to by the Member States of the European Community. In this case, the Council decided that France would have to amend its Constitution before being able to ratify the Treaty.

The Council is composed of nine nominated members, three appointed (each for nine years) by each of the following: the President of the Republic, the President of the Senate, and the President of the National Assembly. In addition, any former President of the Republic is *ex officio* a member for life: at present M. Giscard d'Estaing is the only member in this category but is unable to sit while he holds elected office. The President of the Council is appointed by the President of the Republic from its membership.

Since its creation in 1958 the Council has delivered more than two thousand decisions. Much the largest group relate to references made to the Council under Articles 58, 59, and 60 of the Constitution concerning irregularities in presidential or parliamentary elections or in the conduct of referenda; these decisions, though of much practical (and political) importance, need not concern us here.

The next group of decisions are those under head 4 above and to which we have already alluded in the previous section. Here the Council, under the jurisdiction conferred upon it by Articles 37(2) and 41, has been involved in decisions that parliament has strayed (Article 37(2)) (or is about to stray: Article 41) outside the domain to which *la loi* is confined by Article 34.

Under Article 37(2) the Prime Minister may refer to the Council any *loi* enacted by parliament after 1958 for its decision whether the text in question did properly fall within the legislative domain of Article 34. We have referred above to the process of *délégalisation*.

One of the most striking illustrations of this activity of the Council is the decision of 28 November 1973.[8] Here the government was proposing to reform by decree the *Code rural* (which had been enacted as a *loi*); the decrees would have introduced certain criminal sanctions involving imprisonment. The Council was seised under Article 37(2) with the government's request to declassify the relevant articles of the Code so as to permit their amendment by the decrees. The Council ruled that, even for the lowest category of criminal offences (*contraventions*), a *loi* was required if the offence was to carry any penalty involving loss of liberty (*des mesures privatives de la liberté*). This ruling was not strictly necessary for its decision, as the articles in question envisaged only a fine and so, as the Council decided,

---

[8] See B. Nicholas, 'Fundamental Rights and Judicial Review in France' [1978] PL 82 at pp. 92–7; Bell, *French Constitutional Law*, ch. 2 and Decision 1.

were not of a legislative character and could therefore be 'declassified'.

If (to imagine the unimaginable) the government still persists in making decrees despite an adverse ruling of the Council, then an interested party could challenge the decrees before the Conseil d'Etat by a *recours pour excès de pouvoir* on the ground that they breached the Constitution, as they fall outside the domain of *règlement* (for this *recours*, see Chapter 9); and the Conseil d'Etat would regard itself as bound by the decision of the Constitutional Council on the issue of constitutionality under the doctrine of *res iudicata* (*la règle de la chose jugée*) and would annul the decrees (CE 12 December 1969, CONSEIL NATIONAL DE L'ORDRE DES PHARMACIENS).

The third and last group of decisions (some 300 to date) embrace heads 2 and 3 above, and it is these which have attracted the most attention, since they have gone beyond what was foreseen as the likely role of the Council in 1958. Indeed, until 1971, the Council's functions appeared to be limited to electoral disputes (i.e. head 1) and the policing, as it were, of the border between Articles 34 and 37 (head 4).

On 16 July 1971, however, the Council ruled unconstitutional a government bill (*projet de loi*) which introduced restrictions upon the forming of certain voluntary associations deemed contrary to the public interest. The bill had passed through parliament and was about to be promulgated when the Council was seised of the matter by the President of the Senate pursuant to Article 61 (as to which, see below). The Council ruled that the bill offended the principle of liberty of association as referred to in the Preamble to the Constitution of 1946, itself incorporated by reference in the Preamble to the Constitution of 1958: the allusion in the 1946 Preamble to those 'fundamental principles recognized by the laws of the Republic' was construed by the Council as referring, *inter alia*, to the Law of 1 July 1901 on the right of association.[9]

---

[9] Rivero remarked of this decision that 'The Constitutional Council has taken a step for the protection of the citizen against the arbitrariness of the legislature no less decisive than those by which the Conseil d'Etat has progressively organized the citizen's defence against the arbitrariness of government': *AJDA* 1971, 537. See further, Nicholas, 'Fundamental Rights', 87–92; Bell, *French Constitutional Law*, decision 1.

Since that decision, the Constitutional Council has developed a large case-law in a variety of areas not only in civil liberties, but also governing the conduct of the administration and the legislative process. Some of its decisions have been very sensitive politically. Most prominent among these was the decision of 16 January 1982 concerning the new socialist government's Law to nationalize some leading banks and other industrial concerns.[10] Constitutional principles set out in the Preamble to the 1946 Constitution stated that *de facto* monopolies should become the property of the state, whereas the Declaration of the Rights of Man and of the Citizen of 1789 declared the sanctity of property and that it could only be taken away on grounds of public necessity. The Council had to reconcile these conflicting principles, which the authors of the Constitution had failed to do explicitly. Since the 1946 Preamble affirmed the principles of the 1789 Declaration, the Council held that the latter remained fully in force and were merely qualified by the later constitutional provisions. All the same, the Law was valid, since public necessity in fighting France's economic problems was established (though the original provisions on compensation were held unconstitutional). The Council had, in effect, to put flesh on the skeleton of constitutional principles, and did so in highly charged political circumstances. Similarly, in 1986 it decided that privatization legislation was only valid if it ensured that the state received the real value of the assets sold to the private sector.[11] Perhaps its most outstanding contribution in developing the constitutional provisions on civil liberties has been in media law. Despite the exiguous nature of the texts in the Constitution itself, the Council has developed a body of case-law not only on the press, but also on television, and cable and satellite broadcasting which were barely contemplated when the constitutional principles were written in 1946, let alone in 1789.[12]

The Council has also played an important role in safeguarding the rights of parliament in the face of a strong executive, which

---

[10] See L. Favoreu (ed.), *Nationalisations et Constitution* (1982); Bell, *French Constitutional Law*, ch. 2 and Decision 2.

[11] T. Prosser, 'Constitutions and Political Economy: The Privatisation of Public Enterprises in France and Great Britain' (1990) 53 MLR 304.

[12] L. Favoreu and L. Philip, *Grandes Décisions du Conseil constitutionnel* (5th edn., Paris, 1989) (hereafter *GD*), 596 ff.; Bell, *French Constitutional Law*, ch. 5.

the Constitution endows with powers to ensure that its proposed legislation is enacted. Parliament is very restricted in the timetable for voting finance laws and in their amendment. The Council has been strict in ensuring that the government does not abuse this process by including provisions on non-financial matters in such a law. The government must also respect the procedures for enacting such legislation, and this led in 1979 to the whole Finance Law for 1980 being declared unconstitutional.[13]

In public law, the Council has been concerned, *inter alia*, with the status of judges of the administrative courts and the appropriate jurisdiction (*compétence*) of those courts, on the one hand, and the civil and criminal courts on the other. In a landmark decision of 23 January 1987,[14] it held that fundamental principles recognized by the laws of the Republic established the competence in principle of administrative courts to quash or correct decisions taken by the executive, its agents, or local authorities in the exercise of public power, though this might be modified by the legislator to create 'blocks of competence' (typically in the civil courts) over a particular subject-matter (see Chapter 6). Such a decision has obvious importance for the case-law of the Conseil d'Etat and the Tribunal des Conflits; but, more fundamentally, it places on a constitutional basis the very existence of a separate hierarchy of administrative courts.

The increased activity of the Council since the 1970s was in large measure due to the reform introduced in 1974 by way of a constitutional amendment. Prior to this reform, Article 61, as originally framed, limited the right to seise the Council (other than in electoral disputes) to the four high officers of state (referred to above, p. 14). The reform of 1974 added to these four a new right for the Council to be seised by any group of sixty deputies or sixty senators (as happened for the first time with regard to the controversial abortion bill of 1975). French commentators regard the reform as a vital protection for the opposition parties in parliament, or indeed for any minority group that can enlist the support of the required number from one or other chamber.

[13] CC decision no. 79–110 DC of 24 December 1979, *GD* no. 30; Bell, *French Constitutional Law*, ch. 4 and Decision 15.
[14] CC decision no. 86–224 DC of 23 January 1987, *GD* no. 43.

The new right of the opposition to seise the Council has made it very much a forum for last-ditch resistance to government policies.[15] In such a charged political atmosphere, the Council has inevitably been the subject of criticism from politicians for reaching partisan results. On the whole, however, it has managed to remain aloof from party political conflicts and has gained the respect even of its erstwhile opponents. Indeed, the criticism now made of the Council's impact is that it has led to a degree of legalism in political life, as arguments of constitutional legality become a major feature of parliamentary and political debate.

The Constitutional Council is, as we now see, more than a mere advisory or consultative body. Is it then a court? Certainly, it is a court when called upon the adjudicate upon electoral disputes. In its constitutional functions, although clearly exercising a judicial role, it is not indeed a supreme constitutional court such as that of the United States. It is not supreme because it is in no sense a court of appeal from the Cour de Cassation or the Conseil d'Etat, which remain the highest courts respectively in the judicial and administrative hierarchies (what the French term *l'ordre judiciaire* and *l'ordre administratif*) of the French court system.[16] Nor is the Council a true constitutional court such as those of the United States, Germany, or Italy, because it can only express an opinion on the constitutionality of a measure *before* it comes into effect: as MM. Favoreu and Philip emphasize, it operates a priori, not a posteriori; but that opinion is then final and is binding upon the administration, upon the President of the Republic, upon parliament, and upon the government.

Moreover, on constitutional issues, as distinct from electoral disputes, it can only be seised of a matter upon a reference from the four high officers of state or (since 1974) from a group of sixty deputies or sixty senators.

Unlike the procedure of other French courts, that of the Council is more like an investigation conducted by the French

---

[15] See S. Wright, 'The French Constitutional Council: A Political Weapon in the Amnesty Arena' (1989–90) 14 *Hold. L. Rev.* 41.
[16] But the decision of the Constitutional Council is binding as *res judicata* upon the Cour de Cassation or the Conseil d'Etat if either of these courts is called upon to pronounce on the same text: CE 12 December 1969, CONSEIL NATIONAL DE L'ORDRE DES PHARMACIENS.

Ombudsman (the Médiateur, discussed below, p. 30).[17] A reference to the Council will usually present arguments why certain clauses of a bill are unconstitutional, but it may simply refer the text without arguments. The Council is not limited to the grounds of the reference, but may examine any part of the bill which it considers unconstitutional. For example, seised of a challenge to a bill which proposed to modify the electoral law so as to require at least 25 per cent of candidates on lists for local elections to be women, the Council, of its own motion, declared this provision to be unconstitutional, although it was not challenged in the reference from the sixty deputies.[18] There are no written rules of procedure and there is no oral hearing of the interested parties. Any person, including members of the public, may write to the Council, and it may take such note of the submissions made as it considers appropriate. The member of the Council designated as reporter in the case may take such steps as he thinks fit to inform himself about the issues. This can include interviews with members of parliament, of the government, or with those affected by the proposed legislation. It is this form of procedure which makes the Council appear a non-judicial institution, but its objective is merely to be fully informed when coming to an interpretation of the constitutional texts.

The Council is composed of members with a variety of experience in politics, public life, and the law. No single kind of experience is seen as a necessary or a sufficient qualification. But this does not detract from the essentially judicial character of the Council's decisions. They are legalistic both in style and content. The Council is therefore to be regarded as a court, notwithstanding that its membership is not confined exclusively to lawyers.

The astonishing growth of the influence of the Constitutional Council is relevant to our subject. For there is now placed alongside the Conseil d'Etat, the traditional guardian of the citizen in his dealings with the administration, a powerful and prestigious body to uphold constitutional norms, especially the

---

[17] See D. Turpin, *Contentieux constitutionnel* (Paris 1986), ff.; Bell, *French Constitutional Law*, ch. 1.

[18] CC decision no. 82–146 DC of 22 November 1982, see Bell, *French Constitutional Law*, ch. 6 and Decision 34.

ADMINISTRATIVE BACKGROUND

fundamental rights and liberties of the individual. Thus, the then President of the Republic, M. Giscard d'Estaing, in a speech to mark the twentieth anniversary of the Fifth Republic (on 28 September 1978) could proclaim that, thanks to the Constitutional Council, 'France has become an *Etat de droit*, that is, a State in which each authority, even the highest, operates under the control of a judge.'[19] And the learned commentator Louis Favoreu has said that French administrative law and the activity of the administrative courts have been pushed into the background by the new upsurge of constitutional principles under the guidance of the Constitutional Council: 'public law prior to 1970 will soon be regarded as "ancien droit public" '.[20]

M. Favoreu perhaps exaggerates, but it has to be recognized that the Constitutional Council now takes the lead in defining the rights of the citizen, and the Conseil d'Etat and the Tribunal des Conflits follow its initiatives. In fact, the Constitutional Council often draws on the large body of law developed by the administrative courts, and the relationship between them is one of partnership rather than competition, or, in the phrase of MM. Massot and Marimbert, one of 'harmonious co-existence'.

An example of this harmony is provided by the twin cases before the Conseil d'Etat of ASSOCIATION GENERALE DES ADMINISTRATEURS CIVILS and BLETON (CE 16 December 1988). In France, various administrative bodies are not staffed exclusively by promotion within the civil service but from outside (*au tour extérieur*). We shall see in the next chapter that this is true of the administrative courts themselves. By the Law of 13 September 1984, one third of the senior posts of *inspecteurs généraux* of certain administrative agencies could be recruited externally, without any condition other than that of age. Pursuant to this Law, D was appointed General Inspector of Cultural Affairs in the Ministry of Culture, and S General Inspector of Public Libraries. Both appointments were challenged by the civil service and library associations concerned, as well as by individual librarians. The Law of 1984 had been referred to the Constitutional Council by the parliamentary opposition. The Council

---

[19] Of course, even before the Fifth Republic, France was generally accepted as an 'Etat de droit', that is a country subject to the rule of law.

[20] Favoreu, 'L'apport du Conseil constitutionnel au droit public', *Pouvoirs*, 13 (2nd edn., 1986), 17.

decided not to rule against the Law, but drew attention to Article 6 of the Declaration of the Rights of Man of 1789, whereby 'The law ... must be the same for all ... and all citizens ... are equally eligible for all public ... employment according to their abilities and without distinction other than that of their virtues and talents.' The Council underlined that the government's discretion under the Law of 1984 did not extend to making appointments which would be in breach of Article 6 (CC decision no. 84–179 DC of 12 September 1984). Having regard to the views of the Council,[21] the Conseil d'Etat upheld the appointment of D, but quashed that of S. The former was an appropriate appointment within the Ministry of Culture because he had been an architect, an artistic director of a state theatre in Paris, had written several books, and served for four years within the Ministry preparing architectural and urban plans for the capital. On the other hand, S had served for twenty years in the merchant navy, rising to the rank of captain, and then been active in voluntary associations, some for young people. He obviously did not possess the specialized knowledge and experience required as a General Inspector of Public Libraries, of which there were only four such posts. Hence his appointment did not amount to equal treatment of all appropriately qualified candidates and was vitiated by an *erreur manifeste d'appréciation* (on this concept see p. 245).

4 *DROIT ADMINISTRATIF*

As Ridley and Blondel say in their *Public Administration in France* (p. 125) 'one of the most consistently demanded reforms of the eighteenth century was the reform of a slow and costly system of justice. The demand did not only come from a suffering public but also from the advisers of the monarch.' Political thought in 1789 was also concerned to stop the ordinary courts interfering with the activities of the administration: the Parlements of the Ancien Régime had a bad record of having impeded attempts at administrative reform in the later years of the mon-

---

[21] The decision of the Constitutional Council figures prominently in the conclusions presented for both cases by the Commissaire du gouvernement. The importance of the case is shown by the convening of a plenary formation of the Conseil d'Etat—l'Assemblée du Contentieux (see Ch. 4, p. 71).

archy. The ideas of the original leaders of the Revolution in 1789 and 1790 had been encouraged and inspired by the success and example of the American Revolution and the Constitution of 1787, which provided, following Montesquieu's advice in his famous *Esprit des Lois*, for an elaborate system of checks and balances and a sharp cleavage between the three powers of government; the executive, the judiciary, and the legislature. When Napoleon as First Consul came to build a viable governmental machine on the ruins of the Revolution, restrictions on the jurisdiction of the courts similarly suited him, in his desire to create a strongly centralized autocracy. On the other hand, he also wanted to establish a strong, almost military, control over his administrators, both in the provinces and in the several branches of government. Therefore the administration of post-revolutionary France was thought of as a separate machine, independent of both legislature and the 'ordinary' judiciary. In fact, a hierarchical structure was devised, the early councillors of Napoleon's Conseil d'Etat wearing an elaborate uniform (see Frontispiece), as is worn (in modified form) by the prefects to this day on ceremonial occasions. The surprising feature of French administrative law (described by Professor Weil as a 'miracle' in *Le Droit administratif*, p. 8), is that, in spite of its totalitarian origins, it has survived to provide one of the most systematic guarantees of the liberties of the individual against the state known in today's world. How this came about will be discussed in the next chapter; the extent to which *droit administratif* in practice measures up to this description it is the purpose of this book to describe. One of the reasons for the strength of *droit administratif* throughout the nineteenth century, and down to the present day, has been the relative weakness of the other powers of government. As we shall see, the civil courts were expressly excluded from adjudicating in matters involving the administration, and in the famous *arrêts* BLANCO (TC 8 February 1873) applying to the central government, and TERRIER (CE 6 February 1903) applying to local authorities, the jurisdiction of the administrative courts was confirmed as being exclusive (on the details of these cases, see Chapter 8).

As for the legislature, this has always, since 1789, been highly susceptible to political changes. While the principle of parliamentary sovereignty has remained a constant feature of all regimes since the Revolution, at various periods parliament has surrendered extensive rule-making powers to the executive. The see-saw of power between parliament and executive is a recurrent theme of French politics and partly explains their instability. This political instability, although it can be exaggerated, has contributed to the undoubted stability of the French administrative machine; the Conseil d'Etat, like the prefects and the mayors, has survived two monarchies, two empires, and four republics, to give France very considerable internal stability. It is therefore not surprising that, in the estimation of a traditionally conservative nation, these institutions have acquired very considerable respect, if not veneration.

5 MODERN ADMINISTRATION

It is an outstanding characteristic of modern French administration that discretions should be given to officials rather than politicians. French civil servants have always had a great deal of responsibility, and it is sometimes said in France that 'we are not governed, but administered'. The powers of the administration have grown with the tasks confided to the state. As Vincent Wright has written:[22]

> More than ever, the state has become an employer of men, an owner of property, a protector of the nation's boundaries, a guarantor of its social welfare: its two million agents now constitute more than a tenth of the total workforce and their salaries and social security costs amount (in 1986) to nearly 20% of total public expenditure.

Without significant delegation of power to officials, such a state could not operate. The career structure of the *fonctionnaires* (civil servants) also contributes to the important role played by them (see p. 37). Yet another factor tending in the same direction is the absence under the Fifth Republic of any sharp cleavage between the party politician and the senior departmental administrator. Thus, an impressive number of ministerial posts have been filled by senior officials, and officials have occupied politically sensitive positions as advisers in ministerial offices. In addition, and contrary to the current British position, senior civil servants including *Conseillers d'Etat* frequently occupy important posi-

[22] *The Government and Politics of France* (3rd edn., London, 1989), 100–1.

tions in local government, as mayors or assistant mayors of their local communes, or even as presidents of a region.

The cult of the amateur is alien to the French. Essentially individualist, they dislike government interference in their private affairs, but, provided they are left alone, they are content to leave the details of government to experts. In more recent years, this has been illustrated by the creation of independent and autonomous agencies to regulate or supervise particular policies or policy areas: the *Commission nationale de l'informatique et des libertés*, which regulates the use of computers and databases; the *Conseil de la concurrence*, which monitors monopolies, take-overs, and restrictive practices; the *Conseil supérieur de l'audiovisuel*, which supervises broadcasting; the *Commission des sondages*, which investigates complaints about opinion polls; the *Commission de la consommation*, which advises on consumer affairs; the *Commission des opérations de bourse* which regulates the financial markets; and the *Commission nationale de contrôle des campagnes électorales*, which makes recommendations and monitors the conduct of national electoral campaigns. Control by officials is seen in many cases as a way of resolving problems which divide politicians. The British solution is to invite a senior judge to chair a committee.

It would be a gross over-simplification to see the administration as a monolithic, dominant body. As we will see in discussing the civil service, there is great diversity in the interests and membership of the administration, and this can lead to rivalry and fragmentation, which inevitably weaken its power. In addition, controls on the administration have been reinforced by the strength of governments in the Fifth Republic. Freed from the worry of whether they will still be in office tomorrow, ministers can lead a relatively stable existence and can occupy themselves with the affairs of their departments. All the same, parliamentary control through effective ministerial accountability and workable specialist committees remains to be achieved.

In this context, the administrative courts have an important role. They have long provided an available remedy against administrative action. In essence, administration by bureaucrats is made more palatable because, through the administrative courts, the bureaucrats have established their own control and supervisory machinery.

Although France possesses a separate system of administrative courts having a general jurisdiction, these should not be identified with British administrative tribunals; France has comparatively fewer administrative tribunals of the British kind, that is those having a narrowly limited jurisdiction. (For a list of major bodies of this kind, see Appendix D.) In Britain, on the other hand, most of the functions of the welfare state, such as the health service, industrial insurance, and social security, have led to a proliferation of such tribunals; the pervasive control of land use through legislation on such matters as town and country planning, housing and public health, has similarly led to numberless inquiries of various kinds. The influence of the Franks Committee and of the Council on Tribunals has resulted in some measure of standardization of the procedures governing these tribunals and inquiries, so that they observe at least the Franks 'trinity' of 'openness, fairness and impartiality', but there is no single code of procedure.[23] In Britain much of the machinery of administration has been judicialized in this way; partly no doubt because of the innate British distrust of the official. The jury system, elected representatives on local authorites, and latterly amateurs appointed to rent or national insurance and other tribunals, reflect a desire to control the lawyer, the administrator, and the specialist, by the reasonable man or woman on the Clapham omnibus. The modern concern with public participation, especially in the planning and related administrative processes, is a further example of the same tendencies.

By contrast, in France the general jurisdiction of the administrative courts makes it unnecessary to provide for any special form of appeal to a minister or an administrative tribunal. Thus, if new administrative machinery is created, such as the allocation of dairy quotas, there is no need for express provision to be made in legislation for a complex system of tribunals and appeals to control the discretion of administrative officials, as the normal remedies of *droit administratif* will automatically be available to the aggrieved citizen. In some areas, such as refugees, civil service pensions, and social assistance, a tribunal system does exist in France, but this occurs only because of a political decision

---

[23] But a draft set of rules for tribunals has been prepared by the Council on Tribunals: *Report: Model Rules of Procedure for Tribunals* (Cm 1434, March 1991).

that such issues should be expressly taken out of the normal administrative court system.

The French system of social security does, it is true, more nearly resemble the British pattern, in that specialized tribunals have been established to administer it. However, as we shall see in the next chapter, social security is regarded in France as falling within the province of private, not public, law and therefore is not the concern of *droit administratif*. Consequently the decisions of these special tribunals are subject not to the Conseil d'Etat but to the Cour de Cassation (the supreme court in the French civil law system).

Professor Ganz[24] has emphasized that in England the non-judicial aspects of the administrative process should not be neglected by writers focusing attention upon courts and tribunals. We are conscious that the same distortion is possible in describing the French system of administrative courts. For example, the current congestion of the judicial process in France, especially before the Tribunaux Administratifs, gives a strong inducement to the aggrieved citizen to seek to rectify his grievance without resort to litigation at all. Recently, legislation has been introduced in France to facilitate conciliation in disputes between the citizen and the administration.

The 1987 Law reforming the administrative courts has authorized compulsory conciliation in contractual and non-contractual liability actions, and a 1986 Law made it clear that conciliation is one of the functions of the judge of the Tribunal Administratif.[25] But the experience of legislation earlier in the 1980s was disappointing. In tax cases, by contrast, compulsory appeals within the administrative hierarchy, followed by the compulsory conciliation of a departmental committee, does reduce the case-load significantly. Such conciliation is in some respects reminiscent of the German *Widerspruch* (see Chapter 10), but unlike that procedure not all the new French conciliation committees are compulsory, and their future success remains dubious.[26]

---

[24] *Administrative Procedures* (London, 1974).
[25] See L. N. Brown and J. S. Bell, 'Recent Reforms of French Administrative Justice' (1989) 8 CJQ 71.
[26] A like interest is now being shown in common-law countries in 'alternative dispute resolution' (ADR)—alternative, that is, to both courts and tribunals: see H. Brown, 'Sizing Up Alternative Dispute Resolution' (1989) 46 *Law Society Gazette* 15; id., *Judgement Day: The Case for Alternative Dispute Resolution* (Adam Smith Institute, London, 1992).

Control of the administration also takes place through its openness to public scrutiny and criticism. To promote greater 'openness' in public administration (what the French refer to as 'la transparence de l'administration'), parliament has intervened in recent years on two occasions. First, the Law of 17 July 1978 reaffirms the rule already established by the case-law whereby the citizen is given a general right to be notified of any decisions, individual in character, affecting him (general regulations, of course, have always had to be published); in default of notification, the decision will be of no effect so far as he is concerned. In addition, the citizen is given a new right of general access to administrative documents affecting him individually or of general application; only certain sensitive categories of documents are excluded (e.g. those relating to internal governmental discussions or to national defence, national security, or foreign relations). A special commission has been set up to ensure freedom of access and, in particular, to review refusals to allow access.[27] Secondly, the Law of 11 July 1979 introduces the general principle that any administrative decisions must be accompanied by reasons; this requirement of 'motivation' is discussed at pp. 88 and 218.

There has also been concern in very recent years to make the administration more 'user friendly', as is witnessed by a statement from the Prime Minister to the Council of Ministers in February 1989 (*Le Monde*, 23 February 1989).

The British system also uses the inquiry as part of the process of meeting a citizen's dissatisfaction with a decision of an administrative authority. Inquiries may be held when an individual appeals to a minister against a planning decision, a compulsory purchase order, or a variety of other matters, such as an order to divert a public footpath. The French also know the procedure of *enquête publique*, which may be convened in what may seem similar circumstances. But an *enquête* is held, not because of an appeal or representations by some member of the public, but because the law requires it to be held in every case involving some prescribed issue (such as development in a national park), whether or not any objections have been raised. Proceedings at the *enquête* are relatively informal and some are poorly attended.

---

[27] See R. Errera in N. Marsh (ed.), *Access to Government Documents* (London, 1987), 87.

The inspector is appointed by the authority responsible for making the final decision (often the prefect) and he will not necessarily hold a professional qualification. His report on the *enquête* will be one of the papers inserted in the file leading to the eventual decision, and in many instances it may be of no particular significance. In an *enquête* there is very little of the courtroom atmosphere that is so obvious in many British public inquiries. A *recours en annulation* could be sought from the local Tribunal Administratif if no *enquête* had been held where one was required under the legislation, but there may be less opportunity for a detailed examination of the proceedings by the administrative judge than occurs from time to time under the British system. Nevertheless, challenges are sometimes brought to the court, e.g. as to the independence of the inspector (for instance, if he is in the employ of the administration or owns neighbouring land), or as to a gross irregularity in procedure, or as to a lack of precision in specifying the public works to be undertaken or their cost: see EPOUX NEEL (CE 23 January 1970) and VILLE NOUVELLE EST (CE 28 May 1971, discussed in Chapter 10).

6 *LE MEDIATEUR*

Infected by the 'Ombudsman' fever of the early 1970s, when countries all over the world adopted their own version of this Scandinavian institution, the French established the Médiateur by the Law of 3 January 1973.

A prominent politician, M. Poniatowski, who later became Minister of the Interior in the governments of M. Chirac and M. Barre (1974–7), declared:

> The French administration is often heartless, haughty, and convinced that it embodies the sovereign. Sprung directly from the past and having largely taken shape under the monarchy, it is riddled with monarchical attitudes. It does not regard itself as a public service but as a master ordering subjects about.[28]

M. Poniatowski was a leading voice in promoting the creation of the Médiateur (or 'Mediator') to help restrain the excesses of the administration by providing a remedy which would be

---

[28] Cited in L. N. Brown and P. Lavirotte, 'The Mediator: A French Ombudsman?' (1974) 90 LQR 212.

'simple, free and readily accessible' and would extend beyond the limits of judicial control into the sensitive area of *l'opportunité* (or policy), allowing the public denunciation of the bad decision or the bad rule and so rendering the administration 'less oppressive, more accessible and, above all, more human'.[29]

The Médiateur is an official having much the same investigatory functions as the Scandinavian Ombudsman and was modelled to some extent upon our own Parliamentary Commissioner for Administration, although he is in no sense an officer of the French parliament, nor does he enjoy the vicarious prestige which the epithet 'parliamentary' can still bestow in the United Kingdom. As in the case of the PCA, however, complaints about maladministration (or a breach of *'équité'*—which we may translate loosely as 'fairness') must be routed through a deputy or senator,[30] and his annual reports show that the Médiateur has been busily engaged in reconciling differences between the citizen and the administration. This role of conciliator explains his rather curious name, a *médiateur* being someone who acts as an intermediary or who intervenes to arrange matters (even marriages) between others.

When making his investigations the Médiateur is entitled to call upon the Conseil d'Etat and other state agencies for assistance, but his reports (like his actual investigations) do not amount to *actes administratifs* and so are not subject to supervision or review in the courts: CE 10 July 1981, RETAIL. A complaint to the Médiateur does not stop the running of any period of limitation for bringing an action before the courts. On the other hand, the Médiateur is forbidden to intervene in proceedings already instituted before a court, nor may he call in question the merits of a decision of any court; but the amending Law of 1976 does now permit him to address recommendations to the administration when it is involved in court proceedings, suggesting how the dispute might be settled.

The Médiateur has made a not inconsiderable impact on French public life, and was described by a former holder of the

---

[29] Phrases used in parliament when he was presenting his own private bill for an ombudsman in December 1972; this bill, though not enacted, anticipated and paved the way for the government bill which became the Law of 3 January 1973.

[30] The Law of 24 December 1976 now allows a deputy or senator a 'spontaneous' right to refer any matter to the Médiateur within his jurisdiction.

ADMINISTRATIVE BACKGROUND 31

office as an indispensible cog in French democracy.[31] Certainly he receives more complaints than the PCA or the Commissioners for Local Administration: in 1988 the Médiateur received 3,746 complaints, and his local delegates a further 12,365 complaints. Seventy per cent (2,956) were within his competence: of these, in the event, he rejected 1,668 (56.5 per cent), but for the rest his mediation was successful in 887 (87 per cent) of the cases taken up by him.[32] (By contrast in 1987 the PCA received 719 complaints, of which 24 per cent were within his competence, and the five Commissioners for Local Administration in the UK received 5,596 complaints.) His investigations and annual reports (especially the latter) contain many suggestions for the reform of the law and administrative practice: these are taken seriously by the government.

The advent of the Médiateur, however, has not caused any diminution in the flood of cases to the French administrative courts, especially to those of first instance, the Tribunaux Administratifs; nor, as we shall see later (in Chapter 11), has it in any way called in question the prestige of the Conseil d'Etat.

7 LOCAL GOVERNMENT

In a sense, until very recently France has had little 'local' government, as it is understood in England. For instead of independent bodies exercising functions conferred on them by parliament in their own right, in France there has traditionally been mainly a devolution of power from the centre to local agencies acting on behalf of or under the general supervision of the ministries in Paris, although the *conseil municipal* (even before the reforms of 1982) has had a general jurisdiction (*clause générale de compétence*) over local affairs (*les affaires de la commune*).

[31] M. Robert Fabre, *Rapport du Médiateur au Président et au Parlement, 1981* (Paris), Introduction. Of the four holders of the office to date, M. Pinay had previously been a member of parliament and a Prime Minister under the Fourth Republic; M. Fabre, like his predecessor, M. Paquet, was a member of parliament at the time of his appointment. The present holder, M. Legatte, a member of the Conseil d'Etat, was a member of the Constitutional Council at the time of his appointment.
[32] See Le Médiateur de la République, *Rapport au Président de la République et au Parlement, 1989* (Paris). Unfortunately this report has less statistical analysis than that for 1988 (referred to in our text above) but shows that the Médiateur or his local delegates received in 1989 a total of 17, 758 complaints.

France has three main tiers of local government. Metropolitan France is divided into 13 regions, each of which regroups several of the 96 *départements*. Although the *département* is itself subdivided into *arrondissements*, which are divided into *cantons*, these subdivisions are of little administrative importance. Much more important are *communes*, a subdivision of a *canton*, but the size in terms of area and population of the *communes* varies very considerably, and a majority have under 500 inhabitants. In recent years there have been many groupings of *communes*, co-operating for a variety of administrative reasons and under different laws. Thus, there have been over 700 'fusions' of *communes*, there are nine large *communautés urbaines* (or conurbations) in such areas as Bordeaux and Strasbourg, and there are nearly 1,300 *syndicats*, where two or more *communes* agree to unite for particular limited functions.

Before the significant changes introduced by the Law of 2 March 1982, central government dominated local government. Each *département* had a prefect who held the balance of executive power but who was assisted by an elected *conseil général*. Created by Napoleon, but based on the *intendant* of Richelieu, the prefect is an official of central government and is responsible for most central government functions in the *département*, for example bridges and highways (*ponts et chaussées*), but not education. He is the official chain of communication (in both directions) between the lesser local units and the Ministry of the Interior, he provides advice to local government, and is responsible for internal order and security within his *département*. Before 1982, he also exercised supervision (*tutelle*) over decisions of local authorities and could quash them for illegality on his own authority. Since much of the activity of the *département* and the *commune* involved the provision of public services on behalf of central government, central government, through the prefect, had a very considerable say in the activity of local government.

Political thought in France had hitherto been suspicious of devolving power from the centre, for fear of destroying the unity of the state. But for some years the regional prefects of Corsica, where local patriotism is particularly strong, had enjoyed a considerable measure of freedom from detailed control. In 1964 informal groupings of *départements* were set up as 'regions' under senior prefects with essentially a consultative and deliberative function.

Building on this, one of the principal objectives of the 1982 legislation was to democratize and 'deconcentrate' local government.[33] Considerable powers, especially in the fields of economic planning and regional infrastructure, have been devolved to elected regional councils, to be advised by new-style regional prefects, who retain responsibility for public order. Executive power vests in the president of the Regional Council who has become an important political figure. The importance of the new regional government is shown by the fact that former President of the Republic Giscard d'Estaing chose to retain a seat on a Regional Council rather than one in the National Assembly when he was elected to the European Parliament.

At the level of the *département*, executive power now vests not in the prefect, but in the president of the *conseil général* of the *département*, and the *conseil* itself retains residual competence in social and economic affairs.

The *communes* remain virtually unchanged except in matters of financial control. At the *commune* level, the mayor has always been the representative of central government, although he is elected (usually for a period of six years) by his *conseil municipal*, and he is responsible, through the prefect, to the central government, for the maintenance of law and order within his *commune*. He is much more than a mere chairman of the council like his English namesake. When exercising his 'police powers', an expression which must be understood more widely than in England and much more nearly as used in the United States, neither the prefect nor the mayor is in any way subordinate to the *conseil général* or the *conseil municipal*. The mayor may be disciplined only by the prefect and the prefect only by the minister in Paris; and a citizen aggrieved by a decision (*arrêté*), which may be either a positive action or a refusal of some kind of licence or a failure to take some positive action, cannot complain to the *conseil municipal*, as his appropriate remedy (if any) is an appeal to the local administrative court and, perhaps, from there to the Conseil d'Etat.[34] The mayor is the executive authority in the

---

[33] For an assessment of the effect of the 1982 reforms, see *AJDA* 1992, *numéro spécial, Décentralisation, bilan et perspectives*.

[34] See the hypothetical case in App. I.

commune but he may act in two capacities, as agent of the state, and as agent of his *commune*. In his former capacity, he is the equivalent of the registrar of births, marriages, and deaths, the electoral registration officer, and the conscription officer. In this capacity, he also possesses (as does the prefect in his *département*) a general autonomous power, independent of any express or implied delegation to him by statute, or by decree, to make regulations or take decisions in matters of police or public order. But, in practice, the *Code des communes* (embodied in a Law of 1884) provides for most matters of police. *Conseils municipaux* come into the picture outside the broad field of 'police' or 'public order'.

Autonomy has been enhanced by a new status for officials in local government. Formerly, officials of the *département* were national civil servants, while communal employees formed a true local-government service. However, the reforms of 1982 provided that the staffs of the regions and of the *départements* also are to be employees of their respective *conseils*. The new local and regional autonomy is checked in a number of ways. First, in its decision on the decentralization reforms (25 February 1982), the Constitutional Council insisted that respect for the law should be secured by the practice of transmitting local authority decisions to the prefect before they came into force, so that he had an opportunity to refer them to the Tribunal Administratif to test their legality. The prefect has also been able to insist that relevant ancillary documents are provided before he decides whether to refer a decision to the Tribunal. The procedure of reference by the prefect (the *déféré préfectoral*) is a major safeguard in ensuring that the decisions of local authorities are lawful.[35] Secondly, budgetary control has been restructured. A new body, the *Chambre Régionale des Comptes*, has been established in each region. It performs functions similar to English district auditors in vetting the accounts of local authorities. But it has additional functions. It disciplines local authority treasurers (who are still appointed by central government). It also advises the prefect when a local authority fails to set an annual

---

[35] For a recent example, see CE 26 July 1991, COMMUNE DE SAINTE-MARIE (REUNION), where the prefect of Réunion referred two public works contracts to the Tribunal Administratif of Saint-Denis, and their annulment was upheld by the Conseil d'Etat.

budget in time or makes a legally defective budget. In such cases in England, the district auditor has to proceed to surcharge the councillors (cf. *Lloyd* v. *MacMahon* [1987] AC 625). In France, the prefect simply steps in to impose a budget upon whose content the *Chambre Régionale des Comptes* has advised. From decisions of a Chambre Régionale an appeal lies to the Cour des Comptes,[36] an administrative financial court established before the Revolution, which has functions corresponding to the National Audit Office, together with those of the Audit Commission and the Scottish Commission for Local Authority Accounts. Thirdly, it must be remembered that most of the money for local government is provided by way of grants from central government. The *taxe d'habitation* (a residents' tax) and the property-based *taxe foncière* have both been reformed in recent years to increase local financial autonomy, but central-government influence through the direction of finance remains strong.

In formal terms, *communes* and *départements* may appear to enjoy a wider discretion than local authorities in England. They do not have to be able to point to express statutory authority for each of their acts, since the communal council, like that of the *département* is regarded as possessing a general power over the organization of the public services of the area. It follows that the English doctrine of *ultra vires* is virtually unknown in French local government. This does not mean, however, that a French *commune* can do exactly as it likes, as it is subject to the following important restrictions:

1. As explained above, the mayor, *not* the council, is concerned with public order.
2. Most administrative decisions of importance (and in particular the annual budget) adopted by the municipal council, need the approval of the authorities of the *département* or, in some circumstances, of the *Chambre Régionale des Comptes*. Formerly, all decisions of the *conseil général* of the *département* were subject to the control (*tutelle*) of the central government, but this has been abolished by the Law of 1982. For many activities of

---

[36] On this institution, see J. Magnet, *La Cour des Comptes et les institutions associées* (Paris, 1971); and shorter, but no less authoritative, is J. Raynaud, *La Cour des Comptes* (Paris, 1980).

the *communes*, especially when a *syndicat* of several *communes* is proposed for some particular purpose (such as the provision of an electricity supply for the area), consent at the level of the *département* has to be obtained; for others, such as the formation of a semi-public company (*société d'économie mixte*) in which the *commune* is to hold shares, detailed approval of the administrative sections of the Conseil d'Etat is required.[37] Despite the reforms, this system of supervision remains very detailed and important; in practice it goes much further than any comparable feature in England.

3. Decisions of mayors, municipal, departmental, and regional councils and of prefects are also subject to a *recours en annulation* or *en indemnité* (for these expressions, see Chapter 8) before the local administrative courts and ultimately the Conseil d'Etat. Thus, when a *commune* decided to establish a municipal dental clinic, the Conseil d'Etat, on the application of a ratepayer, quashed the decision on the ground that the *commune* had not adequately established a need, and that such a charge on public funds was improper when there were an adequate number of private dentists in practice in the town (VILLE DE NANTERRE, CE 20 November 1964).

French local authorities—the *communes* in particular—are thus regarded as being part of the administration and susceptible to control by superior authorities in the hierarchy.

Even before 1982, the long-established structures of local government allowed considerable scope for local initiative, as is evident from the successful redevelopment programmes of such energetic municipalities as Lyons and Marseilles. They did not require the special structures created for local government in Paris in 1975, although since 1983 their structures have been more closely aligned to that of the capital. There remains considerable pressure to reform and rationalize the system at communal level. Thus, a Law of 1966 permitted the government to establish the so-called 'urban communities' referred to above, so as to integrate the more important local-government services regardless of communal boundaries. The 'communities' are administered by joint boards which are not directly elected but chosen by the

---

[37] On this subject, see M. Rendel, [1966] PL 213.

conseils municipaux within the conurbation. These and other forms of reorganization are not affected by the Law of 1982.

8 THE CIVIL SERVICE

The standard of behaviour of an administration depends in the last resort upon the quality and traditions of the public officials who compose it rather than upon such sanctions as may be exercised through a system of judicial control. Hence, although the rest of this book will be chiefly concerned with judicial control, this chapter concludes with a few observations about the French civil service.

As in England, the French distinguish between the servants of the central government and those engaged in local government. Both are regarded as constituting *la fonction publique* and fall therefore within the description of *fonctionnaires*. But, as we have seen when discussing local government, civil servants fill many posts which are occupied by local government staff in England; local government officers in France are concerned only with those functions of the mayor which he exercises in his capacity as chairman of the *conseil municipal*. The education service is a notable example of the countrywide activities of the central government, and the schoolmaster of the smallest village is a civil servant. Ridley and Blondel point out in their *Public Administration in France* (p. 29) that:

> The local government service in Britain employs 1,500,000 persons: it is a competitor of the civil service for many important jobs. In France, where it employs less than 400,000 people, it is less attractive and looks more parochial.

It follows that the civil servant is a familiar figure everywhere in France and that a civil service career appears as offering great opportunities to all social classes, brains not birth being the key to appointment and advancement within its ranks. Great stress is laid upon affording all citizens equal access to the service, and this is assured by recruitment through competitive examinations (*concours*) open to all with the required educational attainments (see CE 28 May 1954, BAREL, discussed in Chapter 9). The reality, however, is that few people from poor backgrounds can attain the very high educational qualifications for entry into *les grandes écoles* from which the senior civil servants are recruited.[38]

The civil servant is not in a contractual relationship with the state but enjoys a status determined by the *Statut général des fonctionnaires* (as revised in 1959). This text is supplemented by an extensive and important case-law of the Conseil d'Etat, under whose jurisdiction (at first instance or on cassation)[39] fall all disputes touching recruitment, pay, promotion, duties, and discipline within *la fonction publique*.[40]

The civil service has a tradition, dating back to the Ancien Régime, of active intervention in the economic and social development of the nation. Under the Third and Fourth Republics the chronic instability of French politics meant that senior civil servants often had to assume responsibilities which would be regarded in England as more properly belonging to the politicians, although even in England the initiative in detailed legislation is often really with the civil servants.[41] Before 1958 the distinction was commonly made between 'the government', everchanging and fumbling in policy, and 'the administration', permanent, stable, and forward-looking. After the last war, for example, it was the civil service which initiated the plans for the long-overdue modernization of industry and agriculture. Indeed, much of the credit for France's economic resurgence since 1945 must be attributed to her energetic, brilliant, and far-sighted administrators.

This dynamic outlook and a missionary zeal in public service that transcends politics are fostered by the training which the higher echelons receive in the *grandes écoles*. This term is applied

---

[38] See J. Ardagh, *France Today* (London, 1990), esp. ch. 7. These *grandes écoles* are listed below.

[39] As we shall see (Ch. 3, p. 52), the Conseil d'Etat retains jurisdiction at first (and last) instance in respect of civil servants nominated by decree, but other civil servants have recourse to the Tribunaux Administratifs (especially that of Paris), with appeal now to the Cours Administratives d'Appel and further review (subject to leave) by the Conseil d'Etat *en cassation*. See also App. A.

[40] The 'Eurocrat', a Community civil servant, had, up to 1989, direct access at first instance to the Court of Justice of the European Communities, but since that date 'staff cases' have been transferred to the new Community Court of First Instance, with only a right of appeal to the Court of Justice: see Brown and Jacobs, *The Court of Justice of the European Communities* (3rd edn., London, 1989), ch. 5.

[41] See, e.g., J. T. Craig, 'The Reluctant Executive' [1961] PL 45.

to a number of post-entry training schools, such as the Ecole Polytechnique, Ecole des Mines, Ecole Nationale des Ponts et Chaussées, Ecole Normale Supérieure, and Ecole Nationale d'Administration.[42] These provide the equivalent of an undergraduate, or in some cases postgraduate, course in their particular field. Entry is as fiercely competitive as for an Oxbridge scholarship and carries something of the same cachet, both educationally and socially. Many enter the civil service in order to acquire the status of *ancien élève* of one of these institutions as a stepping-stone to a subsequent career in industry or commerce. This movement from public service into private enterprise (or *pantouflage*, as it is called), which may occur late as well as early in the career of an ambitious civil servant, is an important factor in creating a certain common outlook and a mutual regard between the leaders in the business world and the top levels of the administration.

Among higher civil servants a distinction is made between the 'technical' branches and the general or non-technical administrator. The latter correspond to the administrative class of the British civil service, but the former are in no way regarded as their inferiors. Thus, the civil and mining engineers who make up the famous Corps des Ponts et Chaussées or Corps des Mines rank among the top people in their professions and command that high respect which the French accord to the expert. This notion of 'corps' is encountered no less in the non-technical branches where, traditionally, the three *corps d'élite* are the Conseil d'Etat, the Cour des Comptes, and the Inspection des Finances. It is to these three administrative departments of the civil service that the best graduates of the National School of Administration aspire (of this School more will be said in Chapter 4).

The French civil service has been described as 'one of the main driving forces, if not *the* driving force, in French life'.[43] Wright is more sceptical.[44] He notes that 'certain officials in parts of the

---

[42] 'The principle of the post-entry training school has thus been adopted for almost all branches of the higher civil service. Training plays a much greater part in the recruitment of the French civil service than it does in most other countries' (F. F. Ridley and J. Blondel, *Public Administration in France* (2nd edn., London, 1970), 36).
[43] Ibid. 54.
[44] Wright, *Government and Politics of France*, 131.

administration do enjoy considerable influence but they are by no means alone in the complex fabric of public decision-making'. Ingrained habits and traditions, rather than a dynamic vision of the future, may characterize the activity of many civil servants and reduce their effective power. Though different in structure, there are many similarities with the civil service in Britain and in other western countries in terms of the power effectively wielded by bureaucrats in France. For that reason checks have to be maintained, and important among these is judicial control, to which subject we now proceed.

3

The Administrative Courts

1 INTRODUCTION

Judicial control presupposes the existence of judges who impose upon the administration obedience to the law. The distinctive character of the French system in comparison with our own is apparent once we ask the two questions:

What judges are they who wield this control?
To what law do they constrain obedience?

For the French answer is that the judges are special judges and the law is a special body of law. This and the following four chapters will be concerned with the administrative courts, their personnel, procedure, and jurisdiction. Chapters 8 and 9 will examine the law which they apply.

In France judicial control of the administration is entrusted to a specialist corps of judges who sit in special courts. These courts now form a three-tier hierarchy headed by the Conseil d'Etat in Paris, below which are the five regional Cours Administratives d'Appel and the Tribunaux Administratifs, which number twenty-six in metropolitan France (see Appendix A). In addition to these administrative courts of general jurisdiction, there are a number of other administrative jurisdictions exercising judicial functions in particular spheres. These administrative courts of special jurisdiction much resemble the 'administrative tribunals' of the British legal landscape. But they are less numerous than their British counterparts. Furthermore, these specialized administrative jurisdictions are under the supervision, not of the ordinary civil courts as in this country, but of the Conseil d'Etat as the supreme administrative court.

This complex pattern is the product of developments over the past two centuries which have increased the size and importance of this branch of the French legal system. An appreciation of the

2 THE BIRTH OF AUTONOMOUS ADMINISTRATIVE COURTS

Modern France was born in the Revolution of 1789. Much, however, that was apparently new was in fact a carry-over from the Ancien Régime. The Conseil d'Etat is itself in some measure an example.

In pre-Revolutionary France, the Conseil du Roi advised the King on legal and administrative matters; this council had its origins, similar to those of the English Curia Regis, in the feudal system, but, unlike its English counterpart, it remained primarily a political and arbitral body concerned with the determination of disputes and alleviation of tensions between the great nobles. The power of the lawyers was to be seen rather in the twelve regional royal courts or Parlements (especially the Parlement of Paris), which in the eighteenth century in particular (under Louis XV and Louis XVI) not only interfered to a considerable degree in the executive government but also impeded such reforms as the monarchy sought to introduce.[1]

The Conseil du Roi did hear complaints against the legality of decrees and regulations made by the Conseil itself or complaints on appointments to royal offices. In addition, Richelieu sent out masters of requests (later called *intendants*) from the Conseil to hear complaints on taxation and later on public works. Appeal from these officials lay to the Conseil du Roi. Zealous in their desire to maintain a monopoly of all legal process, the Parlements sought to annul the decisions of their rivals. In 1641, the Edict of Saint-Germain prohibited the Parlements from judging any cases concerning the state, administration, or government and reserved them to the King and his appointees. All the same, the conflicts continued until the Revolution.

Although the Conseil d'Etat initially had certain resemblances to the Conseil du Roi of the Ancien Régime, there is no direct

---

[1] See A. Sheenan, *The Parlement of Paris* (London, 1968) for an authoritative study in English.

THE ADMINISTRATIVE COURTS

link between the two.[2] In both time and character, the Revolution made a major break in the tradition of administrative justice.

The first step taken by the revolutionaries was to break the power of the Parlements. This was done by the famous Law of 16–24 August 1790, which was in part inspired by the Montesquieu theory of the separation of powers, but also owes much to the Edict of Saint-Germain of 1641. Article 13 of the Law, which is still in force, provides:

> Judicial functions are distinct and will always remain separate from administrative functions. It shall be a criminal offence for the judges of the ordinary courts to interfere in any manner whatsoever with the operation of the administration, nor shall they call administrators to account before them in respect of the exercise of their official functions.[3]

This article was confirmed by decree in 1795:

> The prohibition is renewed against the courts taking cognizance of the acts of the administration of whatever kind they may be.

These provisions, which are still in force today, gave complete liberty to the administration; as the power of the King also was curbed and the Conseil du Roi was abolished, there was no one to whom the citizen could appeal for protection against the excesses of the executive. Napoleon, when he assumed power as First Consul, was conscious of the value, for administrative reform, of the information brought by individuals' complaints. In addition, he wanted expert advice on the drafting of laws and regulations, and in his constitution of 'l'an VIII' (1799) he established a Conseil d'Etat, which was to operate under the direction of the three Consuls, but separate from them. The Conseil began its existence on Christmas Day, 1799. It was divided into five sections and presided over by the head of state, at first the First

---

[2] In the 175th Anniversary volume, *Le Conseil d'Etat 1799–1974*, the origins of the Conseil du Roi are traced back to the 13th century, when the Parlement of Paris and the Chambre des Comptes (or 'Exchequer', later the Cour des Comptes) detached themselves from the Curia Regis, leaving the rump as a somewhat amorphous body, designated as the Great Council or Privy Council (*Conseil Privé du Roi*). On the history of administrative law before the Revolution, see J.-L. Mestre, *Introduction historique au droit administratif français* (Paris, 1985).

[3] We are indebted to Martin Weston for correcting a mistranslation of part of this passage in our previous editions: see his invaluable book, *An English Reader's Guide to the French Legal System* (Oxford, 1991), 141.

Consul, and then the Emperor. Initially the work of the five sections (each specializing in particular branches of administration)[4] was to draft new laws and administrative regulations and, perhaps more important in view of later developments, 'to resolve difficulties which might occur in the course of the administration'. It is this last phrase which provided the constitutional basis for the subsequent growth of the judicial activity of the Conseil d'Etat. A Decree of the same year amplified somewhat these terms of reference by giving power to the Conseil to advise the head of state on the setting aside of improper acts of administration authorities and on resolving the jurisdictional conflicts arising between the administration and the civil courts and between the various ministers.

The door being closed by the Law of 1790 to redress in the ordinary courts, some outlet for the aggrieved citizen had obviously to be provided if the new regime were not to be one of administrative tyranny. The solution adopted in 1799 was that the citizen had first to lodge his complaint with the appropriate minister. If still unappeased, he should have a right of appeal from the minister to the Conseil d'Etat.

The Conseil d'Etat had, however, still to build its independence as a true court. This meant the elimination of two doctrines, that of the *justice retenue* and that of the *ministre-juge*.

At this early date, although acting on appeal against the decision of the minister, the Conseil d'Etat had no power actually to decide nor to pronounce judgment; rather, its job was to advise the head of state as the minister's superior in the administrative hierarchy. But in practice its advice was invariably followed, although Napoleon did occasionally refer a case back to the Conseil for a second deliberation. One may compare the way in which a judgment of the Judicial Committee of the Privy Council is cast in the form of humble advice to Her Majesty but always followed. Nevertheless it was of great theoretical importance when the Conseil d'Etat was empowered by the Law of 24 May 1872 to reach decisions in suits against the administration without the formal pretence that it was merely advising the head

---

[4] The original sections reflected, in part, the turbulent times, being: *la section de la guerre*, la *section de la marine*, *la section des finances*, *la section de législation*, and *la section de l'intérieur*. For a full account, see *Le Conseil d'Etat 1799–1974*, ch. 2.

THE ADMINISTRATIVE COURTS 45

of state on a decision which was legally his own. Thus, it is only since the beginning of the Third Republic (1870–1940) that the Conseil d'Etat has had the acknowledged jurisdiction of a court, competent to deliver judgments, not in the name of the head of state, but (like the ordinary courts) in the name of the French people (for such a judgment, see Appendix H). In French parlance, this meant a shift of theory from *justice retenue* to that of *justice déléguée*.

On the other hand, a consequence of the principle according to which the administration should not be under the control of any court persisted in the doctrine of the *ministre-juge*. The Conseil d'Etat did not enjoy immediate jurisdiction in regard to the acts of the administration; complaints had to be brought before the appropriate minister, and only eventually, on appeal from his decision, to the Conseil d'Etat.[5] As a result, it was alleged that all complaints which might compromise high officials or the government's policies were rejected and no leave to bring an action before the Conseil was allowed.[6]

This is why, in 1889, the famous case of CADOT (CE 13 December 1889) marks a decisive stage of the Conseil's evolution. In this case the Conseil cast off the outworn practice that there had first to be a complaint by the aggrieved citizen to the minister. Since CADOT, in any matter involving a decision by the administration, it has been possible to bring the complaint directly before the Conseil. The administration lost its jurisdiction to receive the complaint, as it were, at first instance, and the Conseil d'Etat became, to adopt the French expression, 'the *juge de droit commun* of the acts of the administration'.[7]

Meanwhile, there had occurred changes in the organization of the Conseil which were no less significant of its development as an administrative court. In 1806 there was created within the Conseil a special body, the Commission du Contentieux, to deal with the judicial work of the Conseil as distinct from its

[5] In Germany this doctrine still survives in the procedure of *Widerspruch* (see ch. 10).

[6] See E. Poitou, *La Liberté civile et le pouvoir administratif* (Paris, 1869), ch. 9.

[7] But we shall see that, since 1953, the Conseil d'Etat has become essentially the *juge d'attribution*, and the Tribunaux Administratifs the *juges de droit commun* within the system of administrative courts, i.e. the court which has to be seised of the case when there is no special provision giving competence to another jurisdiction.

advisory work for the various departments of government. This Commission in 1849 changed its name to that of the Section du Contentieux, under which style it has been known to the present day. As the 'contentious section' it co-exists alongside the five other 'administrative sections' into which the rest of the Conseil is subdivided (to which we shall return: below, p. 61).

Another key date is 1831, in which year the Commission du Contentieux began to conduct its judicial business in public, at least to the extent of holding a public sitting at which the parties could be represented by counsel, and after the close of which the Commission had to publish a judgment containing reasons for its decision.

The same year also saw the introduction into the procedural working of the Commission of a new and highly important officer, the Commissaire du gouvernement. Intended originally to present the viewpoint of the government, this officer rapidly arrogated to himself an independent function and began to represent the general public interest rather than the policy of the administration.

The detachment of the Commission du Contentieux from the rest of the Conseil was underlined in 1849 when the decisions of the new-styled Section du Contentieux were no longer required to pass for formal approval through the General Assembly of the Conseil d'Etat, as had been the previous practice (although the General Assembly had long been in the habit of merely rubber-stamping whatever the Commission put before it). A further development of importance was the Decree of 2 November 1864 which exempted the suit to quash an administrative act (*le recours pour excès de pouvoir*) from court fees and from the need for legal representation, thereby encouraging the use of this remedy.

In the result, the Conseil d'Etat emerged by the close of the nineteenth century as a court of first (and last) instance having a general jurisdiction to be seised directly of any complaint or suit against the administration. And in its Section du Contentieux the Conseil d'Etat manifests all the features which the French associate with a court—a public hearing (however perfunctory), the representation of the parties by counsel, a spokesman for the public interest, a collegiate bench, and a published judgment supported by reasons.

It was this court, originating as an offshoot of the administration itself and staffed by high-ranking civil servants, that was to have the vital role of exercising judicial control over the administration from the end of the nineteenth century. In the difficult transition from the *laissez-faire* liberalism of that century to the planned economy and collectivized society of the twentieth, the French were to bless the good fortune which had given them, almost by historical accident, this novel institution before which the humblest citizen could arraign and call to account the all-powerful and interfering state. And once Dicey's distortions had been cleared away, it was to be looked upon with envy by English lawyers. This robust institution has thus survived the vicissitudes of two monarchies, two empires, and five republics, not to mention a foreign occupation.

3 COPING WITH OVERLOAD

Local Administrative Justice

Once established as a court in its own right, the Conseil d'Etat developed an important role as the redresser of grievances against the administration in fields as diverse as the legality of decisions, breaches of contract, liability for torts, and taxation. The ever-widening scope of the activity of the administration in the twentieth century only served to reinforce the importance of the Conseil and to increase its case-load. To meet this demand, increasing importance has been attached to establishing lower-level administrative courts to deal with the bulk of routine litigation.

As has been mentioned, local courts for complaints against the King's administration did exist before the Revolution and were run by the *intendants*. To some extent, these served as models for later local administrative justice. The Conseils de préfecture, like the Conseil d'Etat, were set up by Napoleon in 1799. They were charged in each *département* with affording advice to the prefect as well as a limited administrative jurisdiction, chiefly in connection with direct taxation and public works. From the first they asserted themselves as true courts, although their subordination to the local administration was indicated by the fact that the prefect presided. In the course of the nineteenth century their procedure and organization were improved and their jurisdiction extended.

In 1926 the existing eighty-six Conseils de Préfecture of the *départements* were replaced by twenty-six interdepartmental Conseils, the jurisdiction of each extending over several *départements*. The independence of the Conseils was improved by the prefect's ceasing to be the titular president. In his place, each Conseil was given a president who was to be full-time member of the court. Members of the Conseils were grouped into a separate corps of civil servants with an improved status. Confidence in the reformed Conseils was reflected by extensions made to their jurisdiction in 1934 and 1938, certain categories of matters being transferred to them which had hitherto been within the exclusive jurisdiction of the Conseil d'Etat.

The Reforms of 1953

After the Liberation there was a steep rise in the number of plaints lodged with the Conseil d'Etat. Many stemmed from the process of *épuration* designed to purge the administration of those officials who had collaborated with the enemy during the occupation; the *fonctionnaire* who rightly or wrongly believed himself aggrieved challenged his dismissal or other sanction before the Conseil. Another fertile and more permanent source of litigation was the increasing part played by the state in regulating the social and economic life of the nation.

The Conseil d'Etat gradually fell behind in its race to keep abreast with this spate of litigation. Whilst some 6,000 plaints were lodged in its registry in an average post-war year, it decided only some 4,500, and this despite the most prodigious efforts by every member of the Section du Contentieux to increase the output of decisions. Faced with a backlog of 26,000 cases awaiting decision in 1953, a drastic remedy was needed. This was found in a complete reversal of the pattern of jurisdiction of the administrative courts. In brief, the Conseil d'Etat lost the greater part of its jurisdiction at first instance: the usual court of first instance was to become the Conseil de Préfecture. These local courts, of which there were then twenty-six in metropolitan France[8] were to enjoy a new status more in keeping with the

---

[8] For a list see App. D. Their number was reduced from 26 to 24 by the disappearance of the two which used to serve the Algerian *départements* but then increased to 26 again by the creation of new Tribunaux, those of Amiens and Bastia, in 1967 and 1982 respectively: each is named after the city in which it sits.

dignity of their new role. They were also to receive a new name, that of Tribunaux Administratifs.[9] By this redistribution of administrative jurisdiction it was hoped to relieve the pressure upon the Conseil d'Etat without impairing the quality of justice to which the citizen had become accustomed in his litigation against the state. In the event, these hopes appear to have been fulfilled.

The reform of 1953 made the Tribunaux Administratifs the *juridiction de droit commun* at first instance, leaving only a few (yet important) categories of cases for the Conseil d'Etat as judge of first and last resort. For the most part, the Conseil has acted thereafter as a court of appeal, or a court of cassation for cases coming from special administrative jurisdictions (see Appendix D) and (since 1989) from the new Cours Administratives d'Appel (for the relationship between the various administrative courts, see Appendix A). The problem of the jurisdiction of the Tribunaux Administratifs *inter se* is resolved on a territorial basis, but in order to help limit the burden on the Tribunal Administratif of Paris (which already has seven *sections*, each with two *chambres*), the rules of competence are not based on a simple criterion such as the residence of the plaintiff. Instead, the plaintiff's residence appears only as one of many criteria which vary according to the nature of his complaint. In default of any other rule, the competent Tribunal is the one in whose area the maker of the decision is officially resident. Nevertheless, the various criteria still often lead to Paris, with consequent overloading of that hard-pressed Tribunal.

The Reforms of 1987[10]

Although the 1953 reforms achieved considerable success in reducing the burden on the Conseil d'Etat, it proved only a

---

[9] But, even before 1953, 'Tribunal Administratif' was the title of the Strasbourg administrative court which acted with jurisdiction comparable to a Conseil de Préfecture for the three *départements* of Alsace-Lorraine: the court had been so styled since the return of Alsace-Lorraine to France after 1918. See further L. N. Brown, 'The Reform of the French Administrative Courts' (1959) 22 MLR 357, for a fuller discussion of the reform of 1953; also M. J. Remington, 'The Tribunaux Administratifs: Protectors of the French Citizen' (1976) 51 *Tulane Law Review*, 33.
[10] See L. N. Brown and J. Bell, 'Recent Reforms of French Administrative Justice' (1989) 8 CJQ 71, and for a diagrammatic presentation, App. A.

respite. As the activity of the state increased, so did complaints about its decisions, the modern attitude of the citizen being to expect redress for every grievance. In addition, proceedings on appeal against the Tribunaux Administratifs and on cassation against the administrative courts of special jurisdiction, such as the Commission des Réfugiés, provided a large body of work for the Conseil d'Etat. By 31 December 1987, the Conseil was back in the position of 1953, with a backlog of 25,392 cases awaiting decision, equivalent to over three years' work.

Again the solution adopted was to unburden the Conseil d'Etat of a number of categories of case. Already, improvements in the Conseil's methods of working had been introduced and these had more than doubled its decision-making capacity between 1975 and 1985 (on these see p. 72). All the same, the rate of appeals to the Conseil grew even faster. The Conseil's procedure was unnecessarily painstaking and laborious for the many routine cases which still came before it. A more rational use of resources required some limitation of the use of the Conseil not only at first instance but even on appeals.

The Law of 31 December 1987 created a new tier of courts, the Cours Administratives d'Appel. Five such courts were created in Bordeaux, Lyons, Nancy, Nantes, and Paris. Their jurisdiction covers appeals from the Tribunaux Administratifs on what is known as *le plein contentieux*. This notion of *plein contentieux* is discussed more fully in Chapter 8. It comprises disputes on tax, administrative contracts, public liability generally and civil service matters in all of which issues of fact are particularly important. In *le contentieux de l'annulation* (discussed in Chapter 9), the Law of 1987 envisaged that appeals on questions concerning the legality of administrative decisions, except measures taken under the *pouvoir réglementaire*,[11] should eventually be devolved to the new courts, though this would only be done by a decree made on the advice of the Conseil d'Etat itself: the Decree of 17 March 1992 has now transferred by stages proceedings *pour excès de pouvoir contre les décisions individuelles*, a clear mark of confidence in the Cours Administratives d'Appel.

[11] For this term, see Ch. 2, p. 9. If one takes account of the retained jurisdiction of the Conseil d'Etat at first (and last) instance over the legality of decrees and *actes réglementaires* of ministers (see p. 52), the exception relates mainly to *règlements* of prefects or local authorities.

Appeals on electoral disputes at the municipal or cantonal levels remain with the Conseil d'Etat. Since about 80 percent of all Conseil d'Etat cases have been appeals from the Tribunaux Administratifs, and over half of these now fall within the categories of business transferred to the new courts, a considerable reduction has been made both in the number of cases coming to the Conseil d'Etat and in its backlog of cases awaiting decision.

Two other measures were taken. Firstly, a 'filter' procedure (the *procédure préalable d'admission*) was introduced for all appeals to the Conseil d'Etat *en cassation*, to quash the decisions of an inferior court, be it one of the new Cours Administratives d'Appel or a special jurisdiction. This procedure allows the Conseil to reject a case where it is irregular (e.g. out of time) or where it is not founded on any serious grounds. Unlike the leave stage in common law courts, it is not a matter of unfettered judicial discretion where the judge may choose freely and without explanation whether to allow a case to go forward for a full hearing. The 'filter' procedure is a summary judicial decision for which reasons have to be given and which binds the parties, so that the matter cannot be reopened.

Secondly, the Law enables the lower administrative courts, the Tribunaux Administratifs and the Cours Administratives d'Appel, to submit for the prior opinion (*avis*) of the Conseil d'Etat 'a new question of law, presenting a serious difficulty arising in a number of cases', before it reaches its own decision. This procedure will be discussed in more detail below. It suffices to note here that this procedure enhances the role of the Conseil d'Etat as primarily a body deciding important questions of law. Together with the transfer of appeals on certain matters to the Cours Administratives d'Appel, this reform will have the effect that the Conseil will be primarily a review court, quashing errors of law made by lower courts, providing them with rulings on points of law, and confining its role at first instance essentially to the review of the legality of decrees and similar measures taken by the government.

Summary of the Present Jurisdiction of the Conseil d'Etat

The cumulative effect of the reforms of 1953 and 1987 is to leave the Conseil d'Etat with three heads of jurisdiction. The first is its jurisdiction at first and last instance. At present, this represents some 15 to 20 per cent of the total case-load and includes principally:

1. The *recours pour excès de pouvoir* to annul a governmental decree or *ordonnance*, or an *acte réglementaire* (see p. 9), or other decisions of a minister which have to be taken with the advice of the Conseil d'Etat.

2. Disputes concerning the individual status of those public servants who are appointed to the more important administrative or public bodies by presidential decree.[12]

3. Disputes concerning decisions of collegiate bodies at the national level; this important category includes, for example, decisions of *jurys de concours* (CE 14 November 1980, BRABANT, where the candidate was asked a question outside the published syllabus) or of the *Conseil national de la communication et des libertés*, a body of nine which, *inter alia*, licensed and regulated radio stations (see, e.g. CE 13 February 1991, SOCIETE 'ILE-DE-FRANCE' MEDIA where a decision of this *Conseil* was quashed for inadequacy of reasons).

4. Disputes concerning election to the regional councils in France and to the European Parliament.

5. Proceedings to challenge administrative acts, the application of which extends geographically beyond the area of any single Tribunal Administratif.

The second head is the Conseil's continuing appellate jurisdiction over those appeals from the Tribunaux Administratifs which have not been diverted to the Cours Administratives d'Appel by the 1987 reform. The appeals presently retained, as we have seen, relate to local elections and *le contentieux de l'annulation* in respect of *actes réglementaires*; proceedings to annul *décisions individuelles* have now been transferred to the Cours Administratives d'Appel by stages under the Decree of 17 March 1992, referred to above (p. 50; see also Appendix A).

The third head of jurisdiction retained is that of *cassation*, which presently represents only some 3 per cent of the case-load

---

[12] These are very numerous as they include members of the *grands corps* and other bodies recruited through l'ENA, judges of the ordinary courts, officers of the armed forces, and university professors. (For l'ENA, see Ch. 4, p. 78).

of the Conseil. This permits the Conseil to review the legality of decisions both of the Cours and of a miscellany of specialized administrative jurisdictions. The distinctive nature of the *recours en cassation* is examined in detail later.[13]

4 INTEGRATING ADMINISTRATIVE JUSTICE

To match the increased demands made of them, the judiciary of the Tribunaux Administratifs were endowed with an enhanced status in 1953, bringing improved conditions of service, higher salary, and new avenues of advancement by the provision that several posts in the Conseil d'Etat should be reserved for recruitment from the Tribunaux Administratifs. The Law of 6 January 1986 further enhanced this status by giving these judges (like those of the Cours Administratives d'Appel after 1987) the guarantees of judicial independence and irremovability, similar to those long-enjoyed by the judiciary of the civil and criminal courts. The consequence has been that the lower tiers of administrative courts now enjoy greater formal guarantees of independence than members of the Conseil d'Etat themselves. To this paradox, we return in the next chapter when examining the career structure within the administrative courts (p. 80).

Like those of the Conseil d'Etat, members of the corps of judges of the Tribunaux Administratifs and the Cours Administratifs d'Appel are drawn predominantly from the Ecole Nationale d'Administration. The recent reforms have required the rapid appointment of a large number of new judges, so that recruitment, at least in the short term, has had to extend outside senior civil servants to include others, such as those with legal experience. As a result, the proportion of those administrative judges who have experience of the active administration has been reduced considerably.

A permanent commission of inspection composed of members of the Conseil d'Etat visits each of the Tribunaux and Cours from time to time. The intention is not only to check the technical quality of their operations, but also to bring to those in the relative isolation of the provinces the support of the Conseil's authority. At the same time, the Conseil gets early warning of new

[13] See esp. Ch. 9.

trends or frictions in the relations between the administration and the public, before these appear in actual cases referred to it.

As far as members of the Conseil d'Etat are concerned, they remain very much part of the administration, though at a supervisory level and maintaining an appropriate judicial distance from the politics of active administration. In difficult political circumstances, it has shown a sense of balance—even in the Vichy and Liberation periods—which has protected it from political attack.

Such an equilibrium was easier to attain when judgment was delayed months or sometimes years after the event. This was not possible in 1962, when the Conseil d'Etat was obliged to act with great promptitude in the CANAL case (CE 19 October 1962). In this case the Conseil had to decide on the validity of a decree setting up a military court for the trial of dissident French officers belonging to the proscribed OAS (Organisation de l'armée secrète). The decree was issued pursuant to the Law of 13 April 1962 which had been approved by referendum and authorized the President to take all necessary measures to implement the Evian agreement to end the fighting in Algeria. Canal (nicknamed 'The Monocle') was the treasurer of the OAS: he was condemned to death, and his two accomplices, Robin and Godot, to twenty years' imprisonment. All three appealed to the Conseil d'Etat.

The Conseil accepted jurisdiction and, with remarkable despatch, gave judgment that the military court was illegal a matter of days before the date appointed for Canal's execution.

In Chapter 9 we shall see how the Conseil reached this conclusion by applying its normal principles of judicial control. But to annul such a decree in the political circumstances of the time inevitably produced a reaction from the government.

Apart from saving the life of Canal,[14] the immediate effect of the CANAL judgment was nullified by the passage through parliament of a Law of 15 January 1963 conferring full force of law (*loi*) upon decrees passed pursuant to the Law of 13 April 1962. Of more permanent importance was the setting up by the government of a study group to examine possible changes in the Conseil d'Etat itself. However, all concerned were conscious

---

[14] He was imprisoned at Tulle and then released in 1968 under an amnesty declared in that year.

that any reversal or even serious change of the pattern of judicial control of the administration was inconceivable: the appeal of tradition, the record of the achievements of the Conseil d'Etat, and the support of contemporary public opinion all stood in the way. As we shall see in the next chapter, the measures which were eventually adopted provided a significant, but not revolutionary, reform in the internal structure of the Conseil d'Etat as a whole, bringing closer together its administrative and judicial activities. All the same, the reform has in no way impaired the judicial character of the Conseil's activities in the Section du Contentieux.

5 OTHER ADMINISTRATIVE JURISDICTIONS

Like Britain, France has seen the creation of special jurisdictions, functioning to some extent within the administrative process but applying the essentials of a judicial hearing as a stage in reaching certain administrative decisions. Unlike in Britain, however, this process of what Americans term 'judicialization' of the administrative process[15] has been on a relatively smaller scale, though still considerable having regard to the number of regional jurisdictions to be found often within a single category. The reason for this is not hard to find. Possessing in the Conseil d'Etat and the associated administrative courts an effective system of *general* judicial control over administrative action, the French have felt less need to judicialize their administrative process by a proliferation of *ad hoc* administrative tribunals. On the other hand, the comparative ineffectiveness of general judicial control by the English High Court in the past has had precisely the opposite effect in prompting parliament to incorporate the partial safeguard of an administrative tribunal within the actual process of reaching an administrative decision. Another factor in Britain has been the reluctance to admit within the charmed circle of 'ordinary courts' the various new adjudicating agencies which have been produced by the economic and social developments of the last half-century or so. Debarred from the club, these agencies have all been lumped together as

[15] The French have coined an even uglier word 'juridictionnalisation': see R. Chapus, *Droit du contentieux administratif* (2nd edn., Paris, 1990), 66.

'administrative tribunals'. It is only in recent years, and especially since the Franks Report of 1957, that this exclusive attitude has begun to break down and a less rigid demarcation has been made between 'courts' on the one hand, and 'tribunals' on the other. The different approach in the two countries is demonstrated by both the number and the nature of their 'administrative tribunals'—using this term in its English sense of an adjudicating agency outside the ordinary courts and having a jurisdiction limited to some specific sphere of administrative activity. In England, their number has been put at some two thousand.[16] In France, only some thirty or so categories of such 'administrative tribunals' can be listed: although there may be several tribunals within some of these categories, the total number of individual tribunals is very much smaller than in England.[17] Even so, a leading authority complains that there are too many.[18]

There is also a qualitative difference. Thus, in Britain one thinks naturally of the tribunals associated with social security as typically 'administrative'. In France, most of the tribunals dealing with such questions are regarded as adjudicating matters of private, not public, law, and are therefore subordinated to the supreme civil court, the Cour de Cassation. French eyebrows are also raised at our classification of rent or industrial tribunals as administrative; courts deciding issues between landlords and tenants or employers and employees are, to the French, essentially civil. On the other hand, perhaps the largest single group of French 'administrative tribunals' are ones which we should classify rather as domestic tribunals, namely the disciplinary organs of the various public professions, such as those of medical practitioners, architects, dentists, pharmaceutical chemists, and all levels of the teaching profession. Significantly, the bodies

---

[16] An authoritative, but not exhaustive, list appears each year as App. C to the Annual Report of the Council on Tribunals.

[17] See App. D. But several of the bodies listed there are difficult to compare with the British concept of 'administrative tribunal'. For an exhaustive list and commentary, see C. Même (Conseiller d'Etat), 'Juridictions administratives spécialisées', *Encyclopédie Dalloz, Répertoire de contentieux administratif*, ii (Paris, 1985).

[18] 'C'est beaucoup et c'est trop': R. Chapus, *Droit du contentieux administratif* (2nd edn., Paris, 1990), 61, who likens these specialized administrative jurisdictions to medical specialists, whereas the Conseil d'Etat, the Cours Administratives d'Appel, and the Tribunaux Administratifs are 'the general practitioners'.

controlling the legal profession are subordinated to the civil courts.

One of the oldest and most important jurisdictions is the Cour des Comptes (or Court of Accounts). This body exercises control over the expenditure of public money by governmental bodies (see Chapter 2). Historically, it has affinities with the English Court of Exchequer,[19] but is really equivalent now to the National Audit Office in that it investigates not only the legality but also the wisdom of public expenditure.

Another major jurisdiction is the Commission de recours des réfugiés, which deals with an ever-increasing number of applications. In practical terms, some of its work is equivalent to that dealt with by the Immigration Appeals Tribunal and Immigration Adjudicators in the United Kingdom.[20]

Over all these specialized administrative jurisdictions the Conseil d'Etat exercises supervision by way of cassation. In other words, no appeal may be taken to the Conseil on the merits of a decision; generally, the Conseil may only quash for procedural error or illegality (which may extend even to a mistake of fact) and then refer the case back for a new adjudication.[21] This branch of the work of the Conseil d'Etat in relation to the administrative courts corresponds generally with that exercised by the Cour de Cassation in relation to the civil and criminal courts. Moreover, like the Cour de Cassation, the Conseil d'Etat avoids entrusting the new adjudication to the same body whose original decision has been quashed, at least where there is another body of co-ordinate jurisdiction to which the case may be referred. There are resemblances between cassation and proceedings for *certiorari* before the English Queen's Bench Division, but the parallel is not exact, as we shall see in Chapter 9.

---

[19] 'The *cursus scaccarii* is probably the nearest approach to a body of administrative law that the English legal system has ever known.' (W. Holdsworth, *A History of English Law* (7th edn., London, 1956), i. 329).

[20] The Commission, which is presided over by a *Conseiller d'Etat*, dealt with 20,439 cases in 1989. This compares, to some extent, with the 18,717 cases received in the UK by Immigration Adjudicators and the 799 received by the Immigration Appeals Tribunal in the same year.

[21] In some cases, the Conseil d'Etat, like the Cour de Cassation, may substitute the correct decision without remitting the case to a lower court (the procedure of *cassation sans renvoi*).

As we shall also explain in that chapter, cassation proceedings differ in various respects from ordinary proceedings to annul the decision of a purely administrative organ. In particular, once the deciding body is classified as a jurisdiction, the Conseil d'Etat will impose upon it stringent rules of procedure in the formulation of its decision, but at the same time exercise a more limited control over the decision itself. This is one reason why the creation of such specialized administrative jurisdictions is not considered to provide better guarantees for the citizen. Rather, they come into existence either to meet highly specialized needs (such as the Cour des Comptes) or in response to well-organized pressure groups (as in the case of the disciplinary organs of the older public professions). Very often (though certainly not in the case of the Cour des Comptes) these jurisdictions lack both the expertise and the tradition of the general administrative courts. If their numbers were to multiply (which is not expected), the general pattern of judicial control would be seriously disturbed, without their providing any better way to supervise administrative action.

On the whole, French reformers have sought for improvement in public administration by regulating the actual process of decision-making where it occurs, namely in the office of the bureaucrat. Statute or decree may require discussion or consultation with the persons affected, and access to information. Such prior controls can often be more effective than providing subsequent recourse to courts or tribunals. We return to this point in Chapter 11.

4
The Structure and Membership of the Courts

1 INTRODUCTION

One of the most curious features of the Conseil d'Etat for the British administrative lawyer is its double role as both adviser and judge of the administration. To a much lesser extent, the position is replicated in the Tribunaux Administratifs, which are occasionally invited to give legal advice to the prefect.[1] But, as we have seen, while the functions of the Tribunaux Administratifs have become increasingly judicial, the Conseil has retained its dual activity. By contrast, the Cours Administratives d'Appel have exclusively judicial functions. For these reasons, and because of its seniority in the hierarchy, chief attention will be given to the Conseil d'Etat. Historically too the Conseil d'Etat, as the previous chapter demonstrated, has enjoyed unbroken continuity from its creation by Napoleon, whereas the Tribunaux Administratifs and the Cours Administratives d'Appel only date from 1953[2] and 1987 respectively.

The Conseil d'Etat is a remarkably successful institution. The secret of this success lies in part in its composition. For it is composed of the cream of the French civil service. The prestige which the Conseil enjoys in relation to other public institutions and departments of government can be compared, in England, to the special aura which surrounds members of the Treasury. And not unlike the Treasury, the Conseil d'Etat has something

---

[1] G. Vedel and P. Delvolvé, *Droit administratif* (11th edn., 2 vols.; Paris, 1988), ii. 97.
[2] The Conseils de Préfecture, however, which were the precursor of the Tribunaux Administratifs, had a continuous history from their creation by Napoleon in 1799 (see p. 47).

of the atmosphere of an exclusive club, an impression reinforced by the fact that it is housed in the Palais-Royal, a historic and beautiful building adjoining the Louvre across the rue de Rivoli, or in the adjoining Valois building, which houses a small overflow. The physical concentration of its members within the Palais helps to induce a strong *esprit de corps*—stronger indeed than that which knits together either the superior judiciary or the senior civil servants in England.

The members of the Conseil d'Etat therefore constitute an élite. This élite, moreover, is small in number. There are only some 297 members of the Conseil,[3] but of these as many as 80 or so are usually on detachment or release from the Conseil, serving with government departments, presiding over important commissions, fulfilling judicial functions in international institutions such as the Court of Justice of the European Communities, or holding political office upon election as senator or deputy of the French parliament or as a member of the European Parliament. Those on release (*en disponibilité*) include a growing number working in commerce and industry, in both the public and private sectors. Others, without any formal detachment, serve frequently upon *cabinets ministériels*, where they may come to wield much influence in the corridors of power as the personal and trusted aides of ministers.

Of those actually working within the Conseil, a certain number have particular responsibilities within the Section du Contentieux (e.g. the Président de Section, the ten Présidents de Sous-section, and the twenty Commissaires du gouvernement) and are almost wholly occupied with the judicial business of the Conseil. Similarly, a few key members of the administrative sections have little or no time for work outside those sections. But the majority of those at the Palais-Royal share in the work of both sides of the Conseil: this is the 'mixing' principle (*brassage*) established by the reform of 1963 (see pp. 64 and 77–8).

This internal division within the Conseil is an important feature of its structure. It represents a division of labour and a consequent specialization of function. We have already seen in a previous chapter how the Conseil began its existence as an essentially advisory body, charged with advising ministers and

---

[3] For details of the members on 1 February 1990, see App. B.

STRUCTURE AND MEMBERSHIP OF THE COURTS 61

the head of state on the drafting of legislation and regulations and on administrative problems generally. This role it has retained. But over the intervening period of almost two centuries more than half of its active personnel have been diverted into what at first was only a minor part of its functions, namely the inquiry into complaints against the administration.

2 THE STRUCTURE OF THE CONSEIL D'ETAT

Today, then, this division into consultative and judicial functions is reflected by the internal structure of the Conseil, which is split into five sections—four administrative and one judicial. There is also, since 1985, a Report Section (Section du Rapport et des Etudes), which serves as a centre for research and reflection within the Conseil; to this 'think-tank', which began life in 1963 as the Report Committee (Commission du Rapport) we return later in this chapter, for, although it forms an additional administrative section, its functions are quite distinct from those of the other four.

The Administrative Role of the Conseil d'Etat

As we saw in an earlier chapter, the task of legislating is shared under the Constitution of the Fifth Republic between parliament and the government, although the French reserve the term 'statute' (*loi*) for texts emanating from the former, governmental measures of a legislative character being usually described as *décrets*.

With regard to statutes, it is formally provided that all bills (*projets de loi*) introduced into parliament by the government[4] must have been submitted for the Conseil's advice. This is not to say that the advice must be followed. The government can ignore it. In principle, it can present to parliament a new bill containing provisions which conform neither to its original bill nor to the modifications suggested by the Conseil d'Etat; for, although this appears to frustrate the requirement of consultation, parliament must retain complete freedom to adopt whatever text it pleases. This freedom accords also with the absence in the Constitution, as we have seen (p. 12), of any judicial review of what parliament enacts. In short, once a bill is before parlia-

---

[4] Cf. non-governmental bills introduced by private members; these *propositions de loi* do not have to be submitted to the Conseil d'Etat.

ment, the Conseil's consultative function is exhausted; the eventual enactment merely recites 'Le Conseil d'Etat entendu . . .'

The Conseil's advice on bills is normally given by the General Assembly of the Conseil after an initial examination and report from the appropriate administrative section (or sections). Where the government specifies the matter as one of urgency (and it frequently does) the General Assembly will be replaced by a smaller standing committee (*commission permanente*). More is said of these procedures below.

Where the government uses its *pouvoir réglementaire* to legislate by decree under Article 37 of the Constitution, the text of the decree must be submitted to the Conseil d'Etat. Here too the government is not obliged to heed the Conseil's advice. Unlike, however, what was said above regarding a *projet de loi*, the Conseil, in its judicial capacity, will annul a decree which conforms neither to the original text submitted nor to the modifications suggested by the Conseil (CE 1 June 1962, UNION GENERALE DES HALLES CENTRALES).

The Conseil must also be consulted upon the text of delegated legislation falling into the category of *décrets d'application*, as distinct from *décrets autonomes* made under Article 37. As we saw in Chapter 2, these are decrees expressly authorized by parliament to fill out the details of their parent statute. Here the Conseil acts normally through its General Assembly, again after a report from the appropriate administrative section. Where consultation of the Conseil d'Etat is constitutionally required, the resulting decree is known as a *décret en Conseil d'Etat* and will recite the formula 'Le Conseil d'État entendu . . .'

Where such consultation is not required, the government still remains free to seek the Conseil's advice, and occasionally avails itself of this right. Every year on average a score of texts recite the formula 'Après avis du Conseil d'Etat'; but the final text may, in fact, differ from that submitted to the Conseil, and it may subsequently be modified by a new text upon which the Conseil has not been consulted at all.

Quite apart from the legislative process, the Conseil d'Etat has the duty of acting as general legal adviser to the government and to individual ministers. In some matters indeed its advice must be both sought and followed: for instance, in cases of deprivation of French nationality or the recognition of an association as charitable (*d'utilité publique*). Usually, however, the advice is sought voluntarily, as where a minister wishes to be reassured that he will be acting legally in some matter.[5] This often becomes a formal request for advice where different branches of the administration cannot agree on the interpretation of particular legislation or on the legal solution of a particular difficulty. In a case of this kind it is usually the appropriate administrative section that acts as legal counsel to the administration.[6]

Finally, the Conseil has been charged since the reform of 1963 with the duty of submitting an annual report to the President of the Republic in which, besides reviewing the work of the Conseil over the year, it may indicate what reforms of an administrative or legislative character it deems desirable. One may compare the annual reports of the Council on Tribunals or of the Parliamentary Commissioner for Administration. Unlike such reports, however, the reports of the Conseil are not required to be made public. But from 1988 the Conseil has been authorized to include these annual reports within its own published series *Etudes et Documents* (begun in 1949), which now bears the subtitle *Rapport Public*: the first such combined volume appeared in 1989 as *Etudes et Documents No. 40: Rapport Public 1988*. The only exception to the principle of publication is the exclusion of those *avis* which the government regards as confidential.

President Ducamin (a former Secretary-General of the Conseil) has estimated that, if account be taken of the whole of its consultative activities, nearly one half of the total capacity of the Conseil is absorbed by tasks other than the adjudication of disputes.[7] Accordingly, we must now consider the composition

---

[5] Under Art. 23 of the Ordinance of 31 July 1945, the Conseil d'Etat may be consulted by ministers 'on difficulties which arise in administrative matters', a phrase used in the Constitution of 1799 (Art. 52). By tradition, it will decline to advise if the difficulty is the subject of litigation in any court.

[6] For the years 1986, 1987, 1988, and 1989 such *avis* numbered 40, 38, 29, and 29. These *avis* in the strict sense (unlike advice given in the process of legislative drafting) are usually and increasingly published: the more interesting ones are conveniently reprinted in *Etudes et Documents* each year. The practice of their publication began with the issue for 1976. For examples of such an *avis*, see Bell, (1990) 2 *Education and the Law* 121, and Appendix J (p. 317).

[7] B. Ducamin, 'The Role of the Conseil d'Etat in Drafting Legislation' (1981) 30 ICLQ 882; for two English views of the same subject, see M. Rendel, *The Administrative Functions of the French Conseil d'Etat* (London, 1970) and L. Neville Brown, 'The Participation of the French Conseil d'Etat in Legislation' (1974) 48 *Tulane Law Review*, 796.

and function of those organs of the Conseil concerned with these important activities, that is, the four administrative sections and the General Assembly. The Report Section will be returned to later in this chapter, in view of its distinctive role, being neither adjudicative, nor consultative in the sense applied to the original four administrative sections.

The Administrative Sections

The full-time members of each section consist of a president and six or seven Conseillers d'Etat. In addition, at least one Conseiller from the Section du Contentieux must be attached to each administrative section. This was an innovation of the reform of 1963 and is matched by a corresponding attachment to the Section du Contentieux of Conseillers from the administrative sections, the intention being that the different viewpoints of the two sides of the Conseil should be represented in all its activities. Before the reform, only the lower ranks (namely, Maîtres des requêtes and Auditeurs) were formally attached to 'the other side', and this attachment was, in large measure, notional because of the heavy work-load each had to carry in his or her normal job. The present arrangements reflect the insistence since 1963 on the principle of *brassage* (see p. 60).

The four sections bear their traditional names of finance, home affairs, public works, and social (Section des finances, Section de l'intérieur, Section des travaux publics, Section sociale). Of the four administrative sections at the time of the 1963 reform, the sections for home affairs and finance date back to 1799, the section for public works was set up in 1852, and the social section finally emerged in 1946. Their spheres of interest are wider than these names suggest, as between them they span all the various government departments. In particular, they have wide powers of supervision (*tutelle*) over more important aspects of local administration.[8] Contested expropriation orders (based upon a *déclaration d'utilité publique*), orders changing the boundaries of a commune, the fixing of a *cahier des charges* for a public undertaking established by a commune as a monopoly, or for many types of mixed company in which communes or *départe-*

---

[8] See M. Rendel, 'How the Conseil d'Etat Supervises Local Authorities' [1966] PL 213.

ments may hold shares, are but a few examples of the many matters in this field that may be considered by the administrative sections. They also supervise various aspects of the operation of *fondations* (charities), which in France enjoy important legal advantages if recognised as *d'utilité publique* by a *décret en Conseil d'Etat*. The allocation of work between the administrative sections used to be entirely a matter of internal arrangement, but the reform of 1963 now gives the ultimate control in this matter to the Prime Minister.

Texts of proposed legislation and requests for advice are examined according to a procedure which is the same in all the four sections. Each text is allocated to a member of the section as *rapporteur*. In contrast to the practice of the Section du Contentieux, senior as well as junior members serve as *rapporteurs* in the administrative sections, their role being of capital importance. The duty of the *rapporteur* is to make a preliminary study of the text, paying particular regard to three factors. The first is its legality; in particular, the *rapporteur* will consider whether a text infringes the Constitution as interpreted by the Constitutional Council. The second is its *opportunité*, that is to say, its general merits and suitability as a means of giving legislative expression to what has been for the ministry an *option politique*—the political choice itself not being a matter upon which the Conseil d'Etat may presume to comment. But matters of substance (such as excessive cost, administrative complications, and difficulties of control) will be taken into account by the *rapporteur*. And the third is the actual drafting (*rédaction*) as a question of pure technique and style.

The *rapporteur* will then take up with the ministry any difficulty or objection. The relevant official or officials may be invited to discuss the text with the *rapporteur* at the Palais-Royal. This process is very informal and matters may sometimes be clarified over the telephone. Often *rapporteurs* acquire sufficient authority and experience on certain matters to eliminate most difficulties by persuasion, even before the text comes before the Conseil. The *rapporteur* then prepares a brief comment, to which is attached his own suggested revision of the text, few texts being incapable of improvement at the hands of a diligent *rapporteur*. By dint of hard and concentrated effort, he usually has completed this part of his task within four weeks. The text will then come before the entire relevant section, composed of the president and some twenty to thirty members.[9] Sessions for this purpose are held in the afternoon, usually at weekly intervals, each section having its own day. Thus, the very busy Section de l'Intérieur meets on Tuesday afternoon at 2.30 in a large elegant room of the Palais-Royal with a bust of Bigot de Préameneu gazing down over the president's shoulder.[10] With a short break for tea, the deliberations may extend as late as 7.00 or 8.00 p.m. As each text is reached, the officials concerned with that text are ushered to seats reserved for them in the Salle de Séance. They are then introduced by name as the *commissaires du gouvernment* concerned with the text in question: those spokesmen for the ministry should not be confused with the officers of the same name but of wholly different function who form part of the Section du Contentieux (discussed in Chapter 3, p. 46, and Chapter 5, p. 101). Usually, three or four officials attend, but on complex texts involving numerous ministries the number may range higher. With the twenty or more members of the section itself who are usually present at any one time, the gathering is quite large, but section presidents conduct the business speedily, with much skill and tact.

The *rapporteur* will introduce his or her revision[11] of the text by outlining the legislative background of the measure, together with its social or economic significance, where appropriate. He will draw his colleagues' attention to general problems presented by the government's text, and the section president then invites discussion on the broader aspects of the measure. The discussion, as we indicated above, may extend beyond matters of form and drafting to matters of substance, like excessive cost or administrative complications. And those present, including the officials, engage freely in the debate, which is guided from the chair with a very light rein. After general aspects have been considered, the *rapporteur* is asked to take the section through his detailed amendments and these will, article by article, be agreed

---

[9] In 1990 the Section de l'Intérieur consisted of the Section President, 17 Conseillers d'Etat, and 10 Maîtres des requêtes. Of these, 9 Conseillers and all the Maîtres des requêtes were also members of the Section du Contentieux.

[10] Appropriately, for Bigot de Préameneu was one of the four-man commission charged by Napoleon to draft the French Civil Code.

[11] An increasing number of women are now members of the Conseil d'Etat: see Note 27.

upon, frequently with modifications. With long texts such as a bill, this process may take several days of discussion in the Conseil. An English observer must, at this stage, be impressed by the precision of the French language in the hands of such skilful artists in legal drafting. Drafting in committee is notoriously difficult, but the Conseil d'Etat seems to make light of the task. Where a conflict arises, the art—of which section presidents appear past masters—is to choose the right moment to take a vote. Voting is limited to Conseillers, except a Maître des requêtes or an Auditeur will vote on a matter for which he acts as *rapporteur*.

The General Assembly of the Conseil d'Etat

Unless discussion on a text has to be adjourned for a further round of consultation and reflection between *rapporteur* and *commissaires du gouvernement*, the section has then completed its work, but the more important categories of texts (principally *projets de loi*) have still to go before the General Assembly of the Conseil d'Etat for approval, usually nine days after discussion in section. This Assemblée Générale is designated as 'Ordinaire' so as to distinguish it from the Assemblée Générale Plénière, that is, an assembly of all those members of Conseil d'Etat with the rank of Conseillers d'Etat.[12] Since the reforms of 1963, the latter has shed much of its business in favour of the former and meets only twelve times a year. The Assemblée Générale Ordinaire is a much smaller body (thirty-five in all), consisting of three Conseillers d'Etat drawn from each of the five administrative sections, a strong representation (13) from the Section du Contentieux, the six section presidents, and the vice-president of the Conseil d'Etat, who presides. It usually meets three times a month. The session begins at 2.30 p.m. and may continue until 7.00 or 8.00 p.m., with a short break for tea.

The procedure before the General Assembly resembles that before an administrative section, although there is a greater degree of formality, and interventions in the debate tend to be short and to the point. Proceedings are now recorded, and this

---

[12] A Maître des requêtes or an Auditeur may attend only in a consultative capacity unless he be *rapporteur* for a particular matter (when he may deliberate and vote on that matter); the same rule applies to meetings of the Assemblée Générale Ordinaire.

innovation is said to have encouraged brevity and relevance. Differences of opinions in the Assembly are resolved by voting on a show of hands.

At the end of the process of consultation, the government department will be faced with a text that may differ in important respects from its own submission to the Conseil d'Etat. Where the decree is one for which consultation is legally required, although the government may still reject the text proposed by the Conseil d'Etat in favour of its own original proposition, it may not introduce a third version. Otherwise this would frustrate the legal requirement for the text to be submitted to the Conseil d'Etat. Rather, the new government version ought in turn to be referred to the Conseil d'Etat for a further round of consultation. Moreover, if the government should defy this principle, and (in the case of a decree but not, of course, a *loi*) a challenge should then be made to the legality of the text which is enacted, the Conseil d'Etat will annul the decree, in its capacity as an administrative court in the ensuing litigation.[13]

A certain flexibility, however, is allowed in the application of this principle. Thus, where the text of a decree consists of a number of articles (as most texts do), then the government may properly combine articles that represent its own original proposals with articles emanating from the Conseil d'Etat. But this is only permissible where the articles are sufficiently severable as to be treated as if they were unrelated and self-contained texts in themselves.[14]

It is very seldom in practice that the government elects to submit to the Conseil d'Etat a text which it is not legally required to submit; where it does so, it has, of course, more freedom to follow or disregard the advice it receives. But the government frequently seeks the legal advice of the Conseil not only on a text but also to help resolve an administrative problem: there were, for instance a total of 149 such *demandes d'avis* over the four years 1986–9, spread fairly evenly between the four administrative sections to which they were referred.

---

[13] This principle was firmly established in the leading case of CE 20 January 1926, LACOSTE, and was applied to a decree under Art. 37 of the Constitution in CE 7 February 1968, MINISTRE DE L'ECONOMIE C. BEAU ET GIRARD.
[14] See CE 16 October 1968, UNION DES GRANDES PHARMACIES DE FRANCE.

STRUCTURE AND MEMBERSHIP OF THE COURTS 69

The Commission permanente *or Standing Committee*
The normal procedure is, in its nature, time-consuming. The Standing Committee is a special device (introduced in 1945) to bypass the normal consultative procedure in cases of urgency. It is for the Prime Minister to designate a bill as 'urgent'. This has the effect of placing the text in question on the agenda of the Standing Committee. While the Standing Committee could in theory refer the text, after considering it, to the General Assembly (in the same way as is done by the four administrative sections), in practice this is not done. The single scrutiny of the text by the Standing Committee is deemed sufficient, for, although it is necessarily conducted with dispatch, it is still thorough. The use of the Committee depends very much on political circumstances. In the first year of *cohabitation* (1986) when the non-socialist government, elected in March, wanted to proceed rapidly with the implementation of its programme, thirty texts were designated as urgent, representing 27 percent of all the texts the Conseil considered that year. In less frenetic times, far fewer texts are referred to the Committee. For instance, in 1989 only six texts (6 percent of all texts) were so referred.[15]

The Standing Committee consists of ten Conseillers, who are chosen annually. One of the section presidents is permanently designated as president of the Standing Committee (in practice, the president of the Financial Section). Since 1963, the lower grades of members of the Conseil d'Etat (Maîtres des requêtes and Auditeurs) have also been eligible to be members of the Standing Committee, and some half-dozen attachments at this level are now regularly made, as they can be very helpfully employed as *rapporteurs*; but they may only vote on a matter for which they act as *rapporteur*.

In their consultative role the four sections, the Standing Committee, and the General Assembly will not be confined to giving advice simply on points of law. Often they will offer advice on the merits of the proposed legislation or other administrative action. What if the legal advice given by an administrative section turns out to be wrong in law? In the case of statutes, there may well be a divergence in interpreting the Constitution between the Conseil d'Etat and the Constitutional Council, in

[15] In *Etudes et Documents No. 41, Rapport Public 1989*, 112.

which case the latter's view prevails. This happened in 1981–2 on the issue of the amount of compensation to be paid under the Law on nationalizations. But such divergences of interpretation of the Constitution are rare. In other cases the Section du Contentieux may be called upon to annul the decree or to declare illegal the administration decision to which the administrative section (or General Assembly) wrongly gave its blessing, and it will not hesitate to do so. In other words, no estoppel operates between the two sides of the Conseil d'Etat. The *option politique* (as we have seen) is not a matter within the competence of the Conseil in its advisory capacity; but as a disagreement on administrative rectitude may appear, in the eyes of the public, to carry political implications, it is a rule that the Conseil's advice can only be published by the government or by the minister concerned. For the same reason, the meetings of the administrative sections or of the General Assembly or the Standing Committee are private and confidential.

French and British Drafting Compared

Because of this drafting supervision by the Conseil d'Etat, France has not felt the need for a British-style specialized office of Parliamentary Counsel but leaves the initial drafting to legal staff (which abound) in the various ministries. However, important texts, including all *projets de loi*, have to be filtered through the Secrétariat-Général du Gouvernement, which has been likened to the British Cabinet Office but which has its own legal experts. This will convene interdepartmental meetings to debate the text and to field a team of *commissaires du gouvernement* to attend before the relevant administrative section (as described above).

M. Braibant, president of the Report Section, sees two main differences between the British and French system of drafting legislative texts. First, parliamentary counsel operate 'backstage' and draft the whole text *individually*, whereas the Conseil d'Etat acts *collegiately* and at a later stage, amending or rejecting a text which has already been drafted by officials within the relevant ministry. The second difference is the stress in Britain on parliamentary power to amend, whereas the French place stress on the review power of the courts—Conseil d'Etat, Constitutional Council, and Cour de Cassation. In his view, at either end of the Channel Tunnel, the results are not good: despite the refined procedures and controls, the texts emerge defective—too many, too often changing, too detailed, and too difficult: in brief, he sees more and more texts of poorer and poorer quality.[16]

There is a marked contrast in the output of texts. In 1984, as a typical year, British primary and subordinate legislation amounted to some 2,190,000 words, whereas the French total was 927,000 words: significantly, the *lois* in that year were, in total, between one quarter and one fifth of the length of the British statutes.[17] The explanation lies in the British search, by detailed (even convoluted) drafting, for absolute precision, leaving no ambiguities for the judge, whereas the French pursue, above all, intelligibility with concision. It is, however, worth noting that many detailed provisions, which would be found in a British statute, are contained not in a *loi* but in *décrets d'application*, which are never subjected to parliamentary scrutiny or approval.

The Judicial Role of the Conseil d'Etat: The Section du Contentieux

It is the fifth section of the Conseil d'Etat—the Section du Contentieux—that is responsible for the judicial work of the Conseil.

Consisting of about one hundred members, the Section du Contentieux is subdivided into ten Sous-sections. Since 1980 (reverting to an earlier practice) each of the Sous-sections has been competent to proceed independently to deal with a case and pronounce judgment thereupon in the name of the whole Conseil d'Etat. Normally, as we shall see, two Sous-sections combine in reaching a decision upon a case.[18] For important cases, an ampler tribunal is constituted to represent either the whole Section du Contentieux or, exceptionally, all six sections: the latter is known as 'l'Assemblée du Contentieux' (or simply

---

[16] Observations of M. Guy Braibant in Sir William Dale (ed.), *British and French Statutory Drafting* (IALS, London, 1987), 135–6, which work is based on the proceedings of a Franco-British Conference of April 1986; see review-article Brown, (1988) 37 *ICLQ* 696–701, For a less critical view of the French system, see Ducamin, 'The Role of the Conseil d'Etat in Drafting Legislation'.

[17] Figures provided by Sir William Dale, *British and French Statutory Drafting*, 166–9.

[18] In important tax-cases, all three tax Sous-sections may combine to form the *plénière fiscale*.

'l'Assemblée') and the former as 'la Section'. For the composition of each, see p. 73.

The ten Sous-sections share between them the various categories of case addressed to the Conseil by aggrieved citizens. Where, as is still usual, two Sous-sections combine upon a case, the one acts alone in the preliminary investigation of the matter (which is termed *l'instruction de l'affaire*) being joined by the other only at the final stage of judgment. The allocation of a dossier to a particular Sous-section for *instruction* is the responsibility of the president of the Section du Contentieux. Usually this a routine matter which he delegates to the registry or *greffe* of the Conseil. The seventh, eighth, and ninth Sous-sections deal only with fiscal matters (*le contentieux fiscal*), amounting to 20 to 25 percent of cases, leaving the first six to share what is referred to as *le contentieux général*; the tenth Sous-section (created by decree in 1979) is treated as a 'troupe de choc' available to help any of the others, but also capable of judging alone.

Each Sous-section consists of a president, who is always of the rank of Conseiller, two other *assesseurs-réviseurs* (also Conseillers in rank),[19] and up to six other Conseillers, and some six or seven Maîtres des requêtes or Auditeurs, one of whom will act as *rapporteur* for each case. When two Sous-sections combine as the normal organ of judgment, there are usually present the three Conseillers from each Sous-section, one representative of an administrative section, the *rapporteur* for the case, and the section president (or one of the three vice-presidents). The senior Maître des requêtes of the two Sous-sections may be added, if necessary to obtain the uneven number which the law requires. This would give a deciding body of nine persons. Often, however, a smaller number is present; but this must not be less than five.

Under a Decree of 10 January 1980, it is now more usual in the less important cases for a single Sous-section to proceed to judgment in the name of the Conseil (see p. 106). In 1989 1,845 cases were judged by a single Sous-section, out of a total of 8,237 judgments; one may compare for the same year the figures of 3,410, 55, and 43 as the cases judged by combined Sous-sections, the Section du Contentieux and the Assemblée du Contentieux respectively (see *Etudes et Documents, No. 41* (1990), p. 67).

---

[19] Usually simply styled *réviseurs*.

In addition, some 1,800 cases were determined by decisions of the president of the Section or of a Sous-section; presidents are vested with powers to issue both interim and final orders, e.g. to reject an appeal as clearly outside the Court's jurisdiction ('un recours entaché d'une irrecevabilité manifeste'). In terms of the limited scope of these powers, this represents only a minor departure from the principle of collegiality of decision-making.

Where cases are deemed of sufficient importance to be passed for judgment to the whole Section du Contentieux or to the Assemblée du Contentieux, the composition of the tribunal is altered. Thus, where the Section is convened, it comprises the section president and three vice-presidents of the Section du Contentieux, the ten Sous-section presidents, two Conseillers brought in from the administrative sections, and the *rapporteur*. Where, more rarely, the Assemblée is convened, the number drops from 17 to 12: the president and three vice-presidents of the Section du Contentieux, the president and *rapporteur* from the relevant Sous-section, the five presidents of the administrative sections, and the vice-president of the Conseil d'Etat, the latter presiding and having a casting vote.

A common and essential feature of this organization is that each judgment is the product of a collegiate examination; the joint and several responsibility of each formation in the final decision is enforced by the rule of the *secret du délibéré*: no member of the Conseil may comment on the position taken by himself or his colleagues in reaching any particular decision.

The Section du Rapport et des Etudes

Another object of the reforms of 1963 was to give to the government the benefit each year of the considered and collective views of all the sections of the Conseil d'Etat on current problems of administration. These views were to be the conveyed by the medium of an Annual Report, the drafting of which was entrusted to a standing Report Commission (originally, *la Commission du Rapport*, renamed in 1975 *la Commission du Rapport et des Etudes*, and elevated to *la Section du Rapport et des Etudes* in 1985). Through the Annual Report, the Conseil d'Etat was to add a third function to its traditional consultative and judicial tasks, for it might actively initiate reforms rather than merely react passively to proposed government legislation or exercise judicial control after the event. Informed commentators have referred to the Conseil exercising through the Report a quasi-initiative in legislation.[20]

Thus, the Report Commission in its Annual Report for 1969–70 produced a closely reasoned study of 98 pages on the reform of *établissements publics*, a protean concept in French law that embraces all bodies providing a public service, from a local authority to a learned society. In the same Annual Report there appeared a far-sighted study of the consequences of 'l'informatique' upon the rights of the individual and upon administrative decision-making.[21] In recent years, however, there has been a change in practice: substantial studies, of which there have been many,[22] do not generally appear in the Annual Report but are presented without delay to the government. Instead, especially since 1984, the Report presents the general reflections of the Conseil d'Etat upon imperfections in administrative law-making and organization and practical proposals for their correction. The Report then surveys, in no less reflective fashion, the year's activities of each of the six Sections of the Conseil, complete with statistics.

Since 1988, as we have seen, the Report is made public in the annual number of *Etudes et Documents*. This publication, true to its title, follows the Report with a number of studies on administrative-law topics (often from a comparative viewpoint) and concludes with a selection of documents, e.g. the issues for 1989 and 1990 included several *avis* of the Conseil requested by the government on difficulties arising in administrative matters (for an example see Appendix J).

[20] M. Letourneur, J. Bauchet, and J. Méric, *Le Conseil d'Etat et les Tribunaux Administratifs* (Paris, 1970), 68.
[21] 'L'informatique' is a useful abstract noun to describe generically all processes of accumulating data through the use of computers.
[22] Some 50, according to MM J. Massot and J. Marimbert, *Le Conseil d'Etat* (Paris, 1988), 272, where many of them are listed: e.g. 'Obstacles to Administrative Deconcentration' (1969–70); 'The Impact of New Information Technologies upon Administrative Procedures' (1987); 'Untruthful Advertising in Business' (1974–5); 'Bio-medical Science and the Law' (1988); 'The Penetration of European Community Law into French Law' (1981–2). For such studies, the Report Section may enlist the help of experts from outside the Conseil d'Etat to constitute a working group not unlike a British Royal Commission. The eventual report has to be adopted by the Report Section before transmission by the vice-president of the Conseil to the Prime Minister.

The work of the Report Section (as we will term it) is not subject to formal regulation by law, but it may be treated under three heads. We have already referred to its general studies of pressing problems of administration and government and its responsibility for the Annual Report, which is now published under the title of *Rapport Public* in the series *Etudes et Documents*. Its second, more mundane task is to provide annual statistics of the Conseil d'Etat in considerable detail: these too are reproduced in *Etudes et Documents* and are a mine of information for researchers.

The third and more practical activity of the Report Section, and one placing a heavy burden on its limited resources of personnel, is 'to monitor the execution of judgments' against the administration (Article 58 of the Decree of 10 July 1963). The 1963 reform sought in this indirect way to meet the then common complaint of successful litigants that no modes of execution (*voies d'exécution*) lay against the administration: a merely declaratory judgment in favour of the citizen might prove a Pyrrhic victory if the administration would not, or could not, comply. This deficiency in administrative justice was largely repaired by the Law of 16 July 1980. This introduced a number of remedies to compel compliance with judgments, in those cases (a minority) where the Report Section is unsuccessful in its efforts to produce a voluntary settlement by the administration acceptable to the judgment-creditor, the citizen. This important activity of the Section will be returned to in the next chapter (p. 110).

Viewing its activities as a whole, we can understand why the current president of the Report Section, M. Guy Braibant,[23] has described its work as extending the influence of the Conseil d'Etat in two directions. On the one hand, the Section operates in advance of the consultative process of the administrative sections, for its studies seek to anticipate administrative problems and offer solutions; on the other hand, its monitoring function over the execution of judgments against the administration is subsequent, chronologically, to the work of the Section du Contentieux. We referred earlier to the Report Section as the

---

[23] It was M. Braibant who, in 1961, first drew attention to the practice of the Egyptian Conseil d'Etat (itself of French inspiration) of presenting a report to the head of state on legislative imperfections and problems of executing judgments (see reference in Massot and Marimbert, *Le Conseil d'Etat*, 49.

'think-tank' of the Conseil;[24] it may be that this sixth and latest Section may, in the long term, have the predominant influence upon the development of French administrative law.

3 THE MEMBERSHIP OF THE CONSEIL D'ETAT

It is essential to an understanding of the Conseil d'Etat to appreciate that it is itself part of the French administration and staffed entirely by civil servants. We have already seen how the administrative sections play a central role in the process of legislation and as advisers to government departments.

At the same time, within the Conseil, a separation had grown up between those members involved in advising the administration and those who sat in judgment upon the administration. This separation was reflected by the early division of the Conseil into the administrative sections and the Section du Contentieux. Because of this physical division within the actual membership of the Conseil, the Conseil d'Etat acting in its judicial capacity (*statuant au contentieux*) had been able to overcome the charge of being judge in its own cause. It was this division that made acceptable the paradox of one organ of the administration being singled out to be judge over the administration.

Step by step, from the origins of the Conseil d'Etat down to the reform of 1963, the Section du Contentieux had progressively detached itself both from the administrative sections of the Conseil and from the 'active administration' outside the Conseil. Until 1963 commentators emphasized that, although all members of the Conseil worked alongside each other in the Palais-Royal, sharing the same library and meeting formally together in General Assembly, yet the Section du Contentieux had achieved independence from the rest of the Conseil and the 'active administration' in the field. That this was indeed so could be effectively demonstrated by the body of case-law which the Section du Contentieux had constructed and under which the administration had been made to conform to what the Section considered proper standards of good administration.

---

[24] And of the government: 'le bureau d'études du gouvernement' comments M. Braibant in 'Les nouvelles fonctions du Conseil d'Etat' (1987) *Revue administrative* 415–21, an authoritative account of the Report Section by its president.

The reform of 1963 was undoubtedly inspired by the idea of applying a brake to this process. It was intended by the government to be a reminder to the Conseil d'Etat *statuant au contentieux* that it was not a branch of the judiciary but remained part of the administration. To this end the reform deliberately sought to reinforce collaboration between the administrative sections and the Section du Contentieux. Thus, twelve Conseillers from the Section du Contentieux (as well as its president) must attend the Assemblée Générale Ordinaire (to which the reform gave rise); and all the Conseillers, except those with presidential responsibilities, do double service upon both the Section du Contentieux and one of the administrative sections. Again, at least one Conseiller from an administrative section must attend when two Sous-sections sit to decide a case; this is increased to two when a case is remitted for judgment to the Section du Contentieux, and we have seen above that judgment by the Assemblée du Contentieux brings in all five presidents of the administrative sections as well as the vice-president of the Conseil.

It is possible indeed to exaggerate the degree of separation of the Section du Contentieux which prevailed before 1963. At bottom the Conseil d'Etat *statuant au contentieux* remained and remains part of the administrative machinery of the French state, although a highly specialized part. This very fact has helped to make the judicial control which it exercises more readily acceptable to the official. The Conseil d'Etat commands the general respect of the administrator in action because he knows his judges are fully aware of the special problems besetting public administration.[25] When called to account by the Conseil d'Etat *statuant au contentieux* he has to acknowledge that his judges are not strangers to the administrative process; they are not amateurs throwing legalistic spanners into the administrative works, which is how British ministers have sometimes tended to regard their High Court judges. On the contrary, the French official is well aware that his judges are peculiarly expert in the field of administration.

The administrative expertise derives from a number of sources. In the first place, the juxtaposition of the Section du Contentieux and the administrative sections within the Conseil

---

[25] They are 'immersed (*plongés*) in an administrative atmosphere' in the phrase of Maurice Hauriou (1856–1922).

d'Etat engenders a mutual awareness of the problems on either side. The interchange of views takes place both informally, in the day to day contacts of the members of the different sections, and formally, in the General Assembly of the whole Conseil. This last, as we have seen, meets in most weeks in its smaller, 'ordinary' form, or twelve times a year in its plenary form.

In the second place, the reform of 1963 has deliberately provided for the attachment of personnel from one side to the other side of the Conseil's activities. The admixture of members from the administrative sections cannot but widen the experience of the body sitting in judgment on the administration. Moreover, the active participation of members of the Section du Contentieux in the work of the administrative sections must sharpen their appreciation of administrative problems. Nor is this likely to endanger the traditional independence of the Section du Contentieux; for at the present day members of the Conseil do not see themselves as 'administrative' or 'contentieux', and someone who has spent his early years in an administrative section will react exactly like his colleagues when he comes to sit as a judge in the Section du Contentieux.

In the third place, the Conseil d'Etat has been described as the 'seed-bed' of the administration, by which is meant that many of its members are transplanted or detached to occupy a wide variety of posts in the 'active administration'. In the 'cabinets' of ministers or on governmental missions abroad they gain valuable experience to enrich their work when they return to the Palais-Royal, besides spreading respect for the body from which they come.

The most important factor, however, in ensuring administrative expertise in the Conseil d'Etat is the method by which its members are recruited. This merits consideration in some detail.

4 RECRUITMENT TO THE CONSEIL D'ETAT

There are two distinct avenues of access to the Conseil d'Etat: by examination and by invitation.

Most members are recruited from the National School of Administration (l'Ecole Nationale d'Administration), which was founded by the Provisional Government of General de Gaulle in 1945, to serve as a graduate staff college for the higher ranks of the administration (corresponding to the old administrative class—now 'administrative trainees' of the British Civil Service). Admission to l'ENA (as it is popularly called) is by a stiff *concours*, or open competitive examination, one being conducted for recent graduates of universities and other comparable institutions (only a minority being law graduates), and a second for those who are already members of the civil service. After two years of intensive studies, the outgoing class is arranged in order of merit, according to their performance in the final examination and over the course as a whole. Depending on this placing each successful graduate from l'ENA then chooses from among the administrative posts which happen to be available at the time. It is significant that the Conseil d'Etat shares the honour with the Cour des Comptes and Inspection des Finances of attracting each year those with the highest placings.

The double sieve imposed by the *concours* on entry and the placing at the end of the course guarantees that entrants to the Conseil by way of l'ENA are necessarily of the highest intellectual quality. In addition, the nature and content of their strenuous course at l'ENA ensures that they have a thorough training (both theoretical and practical) in the field of public administration. Once a post in the Conseil d'Etat has been chosen, the newcomer (with the rank of *Auditeur de seconde classe*) will have the help of some personal supervision from a more senior member of the Sous-section of the Section du Contentieux to which he or she is assigned.

The other method of recruitment is by way of the 'active administration'. It is a long-standing practice to recruit about a quarter of the entrants to the Conseil d'Etat 'from outside' (*au tour extérieur*), that is, from the ranks of those who have already distinguished themselves in the practice of public administration. Thus, a prefect or subprefect may be offered the opportunity of quitting the *corps préfectoral* to enter the Conseil. Recruits in this second category will necessarily be considerably older than those in the first and will usually enter at the higher levels of Conseiller or Maître des requêtes. Currently, one Conseiller out of every three and one Maître des requêtes out of every four must be recruited externally.

This mixed system of entry provides the Conseil with a remarkable combination of young intellect and mature experience.

It ensures that the Conseil has within its ranks both theoretical and practical expertise in public administration. And it has proved a highly successful recipe. Not surprisingly, therefore, recruitment to the regional Tribunaux Administratifs and Cours Administratives d'Appel has tried to follow much the same pattern, although the need rapidly to expand their personnel has led in recent years to greater resort to recruitment from beyond the civil service and certainly beyond the ranks of the 'Enarques' (as graduates of l'ENA style themselves)[26].

5 THE CAREER STRUCTURE WITHIN THE ADMINISTRATIVE COURTS

Once admitted to the Conseil d'Etat by one or other of the above routes, the new member has entered upon a career in which the rest of his (or her) working life will be spent.[27] Those recruited *au tour extérieur* do not normally return to their previous vocations but follow the same career pattern as their colleagues.

A special position, however, is held by the twelve 'Conseillers en service extraordinaire', who participate in the activity of the Conseil only for a limited period of time (currently four years). In this way, distinguished jurists, administrators of public enterprises, industrialists, or trade-union leaders have been associated in the advisory tasks of the Conseil: they play no part in the Section du Contentieux.

It remains to say something of the structure of a career, first, within the Conseil d'État and secondly, within the two lower tiers of administrative courts.

Conseil d'Etat

Membership of the Conseil is divided into three basic grades: Conseiller (the highest grade); Maître des requêtes (the intermediate grade); and Auditeur (the lowest grade or 'Auditorat',

---

[26] The indignation of many Enarques has been aroused by the government's decision, in November 1991, for the National School to be moved to Strasbourg from its present prestigious site in the heart of Paris.

[27] Women now represent a significant and growing proportion of the membership of the Conseil: the *Annuaire* for 1990 lists 30 women, (8 Conseillers, 16 Maîtres des requêtes, and 4 Auditeurs); two Sections have women presidents: Mme Questiaux (Section des travaux publics) and Mme Grévisse (Section sociale).

which is in turn subdivided into Auditeur de première classe and Auditeur de seconde classe). There are in addition certain posts of special responsibility, such as Commissaire du gouvernement, Président de Sous-section, Président (and Vice-Président) de Section, and (the supreme position) Vice-Président du Conseil d'Etat. The Président du Conseil d'Etat is the Prime Minister, represented by the Minister of Justice on the rare occasions when his right to preside is invoked.

Most members of the Conseil are at some time or other detached or seconded to posts in the active administration for limited periods of time. In his individual career a member of the Conseil may start as *rapporteur* in the *contentieux* with a part-time participation in an administrative section; next he may become a Commissaire du gouvernement or be called to special duties in, for example, the Prime Minister's office; he may then leave the Conseil to serve as director in a ministry, returning in the end to the normal function of a Conseiller d'Etat. An interesting account of a typical career is contained in the autobiography of M. Georges Maleville: *Conseiller d'État: Témoignagne* (Paris, 1979).

Members of the Conseil are civil servants (*fonctionnaires*) with the usual safeguards which French law confers in matters of promotion and discipline. In matters of discipline, the reform of 1963 has provided a number of new safeguards, but members of the Conseil still lack that status of irremovability (*inamovibilité*) which is the treasured privilege of the French judiciary and, paradoxically, has now been conferred upon the members of the lower tiers of administrative courts (see below). In practice, however, it is unthinkable that a member should be dismissed or otherwise disciplined by reason of political considerations.[28] Indeed, the government of the day had to accept, with such grace as it could muster, decisions as troublesome as TREBES (CE 4 March 1949, where the entire career-structure of a

---

[28] One quasi-exception occured in November 1960, when André Jacomet, a Maître des requêtes on detachment to a senior administrative post in Algeria, criticized the Algerian policy of the French government at a meeting of high-ranking officials. He was permanently relieved of his post and dismissed from the Conseil d'Etat by the Gaullist government—only to be reinstated in the Conseil 8 years later by the decree of 30 July 1968 (see *Le Conseil d'Etat 1799–1974* (Paris, 1974), 900). But in earlier times, compare the extensive purge of the Conseil d'Etat in 1879: p. 5 above.

ministry was pronounced invalid), BAREL (CE 28 May 1954, discussed in Chapter 9), and CANAL (CE 19 October 1962, discussed in Chapter 3). Although this last case did prompt some important reforms in the institution of the Conseil d'Etat as a whole, any action against the individual members would, of course, have been quite unthinkable.

Promotion depends, in practice, upon seniority of service. This principle of automatic promotion is regarded by members of the Conseil as the essential guarantee of their independence.[29] The only exceptions are the six section presidents and, of course, the vice-president of the Conseil: their nominations are truly matters of choice. Promotions (where not automatic), assignments within the Conseil and outside secondments, and other purely administrative matters are discussed by an administrative unit within the Conseil known informally as 'le Bureau' and comprising the vice-president and six section presidents, with the secretary-general in attendance.

The Bureau should be distinguished from the 'Commission consultative'. Joint consultative committees of 'management' and rank-and-file personnel are a general feature of the different branches of the French public service. This 'paritary' principle, as applied to the Conseil d'Etat, has produced a committee of twelve, consisting, on the one side of the six section presidents and, on the other, of six elected members (two Conseillers d'Etat, two Maîtres des requêtes, and two Auditeurs). In addition, the vice-president presides. Under the relevant texts, this committee 'may be asked for advice on all problems concerning the organization and functioning of the Conseil'; 'may be consulted on all questions concerning the status of members of the Conseil'; and 'must give its advice on individual measures concerning the discipline and promotion of members' as well as in certain other cases specified in the texts.

It is always dangerous to compare salaries of different countries, and this is especially so in the case of France with its (by British standards) massive family-allowances. Nevertheless, it is perhaps worth noting that the average salary of a Conseiller d'Etat is about the same as the highest judicial salaries, namely

[29] For the text of the relevant legislative provision, see Ordonnance 45–1700 of 31 July 1945, printed in *Les grands textes administratifs* (Sirey, Paris, 1970), 524.

STRUCTURE AND MEMBERSHIP OF THE COURTS 83

those of the judges of the Cour de Cassation. At the other end of the scale, a Conseiller of a regional Tribunal Administratif commences with a salary equivalent, roughly, to that of a judge of a Tribunal de Grande Instance, the usual civil court of first instance. It would be misleading to compare these salaries with those (much higher) of the British judiciary. Rather one should equate them with those paid to senior British civil servants. By this standard they appear rather low. On the other hand, a Conseiller d'Etat has a much greater social and intellectual prestige than a high-ranking British civil servant. Nor is he so anonymous a figure. And many of the members of the Conseil are active also in writing and teaching and perhaps as mayors of their local communes, while a chosen few have achieved the highest political offices of state: thus, Georges Pompidou (Maître des requêtes) became President of the Republic, having previously held the office of Prime Minister; and Léon Blum (Commissaire du gouvernement), as well as Michel Debré and Laurent Fabius (Maîtres des requêtes) went on to become Prime Ministers.

Cours Administratives d'Appel and Tribunaux Administratifs

The reform of 1953 gave importance to the members of the Tribunaux Administratifs by making them the normal judges of first instance in administrative disputes. At the same time, from 1953, an enhanced career became possible for members of the Tribunaux. Pay scales were improved; and promotions, at least for a few, could give access to the Conseil d'Etat: one in six Conseillers d'Etat and one in four Maîtres des requêtes were to be recruited *au tour extérieur* from the Tribunaux.

Recruitment to the lower courts resembles that for the Conseil d'Etat: members are recruited from l'ENA, *au tour extérieur*, or by special *concours*. The urgent need for rapid expansion, especially upon the creation in 1987 of the five appellate courts, prompted the holding of special *concours* as well as the use of an *ad hoc* committee to select members for the new courts among civil servants, magistrates,[30] and members of the bar. In the result, by 1990 the judiciary of the lower courts numbered over

---

[30] *Magistrats*, the judicial members of the civil and criminal courts or the prosecution service.

500. But such a rapid enlargement and consequent recruitment beyond the ranks of the civil service has reduced considerably the proportion of those administrative judges who have experience of the active administration.

The three tiers of administrative courts now in place appear to offer good opportunities for a judicial career within this hierarchy. The highest posts in the five new Cours Administratives d'Appel, those of president, carry the rank of Conseiller d'Etat: of the original five presidents of the new courts, two were already members of the Conseil d'Etat; the other three were presidents of important Tribunaux Administratifs. This ranking as Conseiller d'Etat was no doubt intended to boost the prestige of the new courts and to attract outstanding candidates. It will not, however, do anything to counterbalance the worrying trend in recent years for too many of the brightest recruits even to the Conseil d'Etat[31] to depart *en disponibilité* for more lucrative (or more exciting) posts in the French public or private sectors of commerce and industry or in the international job-market.

Members of the Tribunaux Administratifs remained, however, civil servants, under the control of the Minister of the Interior, a legacy of their origin in the Conseils de Préfecture, despite their ambition to be placed under the authority of the Minister of Justice as more becoming their purely judicial functions.

In 1986 their status was transformed and their judicial independence guaranteed. By the Law of 6 January 1986 (Article 1) they achieved the status of irremovability; they cannot be transferred to a new post, without their consent, even by way of promotion. By the same law, a Conseil Supérieur des Tribunaux Administratifs was established; this in turn was extended into a Conseil Supérieur des Tribunaux Administratifs et des Cours Administratives d'Appel by the Law of 31 December 1987, which created the new appellate jurisdiction.

As a result, the members of both the Tribunaux Administratifs and the Cours Administratives[32] have become a single body or

---

[31] The *Annuaire du Conseil d'Etat* for 1982, cited in our 3rd edition (p. 41, no. 1), showed only 7 members *en disponibilité*; the *Annuaire* for 1990 lists 30 (see also App. B). The number *en détachement* remains much the same (49 in 1982, 52 in 1990).

[32] In the Cours Administratives, all members, other than presidents and vice-presidents, are known as 'Conseillers'.

corps under the management, no longer of the Minister of the Interior, but of the Secretary-General of the Conseil d'Etat. The latter sits, *ex officio*, on the Conseil Supérieur. This Council is presided over by the vice-president of the Conseil d'Etat and consists of twelve members: these include three representatives of the administration (including the director-general of the civil service), five elected representatives from members of the Tribunaux and Cours, the Conseiller d'Etat in charge of the permanent commission of inspection of the lower courts (see p. 53), and three persons nominated respectively by the President of the Republic, the President of the National Assembly, and the President of the Senate. The elected and nominated members serve for three years, with possibility of one further renewal.

This Council has considerable powers. First, it makes proposals concerning promotions, nominations as Commissaire du gouvernment, recruitment *au tour extérieur*, transfers, and detachments. Secondly, it must be consulted upon all matters affecting the status of members of the two lower tiers of courts. Thirdly, it acts as the disciplinary body for such members.

The Secretary-General of the Conseil d'Etat, besides providing the secretariat for the Council, is responsible now for the management (or 'judicial administration') of the lower courts; thus, he organizes their registries, staff training, and funding; he also undertakes for the Council studies of the functioning and procedure of the lower courts; and from 1989, he prepares the budget for these courts with the help of the Ministry of the Interior. This budget, like that of the Conseil d'Etat itself, is then incorporated as a separate chapter in the fiscal estimates of the Ministry of Justice, which (with those for the other ministries) are consolidated into the draft finance-bill for presentation each year to parliament. In this respect, the administrative courts have now come to be formally treated in the same way as the ordinary courts and with the same guarantees of independence.

5

The Procedure of the Courts

1 INTRODUCTION

The procedure before the French administrative courts is quite unlike any court procedure in England and differs also from that before the civil courts in France. This is principally because it shares the characteristic of French criminal procedure in being 'inquisitorial' rather than adopting the 'accusatorial' method of the English criminal process or the 'adversary' procedure of our civil trial. In other words, the court takes upon itself the task of finding out the facts, not being content to decide the case on the facts as established by the parties. Indeed, the court will also pursue an independent investigation of the law, again not being content or constrained (as in England) to rely upon the investigations and arguments of counsel.

In a sense, however, the procedure is also adversary. For it is, in French terminology, *contradictoire*, a characteristic common to French civil procedure and denoting that each side must be given an opportunity of contradicting what the other party has said. One may compare, in England, the exchange of pleadings before trial or the general principle of natural justice that both sides should be heard.

To these characteristics of 'inquisitorial' and *contradictoire* must be added a third. The procedure is essentially a written one. Again, this contrasts sharply with the oral nature of English procedure, which is carried to its extreme in a criminal trial. In France, both the civil and criminal process make much greater use of written procedures, so that the importance of the *dossier* before the administrative courts is nothing strange to French eyes. What does appear unusual, however, is the almost complete absence, at least in the Conseil d'Etat, of oral argument by counsel. Before the lower administrative courts (Tribunaux Administratifs and Cours Administratives d'Appel), the rules of procedure provide that the parties may, either in person or by counsel, submit oral observations confined to matters already raised in the written stage of the procedure. The observations are usually short but may sometimes clarify a point of fact, especially if the president of the court questions the parties. It is in the administrative courts, unlike the civil courts, that one encounters the written 'brief', or summary of authorities and argument, on the lines extensively developed in the United States but almost unknown in England.[1]

The fourth characteristic of administrative procedure is the importance of the role played by the Commissaire du gouvernement. Where both the administration and the complainant are represented by counsel, the case assumes the aspect of a three-cornered debate quite unlike the duel-like confrontation of English proceedings—an aspect found also, of course, in the two courts of the European Communities in Luxembourg.[2] Only in those rare cases in England where counsel appears for the Queen's Proctor or for the Official Solicitor, or as amicus curiae, is there a pale shadow of this 'triangular' characteristic of French administrative procedure.

It is now proposed to examine chronologically the normal sequence of proceedings in an administrative case. Although the vast majority of proceedings begin in the Tribunaux Administratifs, the focus will be mainly on the procedure in the Conseil d'Etat, which provides both the pattern and the most developed form of administrative court procedure.[3]

Proceedings in the administrative courts may be divided into four stages: (*a*) commencement of proceedings; (*b*) *instruction*; (*c*) judgment; (*d*) execution.

2 COMMENCEMENT OF PROCEEDINGS

Let us take for example an application to a Tribunal Administratif to quash the refusal by a local commune to grant planning

---

[1] Skeleton arguments used in English civil appeals are more schematic versions of what would be contained in a written brief. On the whole, French written submissions are less lengthy than American briefs.

[2] But, as we shall see, the Commissaire du gouvernement, like the Luxembourg Advocate-General, is much more than a third counsel in the case.

[3] Yet, paradoxically, the new Code des Tribunaux Administratifs et des Cours Administratives d'Appel (Code des TA-CAA) (1989) now provides an elaborate code of procedure for the lower administrative courts, whereas the Conseil d'Etat lacks such a single text for its own procedure.

permission to Mme Dupont for the construction of a garage adjoining her house.

The applicant's success in mounting a case in the first place depends on her ability to specify the mistake made by the commune. To this end, two features of French law are of significance. In the first place, there is often a legal duty to give reasons. Although the French, like the English, have long accepted the principle of *pas de motivation sans texte*, the practical situation has been altered profoundly by the Law of 11 July 1979. Article 1 states that:

> Individuals or legal persons have the right to be informed without delay of the reasons for individual administrative decisions which affect them unfavourably. To this end reasons must be given for decisions which: restrain public liberties or constitute a regulatory measure; impose a penalty; subordinate the grant of an authorization to restrictive conditions . . .; withdraw or restrict rights; set a time limit or foreclosure; refuse a benefit to which a person has a right if the legal conditions are met.

Thus, where legal rights are infringed, there will always be a duty to give reasons. In other cases, there may be other provisions requiring reasons to be given, as in the *Code de l'Urbanisme*. Where the administration provides explicit reasons for a decision, the citizen is better placed to challenge it.

In the second place, there is a general right of access to administrative documents. A wider access to official documents can ensure that citizens know the general basis on which the administration ought to proceed or can fill in the background of individual decisions. The Law of 17 July 1978 gives a right of access to any person, on demand, to files, reports, studies, minutes, statistics, directives, circulars, and so on, held by the administration. In addition, individuals may request documents which concern them personally. If the administration refuses the request, they may refer the matter to the *Commission d'accès aux documents administratifs* (CADA), which will then give its advice upon whether the document is, or is not, one to which they may have access. Thus our plaintiff might obtain copies of the district plan (the *plan d'occupation des sols*) and any special reports on her zone within it in order to establish that no possible harm to the environment would be caused by her building a garage.

THE PROCEDURE OF THE COURTS 89

With the help of such information, and any further information volunteered by the administration, our plaintiff can formulate a *recours*. She must lodge it with the court (in this case the local Tribunal Administratif) in a letter specifying the basic grounds of complaint. The *recours* must be written or typed. Lodging a *recours* by telegram, telex, or FAX (but not by telephone or Minitel) is also permitted.

The complaint itself does not have to be formulated in any precise manner, but it must at least contain a short statement of the facts, of the *moyens* or legal grounds on which the case is based, and of the *conclusions* or the actual relief which is being sought, e.g. for annulment of the administrative decision. The *recours* must then be signed by the plaintiff or her lawyer.[4]

In principle, a plaintiff must be legally represented in presenting a *recours*.[5] Given the written nature of proceedings, it is argued that presentation of the case by a lawyer will make the job of the court easier. All the same, this argument of the court's convenience gives way to the concern to ensure that litigants are not seriously impeded in presenting major complaints against the administration. Thus all *recours* based on grounds of *excès de pouvoir*, and all proceedings concerning elections, pensions, or taxation, as well as most highway matters in the lower courts, do not require legal representation. Indeed, it might almost be said that in the lower administrative courts (the Tribunaux Administratifs and the Cours Administratives d'Appel), legal representation is only required when a pecuniary claim is made against the state or any public body or local authority based on contractual or non-contractual liability. Before the Conseil d'Etat, the requirement of legal representation will apply mainly to pecuniary claims and *recours en cassation*. Many plaintiffs do, of course, elect to employ counsel, but a substantial minority are content to leave the case to be investigated by the administrative courts.

---

[4] Strictly, the document lodged by the *requérant* to commence proceedings is called a *requête*, except where a minister initiates proceedings (as, for example, by way of appeal from an adverse decision in a Tribunal Administratif), in which case the document is correctly termed a *recours*. But *recours* is commonly, if loosely, used to describe any document to begin proceedings, perhaps because it also denotes the actual process itself. Some examples of typical *requêtes* are included in Appendix H.
[5] Code des TA-CAA, Arts. R108 and R116; Ordonnance 31 July 1945, Art. 41.

It should be noted that the state is not required to be legally represented. Its officials, it is assumed, can be trusted to present a case in an organized way. On the other hand, legal representation is required for *départements*, regions, and other local authorities and *établissements publics* (for this term, see p. 74).

Before the Tribunaux Administratifs and Cours Administratives d'Appel legal representation is either by an *avocat*, who is not necessarily, and in practice rarely, an *avocat aux Conseils*. Where counsel is employed before the Conseil d'Etat, he must belong to the specialist bar of sixty members (or partnerships[6]) who constitute the Ordre des avocats aux Conseils and form a closed corporation with a monopoly of legal representation before the Conseil d'Etat and the Cour de Cassation, the two supreme courts of the French legal system, as well as before the Tribunal des Conflits, the Cour des Comptes, and the Prize Court (Conseil des Prises). Like the notary and certain other specialized French lawyers (but unlike ordinary *avocats*), they hold an 'office ministériel', by which is meant that they are entrusted by the state with a public office (their formal appointment is by the Minister of Justice), although they practise without interference and are remunerated by their professional fees.

Where legal representation is not obligatory, the *recours* and submissions may be drafted (but not signed) by either an in-house lawyer not able to practise in the courts or by another professional adviser (e.g. an accountant in tax cases). In this respect, the procedure is closest to that before the French commercial courts.

No court fees are payable on the lodging of a *recours*, and therefore the plaintiff in a *recours pour excès de pouvoir*, not being required to employ an advocate, is not obliged to incur any substantial legal expenses. A system of legal aid is available for poor litigants before all the administrative courts.[7]

---

[6] The right of lawyers to form partnerships was introduced by the Law of 16 September 1972. It increased the number of *avocats aux Conseils* practising on their own account or in partnership to 79 in 1989; it will be seen therefore that the limit of 60 applies to practices. See P. Herzog and B. E. Herzog, 'The Reform of the Legal Professions and of Legal Aid in France' (1973) 22 ICLQ 462.

[7] A major reform of legal aid and access to justice has been effected by the Law of 10 July 1991. Income limits for full and partial legal aid are respectively 4,400 and 6,600 francs per month (Art. 4). Various *décrets d'application* will improve access to justice before all French courts, including the administrative courts (at all levels).

THE PROCEDURE OF THE COURTS 91

The *recours* must be accompanied by a certified copy or copies, the number of copies being determined by the *greffier* (registrar): usually a single copy suffices. The copy will be needed to start the duplicate dossier of the case required for the 'contradictory' procedure, the original dossier being made available to the parties or their lawyers, while the duplicate remains in the custody of the court. The plaintiff must also annex to his *recours* a certified copy of the administrative decision attacked (such as a letter from the administration rejecting some request on her part); this accords with the principle of *la décision préalable*, to be discussed in Chapter 7. There is no question of any affidavit being required in support of statements made in the *recours*.

Finally, the case must be commenced, as a general rule, within a period of two months from the notification to the *requérant* (as the plaintiff is called) of the administrative decision complained of (see *délai*, Chapter 7, p. 161).

3 *INSTRUCTION*

When the *recours* is received by the court office (*greffe*) of a Tribunal Administratif or, on appeal, of the Cour Administrative d'Appel or the Conseil d'Etat, it will be registered and given a number. In the Conseil d'Etat, the *greffe* will analyse the file in terms of its basic component issues, to help with computerized recording and following-up of cases. Depending on the size of the Tribunal Administratif, the president of the court or the president of a section will receive the dossier from the *greffe* and assign it to a particular member as a *rapporteur*. In the Conseil d'Etat, the *greffe* allocates the file to one of the ten Sous-sections, whose president will assign it in turn to one of its members as *rapporteur* unless he can deal with it himself.

The case will now undergo *instruction*, that is, the preparation of the case for eventual judgment—a process that may take several months, or even years if there is stalling or dilatoriness on the part of the administration or of the plaintiff (or his advocate). Powers do exist for sanctions on excessive delay. Where the *recours* must clearly be rejected (as *manifestement irrecevable*), e.g. because it is presented to the wrong order of courts, or the plaintiff has no standing, or it has been presented out of time, all administrative courts have power to reject the *recours* by a judicial decision without *instruction*. In the Conseil d'Etat, this judicial decision can be taken by *ordonnance* of the president of the Sous-section alone under powers granted in 1984. When the Sous-section receives the dossier, it will contain the *recours* itself (with annexed administrative decision) and, perhaps, a *mémoire ampliatif*, especially where an advocate is acting for the *requérant*. Each Sous-section has a *secrétaire de Sous-section*, who is a young civil servant with a law degree and who acts as a kind of 'progress-clerk' in the process of *instruction*. The *rapporteur* has to decide what measures of *instruction* are called for, see that those measures are carried out by the *secrétaire*, and at the end of the *instruction* draft his report.

The purpose of the *instruction* is to elucidate the facts and to research the law. As Chapus suggests, 'there is certainly a natural and close connection between the notion of an administrative court and the written character of procedure, which does not have its equivalent in civil matters.'[8] The written character of the procedure is not unlike the typical form in the United Kingdom of a judicial review hearing: facts are not found by hearing witnesses at trial. However, the written character of pleadings is more radical than anything in British procedures, since even disputes over facts are resolved typically without any oral hearing of witnesses.

The written procedure may involve a number of elements: requests for additional information from the parties, expert reports, site visits, *enquêtes* and further questioning.

(a) Requests for Information

The first step in putting forward a request for information (*demande d'explication*) will be to write to the government department or local authority concerned, inviting them to comment upon the plaintiff's *recours* and other documents within a time-limit prescribed by the *rapporteur*; usually this will be two months, but it may be as short as two weeks. When the administration has submitted its *observations en réponse*, these must be made available for inspection by the plaintiff, who can put in a *mémoire en réplique*. There the exchanges usually end, unless the *réplique* sets up new *moyens* attacking the administrative decision,

---

[8] R. Chapus, *Droit du contentieux administratif* (2nd edn., Paris 1990), no. 716.

when the administration may rejoin with further observations, giving a further right of reply by the plaintiff.

This power to require further explanation or documents from the administration (or, equally, from private parties to the litigation) demonstrates the inquisitorial aspect of the proceedings. The power is based simply on long-standing judicial practice: see the leading case of COUESPEL DU MESNIL (CE 1 May 1936), which declared the right of the court 'to require the relevant authority to produce all documents needed to convince the court and to prove the allegations of the plaintiff'.

If there is any resistance to a request for information, the *rapporteur* may remit the matter for a decision of the court ordering the production of documents. Thus in BLANKAERT (CE 26 September 1986) a physician received a telephone bill more than eight times higher than usual. It related to his private line and he had been absent for a fortnight during the accounting period. The Ministry of Posts and Telecommunications claimed to have checked the line and that the burden of proof was on the plaintiff. The Tribunal Administratif, however, ordered the ministry to produce 'all elements and documents that have been used in arriving at the amount in dispute' and this was upheld by the Conseil d'Etat as a legitimate way of obtaining evidence. If the documents were not sufficient the case should go against the Ministry.

In the course of his correspondence with government departments, the administrative judge cannot be met with any plea corresponding to the English claim of 'public interest immunity'; as in England since *Conway* v. *Rimmer* [1968] AC 910, the court can be trusted to decide for itself whether any question of state secrecy is involved, and in that case the document would not be disclosed to the *requérant*. If the administration refuses to answer a specific question, the court is entitled to draw its own conclusions (as in BAREL, discussed below). In an exceptional case (like *Duncan* v. *Cammel Laird & Co. Ltd.* [1942] 1 All ER 587), the administrative judge may not be permitted to see secret documents, but he will still expect an answer to the plaintiff's allegations: COULON (CE 11 March 1955). Here an employee in a state munitions-factory challenged his dismissal by the Ministry of Defence, alleging that it was due to his political and union activities and claiming production of his personal file; the Conseil upheld the Ministry's plea of state secrecy but required it to justify the dismissal.

(b) Expert Evidence

If more formal investigation is needed, the court may order an *expertise*, that is, a report to be made by an expert either on the whole or on some particular facet of the case; for example, a medical report may be called for as to the extent of the injuries sustained by the plaintiff who is bringing a *recours en indemnité*. Thus, where the local prefect and the Tribunal Administratif (on appeal from him) had ordered the demolition of buildings said to be dangerous, the Conseil d'Etat ordered an *expertise* because the dossier did not disclose exactly what works would be necessary to remove the dangerous nature of the buildings: MANOUVRIER (CE 23 February 1968). But an *expertise* is very unusual at the Conseil d'Etat,[9] although a common procedure in a Tribunal Administratif.

(c) Site Visits

In the case of a planning application, it may be more appropriate for the court to order that one of its members (or even the whole Sous-section) conduct a site visit (*visite des lieux*). Although the *mémoires* of the parties in planning cases are usually replete with photographs of the location of the disputed planning application, a site visit may help further to form a judgment about the validity of the reasons given by the administration, especially where the ground of the *recours* is a 'manifest error in the evaluation of the facts' (see below, p. 245). Site visits are not however limited to planning matters. Thus, in ELECTRICITE DE FRANCE c. SPIRE (CE 2 October 1987) the Tribunal Administratif of Orleans inspected the site of Mme Spire's house in connection with her claim for damages through the close proximity of a nuclear reactor (see further Chapter 8).

Generally, the court is entitled to order all such measures necessary for the information of the court; for example, in a case involving the banning of a film, the completion of the *instruction* was delayed to enable members of the Sous-section to attend a showing of the film in question: SOCIETE 'LES FILMS MARCEAU' (CE 14 October 1960).

---

[9] And increasingly so, since the 1987 reforms have made the Conseil d'Etat primarily a court of *cassation*.

(d) Inquiry into the Facts

The court may also order the holding of an *enquête*, or inquiry into the facts. This is the nearest the administrative courts come to hearing witnesses. At a separate hearing, witnesses will be called to answer questions from the *rapporteur* or another nominated judge. The answers are recorded in writing and are available to the parties. The parties' advocates may be present at the *enquête*, but they do not question the witnesses directly. They may suggest issues that the *rapporteur* may wish to raise and they can make observations on what a witness has said.

Such an inquiry is rarely encountered in practice, especially before the Conseil d'Etat. An illustration of the sort of situation where an *enquête* might be held is PONCIN (CE 22 June 1963 and 17 June 1964), where an *enquête* was held as to what precisely were the determinations reached by members of a *jury de concours* responsible for an appointment to a public office, since it was impossible to sort out the conflicting accounts of their deliberations from the documents in the dossier. *Enquêtes* are ordered more commonly by the Tribunaux Administratifs, who do not enjoy such ready contact with the administration as the *rapporteurs* in the Conseil d'Etat.

(e) Further Questions

It is in general the responsibility of the *rapporteur* to satisfy himself that all the necessary facts are established by documents in the dossier; but an oath[10] or an admission is an acceptable means known to administrative procedure of establishing such facts. In addition, in the lower administrative courts the president may put formal questions to the parties if they are present at the public hearing in order to clarify aspects of the case: commentators refer to this as *l'interrogatoire*[11] but it should in no way be confused with the interrogatories used in English civil procedure. Also apt to be confused with this English procedure is the use of *commissions rogatoires*: this practice of French civil procedure is now made available to the lower administrative courts[12] and enables the court seised of the case to delegate to

---

[10] As in PONCIN (CE 22 June 1963 and 17 June 1964).
[11] For further discussion, see Chapus, *Droit du contentieux administratif*, no. 742, commenting on Code des TA-CAA, Art. R. 196.
[12] Code des TA-CAA, Art. R. 184.

another court some measure of *instruction* which the latter can more conveniently undertake, usually because of the geographical proximity of the parties or of the site involved in the case.

(f) Interlocutory Orders

Expertises, site visits, *enquêtes*, and *commissions rogatoires* are measures of instruction which require an interlocutory order of the court (*un jugement d'avant dire droit*). Additional documents, however, may be asked for without formal order.

This stage of the proceeding, which corresponds only approximately to interlocutory proceedings in an English civil action, may be concluded quickly or may take a long time, according to the diligence of the parties and the complexity of the matter at issue. The secretary of the Sous-section will periodically remind outside correspondents if there are delays, but the initiative is primarily that of the *rapporteur*, although the parties themselves will often suggest a new line of inquiry.

How Inquisitorial is the *instruction*?

The procedure followed in the *instruction* will appear much more inquisitorial than in common law procedure. But, as Chapus observes, 'to say that the procedure is inquisitorial is (by a traditional and customary term) to express that it is under the control not of the parties, but of the judge'.[13] While it is true, as we have seen, that the judge plays an interventionist role, this should not be exaggerated.

In the first place, the administrative court may not act *ultra petita*, that is, it cannot introduce a head of claim not included in the plaintiff's own submissions. The courts are generous, however, in their reading of the submissions of legally unassisted parties and may be able to 'interpret' the submissions of the plaintiff as raising a good legal ground. But such generosity is not shown where the litigant in question is the administration or is legally represented. The attitude is well expressed by Massot and Marimbert:

> the *rapporteur* may be faced with applications coming from ordinary citizens little versed in rules of law in which the subject-matter or legal basis remains concealed rather than being made clearly explicit. In

[13] Chapus, *Droit du contentieux administratif*, no. 718.

these cases, the *rapporteur* is forced to interpret the application in a constructive way, not sticking to the letter of the appeal while not contradicting what it says. This effort to reclassify applications is made most often for the benefit of the applicant. It constitutes, in our view, one of the most attractive features of the tradition of the Conseil d'Etat.[14]

In all cases, however, the court may raise a *moyen d'ordre public* of its own initiative (see Chapter 9, p. 224).

A second constraint is that the procedure is *contradictoire*. This means that both sides must be given a chance to see the evidence or arguments raised by the other side and to make observations on them. Even where the court takes the initiative of commissioning an *expertise* or holding an *enquête*, the parties' lawyers will be entitled to be present and to see the report placed on file. The judge is not allowed to base his decision on documents which the parties have not seen (CE 13 December 1968, ASSOCIATION DES PROPRIETAIRES DE CHAMPIGNY-SUR-MARNE), unless the very subject-matter of the litigation is whether the applicant is entitled to access to a document which the court will then examine before giving judgment on the issue (CE 23 December 1988, HUBERSCHWILLER). Again, *moyens d'ordre public* used to be an exception to this rule: (e.g. CE 20 January 1988, BRUNAUD—but now the Decree of 22 January 1992 requires that grounds so raised by the court must be put to the parties for their observations.

Thirdly, it remains for the parties to prove their case on the basis of the materials found on file. Of course there will be a burden of proof on one party, normally the plaintiff. But a failure to reply, whether to the *rapporteur*'s questions or to the submissions of the other side, allows the court to draw the conclusion that the party in default has no case to make in answer to that put to it (for a famous example, see CE 28 May 1954, BAREL, discussed in Chapter 9).

Thus, although the administrative courts are indulgent towards an unassisted party and will draw inferences from evidence submitted or gathered, the parties do have an important

---

[14] J. Massot and J. Marimbert, *Le Conseil d'Etat* (Paris, 1988) 153; see further J. Bell, 'Reflections on the Procedure of the Conseil d'Etat', in G. Hand and J. McBride, *Droit sans frontières: Essays in Honour of L. Neville Brown* (Birmingham, 1991), 211 at pp. 213–4.

role in ensuring a favourable outcome for themselves. The court may take steps to find out the relative merits of each side's case, but it is not engaged in a fully fledged administrative inquiry into defects in the administration. Its task remains that of resolving a dispute.

Instruction is concerned essentially with establishing the facts. Questions of law, however, are researched by the court, following the principle *curia novit legem*, the court being assisted by the arguments of the parties and (later) by the conclusions of the Commissaire du gouvernement.[15] The dossier prepared for judgment will thus contain the *rapporteur*'s photocopies of case-law and legislation relevant to the case. There is no need for counsel to come to court laden with authorities as in Britain.

Formality and Informality

In the course of his preparation of the case in the Conseil d'Etat the *rapporteur*, who may well (especially in a comparatively simple case) be a young Auditeur with perhaps only one or two years' experience in the Palais-Royal, will discuss any difficulties with the president of the Sous-section and perhaps also with his other colleagues. Indeed, the atmosphere at the Palais-Royal in the offices, corridors, excellent law library, and friendly canteen (with its view of the Eiffel Tower) is very much nearer to that prevailing in a large law school in a British or American university than to the formal dignity of the Royal Courts of Justice in the Strand. On an ordinary working day one will find small groups of men and women discussing legal problems with animation almost wherever one turns. Each decision is a decision of the whole Conseil d'Etat, and it is therefore natural that interesting or difficult problems should be discussed by as many members of the Conseil as is practicable. A similar picture of informal consultation and collaboration can be given of the other administrative courts, where the atmosphere behind the scenes is very domestic. Discussions in rooms of colleagues, in corridors, or over coffee are frequently the way in which cases are prepared for judgment by judges and Commissaires du gouvernement in all administrative courts.

[15] See Bell, 'Procedure of the Conseil d'Etat', 216–7.

THE PROCEDURE OF THE COURTS 99

The *rapporteur* is able to pace cases according to their difficulty, which allows a flexibility in working which is not comparable to the listing of cases in British courts.

4 RAPPORT

The exchanges mentioned above cannot be prolonged indefinitely, even if the parties should seek to do this. For the *rapporteur* has the right to end the exchanges when he thinks the issues of fact and law have been sufficiently ventilated. The dossier, now much fatter, is re-examined and, if he believes (but not until) the case to be *en état (d'être jugé)*, the *rapporteur* proceeds to draft his report.

The report consists of a formal report on the facts and the law (*la note*) and a draft judgment (*le projet d'arrêt*). In the *note*, the *rapporteur* will set out the facts, summarize the views of both parties, and then give his view of the law relevant to the case. He will annex extracts from relevant statutes, cases, law-review articles, and text-books. The aim here is to lighten the task, in the first place, of the senior colleague of the *rapporteur* assigned to act as *réviseur*, who will be the next person to see the report, and, in the second place, of the Commissaire du gouvernement, who will eventually take over the dossier to compose his 'conclusions' (see below). Finally, the *rapporteur* will draw a draft order in the form in which he believes the Conseil d'Etat should issue its judgment: this *projet d'arrêt* must deal expressly with all the *moyens* (arguments of law) of the parties.

Both parts of the report will be seen and commented upon by the *réviseur*. In the Tribunal Administratif or the Cour Administrative d'Appel, this person will be the president of the section of the court which will judge the case. In the Conseil d'Etat, he or she will be president of the *rapporteur*'s Sous-section or one of the two other Conseillers deputed to act as *assesseurs-réviseurs* in the Sous-section. The task of the *réviseur* is to reread the dossier and to suggest corrections to the *projet* or even to propose new lines of inquiry. Given that the dossier will be reviewed also by the Commissaire du gouvernement, the *rapporteur*'s efforts are subjected to serious scrutiny.

In the Tribunal Administratif, once the *rapporteur* has finished his report and it has been reviewed by the *réviseur* and the

Commissaire du gouvernement, the case is ready for judgment. In the Conseil d'Etat and in the Cour Administrative d'Appel, the preparatory stages are more elaborate and, unlike in the Tribunal Administratif,[16] there is a formal *séance d'instruction* before the whole Sous-section or chamber of the Cour Administrative d'Appel to whom the *rapporteur* and the *réviseur* report: this *séance* warrants the following description.

5 SÉANCE D'INSTRUCTION AUPRÈS DU CONSEIL D'ETAT

The report completed, the case will be put in the list for an early *séance d'instruction*. This is a session of the Sous-section to which the case has been referred, traditionally held (by all Sous-sections) on Monday afternoons, commencing at 2.15 p.m. and concluding when the business has been completed, at 7.00 or 8.00 p.m., or even later. All the members of the Sous-section, but no one else, and certainly not the legal representatives of the parties, will be present at the session. Some ten or fifteen cases will be in the list for a particular session, and similar cases are often taken together. When the session opens, the only members of the Sous-section who will know anything at all about the cases coming before it will be the respective *rapporteurs* and the president or other senior members of the Sous-section who have supervised their reports; the president will have read the dossiers and reports of all the cases and made his own notes and formed his preliminary opinions about them.

When a particular case is called on at the session (by the president), the *rapporteur* will read his report; this will take anything from twenty minutes or so to an hour or more, according to the complexity of the case and the prolixity of the *rapporteur*. The *réviseur* will then be invited to add his comments or to outline disagreements with the *rapporteur*. The report will then be discussed by the members of the Sous-section. The president will probably make the first contribution, but he will rarely force his views on the Sous-section. The position of the president, especially at this stage of the proceedings, is an important one;

---

[16] Members of a Tribunal Administratif may discuss the cases to be heard a week or so later at an informal *pré-délibéré*.

he is something more than a chairman of the session, as he is also responsible for the proper conduct of the case, and most presidents recognize a responsibility to train the younger members of their Sous-sections. The president will therefore frequently invite contributions, especially from the younger members of their Sous-sections. Throughout the procedure, there is an emphasis on the collegiate nature of the decision: all members of the Conseil d'Etat are equally responsible.

It may seem strange that those who have never read the file and who have only just heard an oral report should presume to discuss a case. The role of the Sous-section is not to consider matters of detail, but to be convinced that the *rapporteur* has examined the problem in an appropriate manner and has come to a solution consistent with general principles of law. The role is one of review rather than detailed analysis.

When the discussion is concluded, the president will ask the Sous-section whether they agree that the conclusions of the *rapporteur* may be accepted on behalf of the Sous-section; if not, the case may be remitted to the *rapporteur* for him to present a revised report on another occasion, but this is rare in practice. Should a vote be necessary, the only persons entitled to vote are the president, his two Conseiller colleagues, the *rapporteur* for the case, and, in case of a tie, the next most senior member of the Sous-section. Finally, where the draft decision is accepted by the Sous-section (perhaps with agreed amendments), the president will hand on the dossier to the member of the Sous-section who is acting as Commissaire du gouvernement for the particular case.

6 THE COMMISSAIRE DU GOUVERNEMENT

As we have seen, the office of Commissaire du gouvernement dates from 1831, but as it has developed he (or she) is in no sense a representative of the government. The Commissaire is fully independent and gives an opinion in the interests not of the parties but of the law. The Commissaires are appointed by a decree of the Prime Minister in all administrative courts.[17] Their workload is heavy, especially as the Commissaire has a duty to

---

[17] At the Conseil d'Etat, they are chosen from among the Maîtres des requêtes.

write conclusions for every case.[18] Between 1980 and 1986, an attempt was made to lessen the load by permitting courts to dispense the Commissaire from writing conclusions in some cases. But the Law of 6 January 1986 restored the obligation as an integral part of the French conception of administrative justice.

The best description of the role of the Commissaire comes from the Conseil d'Etat in its decision of 10 July 1957, GERVAISE:

the Commissaire du gouvernement is not the representative of the Government; as far as the internal operation of the court is concerned, he is under the sole authority of the latter's president, his mission is to explain . . . the questions which each *recours* before the court raises for decision, and to make known, in formulating his conclusions in total independence, his assessment (which should be impartial) of the circumstances of fact of the case and the applicable rules of law, as well as his opinion on the solutions which, according to his conscience, the dispute submitted to the court requires.[19]

The Commissaire's first function will be to read through the dossier when it is handed to him at the end of the *séance d'instruction*. He will then undertake any further inquiries that he may consider to be necessary, in particular into any matters that he considers the *rapporteur* has overlooked. If he finds himself in disagreement with the draft judgment (*projet d'arrêt*), it is customary for him to inform the *rapporteur* and, in the Conseil d'Etat, the Sous-section president: the latter may then reconvene the Sous-section to reconsider the *projet*, at least if a major—rather than a mere drafting—modification appears desirable. When the Commissaire is ready, he will prepare his own report or 'conclusions'. The conclusions will be drawn up with two purposes. The first is to sum up the case as clearly as possible for the information of those members of the court who have not participated in the *instruction* or have not been *réviseur*

---

[18] The French term *conclusions* has been rendered by us 'conclusions' in English. Of various possible translations (e.g. 'submissions', 'recommendations', or 'opinion') none are wholly satisfactory, we feel, although 'opinion' is the term that was adopted in the Court of Justice of the European Communities, following the accession of the United Kingdom and the Republic of Ireland.

[19] See also the conclusions of Mme Grévisse (now president of the Social Section) in CE 13 June 1975, ADRASSE, where the judgment of the Tribunal Administratif of Martinique was quashed for procedural irregularity, because the Commissaire du gouvernement in that lower court simply referred the case to 'la sagesse du tribunal', without giving any conclusions.

of the report, e.g. (in the Conseil d'Etat) because they are not members of the Sous-section from which the case originated. The second is to relate the proposed solution to the general pattern of the case-law and, if possible, to foreshadow its future development. Of course, the chance to achieve these broader aims varies with the nature of the particular case. In that the Commissaire may bring together a number of cases on related issues for hearing at one *séance de jugement*, this does allow him to shape the development of case-law on a particular topic. But since the Commissaire may well have twenty or thirty cases a month to deal with, he may lack adequate time to make the kind of mark on the case-law which is necessary or which he desires.[20] Nevertheless, despite the increased pressure of case-load at the Conseil d'Etat, some Commissaires still succeed in influencing the development of the law in a way comparable to such brilliant predecessors as David in the last century, Romieu and Blum in the early part of this century,[21] or Latournerie and Letourneur in the years before and after the Second World War.

7 JUDGMENT

When the conclusions of the Commissaire du gouvernement are ready, the case will be put in the list for the *séance de jugement*, which consists of two stages, the *audience publique* and the *audience privée* (or the *délibéré*). The courts may sit in various formations for the *séance de jugement*. In the Tribunal Administratif or the Cour Administrative d'Appel, the court will usually consist of three members: the *rapporteur*, the president (who has also been *réviseur*) and one other judge. In more important cases, a full court of up to seven judges may sit. In the Conseil d'Etat, the *séance de jugement* will consist of the original Sous-section in the less important cases, under the Decree of 1980, or, more commonly, of two Sous-sections combined (*réunies*), one of which will be the Sous-section before which the case was heard

---

[20] Massot and Marimbert, *Le Conseil d'Etat*, 161, note that in the Conseil d'Etat, a Commissaire du gouvernement has over 300 dossiers to deal with a year.
[21] On Jean Romieu, see G. Cahen-Salvador, and on Léon Blum, see P. Juvigny, in *Le Conseil d'Etat—Libre Jubilaire* (Paris, 1952) at 323–36 and 337–40 respectively.

at the *séance d'instruction*, or of the full Section du Contentieux, or even of the Assemblée du Contentieux, according to the complexity and importance of the issues in question; matters having political overtones, for example, tend to be heard by the Assemblée (for these different compositions of the court, see pp. 105–6).

The question as to which of these panels of judges within the Conseil d'Etat is to be seised of a particular matter may be determined by any one of the following: the vice-president of the Conseil d'Etat, the president of the Section du Contentieux, the president of the Sous-section d'instruction, or the Commissaire du government charged with the case.[22] In the other courts, the decision lies with the president of the court, the president of the section charged with the instruction, or the Commissaire du gouvernement.

The Audience publique

The first part of the *séance de jugement* in point of time is the *audience publique*. The proceedings in the Conseil d'Etat and in the Cours Administratives d'Appel are more formal than in a Tribunal Administratif, but they all differ substantially from the oral hearing in a British court. Typically, a large number of cases will be scheduled for hearing at one session, perhaps as many as twenty to thirty for one morning in a Tribunal Administratif. The oral hearing provides an opportunity for a few points of clarification, rather than for rehearsing the full arguments of each side.

It is in the Tribunal Administratif that the oral hearing has most importance. Parties frequently appear in person. Since they may have been legally unassisted, there may be points on which the judges want clarification. The atmosphere is relatively informal, even if the judges do sit on a bench,[23] being closer to that before a British social security appeal tribunal. The plaintiff or his lawyer may make a short speech to the court for five or ten

[22] The last word being with the vice-president in the unlikely event of disagreement.

[23] Prior to 1990, in certain cases (and sometimes only if the parties consented) a member of the Tribunal might sit alone to try the matter; such a *conseiller délégué* was permitted to sit in a town which is not the normal seat of the Tribunal. This very rare exception to the collegial character of the administrative courts was abolished by the Law of 25 June 1990.

minutes, outlining the major points of the case, and then the judges may ask questions. The administration may reply briefly and the court moves on to hear the next case.

The Cours Administratives d'Appel and the Conseil d'Etat are geographically more remote from litigants and, especially in the Cour Administrative d'Appel, they are unlikely to be present. Indeed, before the Conseil d'Etat, a litigant in person has no right to make any representations. However, before the Tribunaux Administratifs and Cours Administratives d'Appel, oral arguments from counsel are frequent.

The *séance de jugement* in the Conseil d'Etat is held traditionally on Monday and Wednesday afternoons in the case of the single Sous-section or two Sous-sections combined, and on most Friday afternoons in the case of the full Section du Contentieux; the difference between the two is also marked by the length of their respective cause lists, although both follow the same procedure. Whereas the single Sous-section or two Sous-sections combined may undertake twenty cases in an afternoon (and the session will continue until all the cases in the cause list have been disposed of), the full section will not usually have more than five or six cases in its cause list. The Assemblée du Contentieux also will not handle many cases at a sitting; this is convened normally only once in two months or so.

At the *séance de jugement*, in both its public and its private stages, the court will consist of the following:

1. *Two Sous-sections réunies or single Sous-section.* Where (as is common) two Sous-sections sit in judgment together, the president will be one of the three vice-presidents of the Section du Contentieux or sometimes even the president of the Section, and there will also be present three Conseillers from each of the two Sous-sections concerned, one representative of an administrative section of the Conseil d'Etat, and the *rapporteur*. The Commissaire du gouvernement is also present but only at the *délibéré*, in a consultative capacity, not as a member of the court: his function is to provide such further explanations and clarifications as members of the court may require.[24]

There is an internal rule about groupings of pairs of Sous-sections, whereby in practice Sous-sections 1 and 4 hold joint

---

[24] See further Chapus, *Droit du contentieux administratif*, no. 851.

sessions, as do 2 and 6 and 3 and 5. Sous-sections 7, 8, and 9, which are concerned exclusively with fiscal matters, unite either all three together,[25] or in pairs. Sous-section 10 was added in 1980 and is available to pair with any other Sous-section, as need arises; or (like the other nine) it may sit in judgment as a single Sous-section in simple cases. It should be noted that this procedure of the Sous-section sitting in judgment alone (*la Sous-section jugeant seule*) is of increasing importance in the day-to-day operation of the Conseil, at least for dealing with routine cases.[26]

2. *Section du Contentieux.* The president of the Section will preside, and there will also be present the three vice-presidents, the presidents of the ten Sous-sections, together with two Conseillers from the administrative sections of the Conseil d'Etat, and the *rapporteur*. As before, the Commissaire du gouvernement may (and does) attend the *délibéré* in a consultative capacity. It is customary for the Section at a particular session to take cases only from one Sous-section d'instruction.

3. *Assemblée du Contentieux.* As we saw in Chapter 4, since the reform of 1963 the Assemblée du Contentieux is composed of the vice-president of the Conseil d'Etat (who presides), the six section presidents, two of the three vice-presidents of the Section du Contentieux, the president of the Sous-section concerned, and the *rapporteur* for the case. The Commissaire du gouvernement must, of course, attend the *audience publique* to present his conclusions and may attend the *délibéré* as before.

By way of comparison, in 1990, 3,044 cases were dealt with by Sous-sections réunies, 2,618 by a single Sous-section, 57 by the Section du Contentieux, and 25 by the Assemblée du Contentieux.

Whatever the composition of the court, the public may be present at the *audience publique*—those who are sufficiently interested to come. The few members of the public who are present at the Palais-Royal will probably be mostly students from l'Ecole Nationale d'Administration, or possibly from one of the Paris Facultés de Droit. The police, who are usually so much in evidence in public buildings in France, are conspicuous by their absence, the Conseil d'Etat being content to rely on the services

---

[25] Then referred to as *la plénière fiscale*. [26] See statistics in App. E.

of one uniformed *huissier* or usher. The atmosphere in the *audience publique* is of course much more formal than at the *séance d'instruction*, but there is no exact parallel with British court or tribunal proceedings, as there are no witnesses.

In addition to the public, the plaintiff's legal representatives—the *avocats*—are entitled to be present, and they will normally attend, robed, although the members of the court are not robed, nor indeed do they have the right of the French judiciary to 'porter la robe', being civil servants and not *magistrats*. Before the Conseil d'Etat the parties themselves do not attend (except, rarely, as members of the public), but parties are quite commonly present before the Tribunaux Administratifs and sometimes before the Cours Administratives d'Appel. The case for the administration will have been conducted, during the process of *instruction*, either by a member of the legal staff of the government department concerned or by an *avocat aux Conseils* briefed for the purpose: the administration's representative may attend the *audience publique*, and commonly does so before the Tribunaux Administratifs. Occasionally a member of the press may be present, but journalists (even those representing *Le Monde*, which gives a good coverage to the Conseil d'Etat) will normally be content to get their material from the official copies of the conclusions of the Commissaire du gouvernement and from the *arrêt* itself (see below). At the Conseil d'Etat there is a press office (*service de presse*) to assist this liaison.

A case will be called on by the clerk to the court (by number and title), and the *rapporteur* will then read very rapidly, and usually in a low voice, a summary he will have prepared of the submissions (*mémoires*) put in by the lawyers for both sides. The president will ask the advocates present whether they wish to plead, that is, to add anything to the *mémoires*. This he will normally do by saying the single word 'Maître?' The *maître* so addressed will rise and may merely say that he relies on the contents of the *mémoires*. Of course he *may* plead if he wishes to do so, but this is rather the exception; it seems to have become the practice for an advocate to plead only in cases where he knows (by prior inquiry) that the Commissaire du government is proposing to recommend conclusions unfavourable to his client, and even then only in an unusual case. Or he may plead in a case attracting great publicity, such as the decision on the validity of the franchise granted to a commercial television station (CE 16 April 1986, COMPAGNIE LUXEMBOURGEOISE DE TELEVISION).

Generally, as the client is seldom present before the Conseil, the advocate does not feel obliged to make a good impression by addressing the court. Where the advocate does decide to plead he is expected to alert the president beforehand (so that the list for that session may be abridged); in his remarks he is allowed only to expand and comment on the points made in his written submissions.

A more important role for the *avocat* at the *audience publique* is to check, by listening to the conclusions, that no fact or argument has been omitted or misinterpreted to the court. Any apparent mistake or misinterpretation can be corrected by the immediate drafting of a *note en délibéré*, which is handed to the court so that it might be considered at the subsequent *délibéré*.

The president will then call on the Commissaire du gouvernement (who will be sitting on the bench with the members of the court, but at a special desk) to read his conclusions. He then stands to read his text. The court will not discuss the case at all at this stage, but at the end of the reading of the conclusions the president will remit the case to the *délibéré* and the next case will be called with virtually no break in time, and the procedure will be repeated. When the cause list for the day has been completed to this extent, the *audience publique* is concluded, and the members of the public are requested to leave the chamber.

Before the Tribunaux Administratifs oral pleading is much more common, partly because the client is more commonly present in court, and partly because the lawyers who appear are not usually specialists in the written administrative procedure but are accustomed to civil procedure with its greater oral content.

The *délibéré*

The second stage of the *séance de jugement* is the *délibéré*, or *audience privée*, which immediately follows the conclusion of the *audience publique*.[27] At this stage, the cases in which the conclu-

---

[27] Compare the practice in the European Court of Justice: at Luxembourg the conclusions of the Advocate General do not immediately follow the oral pleadings but are delivered at a reconvened hearing a week or more later. From the subsequent *délibéré* the Advocate General is excluded; but a written copy of his conclusions is in the hands of the court—which is not the practice of the French administrative courts, even in the Conseil d'Etat.

THE PROCEDURE OF THE COURTS 109

sions of the Commissaire du gouvernement have just been pronounced are discussed by the members of the court. The Commissaire du gouvernement, although he will be present throughout the *délibéré*, is not permitted to take part in the discussion or eventual decision as he has already expressed his opinion in his conclusions. At the most he may be asked to clarify a point in the latter. The court will eventually arrive at a decision, usually (but by no means necessarily) that recommended by the Commissaire du gouvernement, and this will then be put into the form of a very concise judgment, or *arrêt*,[28] which will within the next few days be drafted by the *rapporteur*, in consultation with, and with the approval of, the president of the court. Depending on whether the judgment follows or runs contrary to the views of the Commissaire du government, the latter's conclusions are then said by commentators to be *conformes* or *contraires*, although, of course, the decision itself does not state whether it is *conforme* or *contraire* to the conclusions.

In addition, a note of the deliberation is taken by a member of the documentation service who is in attendance: such notes help provide fuller information on the reasoning if the case is being relied upon in future litigation.[29]

It is a feature of an *audience privée* held by the full Section, or by the Assemblée, that any members of the Conseil d'Etat may attend as observers, and in practice many of them, especially the younger Auditeurs, take advantage of this opportunity of hearing the arguments put forward by their senior colleagues.

8 THE DECISION

The decision itself, in accordance with the normal French preference for public participation in the judicial process, has to be

---

[28] Strictly the Conseil d'Etat reaches 'une décision', the Cours Administratives d'Appel 'un arrêt', and the Tribunaux Administratifs 'un jugement'.
[29] See Bell, 'Procedure of the Conseil d'Etat', 224. The privacy of the deliberation (*le secret du délibéré*) would not allow the note to divulge how members of the court voted or to ascribe a particular expression of view to a member by name; nevertheless, the attitude to the secrecy principle is much more relaxed at the Palais-Royal than before the European Court of Justice in Luxembourg, where the *délibéré* is attended only by the judges concerned, unaccompanied by the Advocate General and registrar and without interpreters.

given in public. For some years now, however, this has become a mere formality. At the commencement of the *audience publique* held next after a *délibéré* (usually the following fortnight), the president will announce that 'Les décisions sont lues' thereby giving publicity to the *arrêts* and conclusions of the Commissaire du gouvernement, which by this time have been made available to the parties and those members of the public and the press who may ask for them.

As will be seen from the specimen reproduced in Appendix H, the *arrêt* is a very bald and somewhat uninformative document; it will satisfy the general principle of *droit administratif* that any decision must be *motivé* and must deal with each *moyen*, but it will not detail the legal principles on which it is based nor discuss the relevant case-law, in the manner of an English judgment. For these matters the French administrative lawyer must have recourse to the conclusions of the Commissaire du gouvernement, and as a consequence, some of these conclusions (especially those of the great Commissaires du gouvernement of the formative period of the jurisprudence of the Conseil d'Etat, such as David, Romieu, and Léon Blum) have become classic repositories of the law, in much the same way as the judgments in the House of Lords of such judges as Lord Atkin and Lord Diplock, or of Lord Denning in the Court of Appeal. It should also be noted that in this procedure there is no room for any dissenting judgments; such would be quite alien to the whole spirit of the French court as a collegiate body. On the contrary the corporate character of the decision is enforced by the rule of 'the secrecy of the *délibéré*', under which there can be no disclosure of what transpired in the course of the discussion or the voting.

It will also be noted that each *arrêt* commences with the formal words 'Au nom du Peuple Français'; this is because the Conseil d'Etat is exercising *la justice déléguée*, and pronouncing judgment in the name of the people, a judgment having the force of *res iudicata*, or what the French term *l'autorité de la chose jugée*.

9 EXECUTION

It is in the matter of the execution of their judgments that the French administrative courts—in English eyes, at least—have previously been open to criticism. Being essentially part of the administration and not one of the judicial organs of the state, the Conseil d'Etat has no officers charged with the duty of enforcing its judgments; in some measure, the Conseil d'Etat has been in a similar position to that of the Queen's Bench Division of the High Court in England, which cannot issue an order against the Crown as such.[30] In France this procedural difficulty (avoided in England in some degree by the declaration, or by issuing a mandamus against a minister) used to apply in all types of proceedings against the administration. But, as we see below, in 1963 the Report Commission (now the Section du Rapport et des Etudes) was introduced within the Conseil d'Etat as an extra-judicial means of improving the execution of judgments against the administration, and in 1980 parliament added new and important judicial remedies of enforcement. Nevertheless, it is a remarkable feature of *droit administratif* that the decisions of the administrative courts are commonly obeyed without the need for any threat of enforcement.

Exceptionally, however, although the members of the Conseil d'Etat are members of the administration—perhaps because of this fact—the administration does not always consider itself as being even morally bound to implement the decisions of the Conseil *statuant au contentieux*. Where there was such a refusal, it was formerly always left to the person affected (normally the successful plaintiff) to raise a fresh *recours* if he saw fit, and thus to bring the matter before the court afresh. This happened in RODIERE (CE 26 December 1925). The plaintiff had objected to being passed over in the system of promotions within the government department in which he was employed, the minister having made a number of appointments over his head. The Conseil d'Etat declared these appointments void, but the minister made new appointments by creating informal posts within the department in the same grades and for the same individuals as before. In further proceedings taken by the same plaintiff the Conseil d'Etat then had to annul these appointments also. This case established the important principle that the court has absolute control over the manner in which its decisions are executed;

---

[30] But subject to the important exception that our courts may now issue an interim order against the Crown (even to disapply a statute), where necessary to safeguard rights under Community law: *R. v. Secretary of State for Transport, ex parte Factortame (No. 2)* [1991] 1 All ER 70, HL

thus, the court would not normally go so far as to insist on the right of an individual civil servant to a particular appointment, but it may do so when determining the consequences of an illegal dismissal from the service, and re-creating the plaintiff's position as it would have been if he had not been the victim of an illegal decision.

The reforms of 1963 and further legislation in 1980–1 and 1988 have effected considerable improvements in this matter of execution. In the first place, the government department whose decision has been the subject of proceedings before an administrative court may ask the Conseil d'Etat to explain the implications of the judgment and how the department should ensure that it is observed ('d'éclairer l'administration sur les modalités d'exécution de la décision');[31] in such a case, a *rapporteur* is appointed by the court, and he will discuss the matter with representatives of the department. There were only fifteen such requests for *éclaircissement* in 1990. Alternatively, either the vice-president of the Conseil d'Etat or the president of the Section du Contentieux may take the initiative and ask the Section du Rapport to draw the attention of the appropriate government department to the implications of the decision. As a further alternative, a plaintiff may himself, within three months of the decision of the administrative court, refer the matter to the Section du Rapport for investigation.

After a hesitant start, this machinery is increasingly set in motion by dissatisfied plaintiffs. From an average of only 20 or so complaints (*réclamations*) in the years 1964–8, the figure rose to over 745 in 1990. The vast majority of complaints relate to the execution of decisions of the Tribunaux Administratifs (81 percent of complaints in 1990), and the favoured subject-matter is public service employment (42 percent in 1990). In most cases, the intervention of the Section du Rapport brings results, albeit slowly. Experience has satisfied the Section du Rapport that the administration is very rarely guilty of wilfully obstructing a decision; usually, it simply does not know how to execute it, and here the mediation of the Section du Rapport, often by telephone,[32] can point the right road. For valuable analysis, the

---

[31] Since 1969, a judgment of a Tribunal Administratif may also be the subject of a *demande d'éclaircissement*.
[32] *Etudes et Documents* No. 38 (1987), 194.

reader is encouraged to consult the annual report of the Section in *Etudes et Documents du Conseil d'Etat*.[33]

As will now be evident, the Section du Rapport provides only a form of conciliating machinery. The *rapporteur*, with the assistance of the Section, may be able to iron out difficulties, but if the administration does not want to implement the decision, he will be powerless. Like the Médiateur (who may also, upon complaint, enjoin the administration to respect *l'autorité de la chose jugée*), the Section du Rapport is not equipped with any real powers of enforcement. Both can only note and report non-compliance.

To reinforce these extra-judicial remedies, the Law of 16 July 1980 introduced two important judicial means to enforce judgments against the administration. First, where the administration has been found liable, whether by the civil or the administrative courts, to pay a fixed sum of money (e.g. as damages in contract or tort) and any right of appeal has been exhausted, payment of the sum must be authorized by the administration within four months. In default of payment by the end of this period, the judgment creditor may proceed to enforce the debt against the public authority responsible. Secondly, the Law borrows from the procedure of the civil courts the remedy of *astreinte* and permits its use by the Conseil d'Etat (but not the Cours Administratives d'Appel nor the Tribunaux Administratifs) in order to enforce a judgment of any administrative jurisdiction. The *astreinte*, an innovation of the judicial practice of the civil courts, consists in a court order requiring payment of a certain sum of money to the judgment creditor for every day for which the original judgment remains unexecuted. The civil courts have found such a cumulative penalty an effective means to secure compliance with their judgments. Use of the *astreinte* by the Conseil d'Etat applies under the Law of 1980 to judgments for sums of money (although, in practice, the provisions just referred to should suffice to guarantee payment); but it may also be used (which would doubtless be the chief value of the *astreinte*) in order to ensure the execution of a judgment quashing an illegal decision following a successful *recours pour excès de pouvoir*.

---

[33] Ibid. 193–209 (by way of example).

The *astreinte* may be requested from the Conseil d'Etat either by the successful plaintiff in the original suit or by the Section du Rapport acting on the plaintiff's behalf. But the *astreinte* cannot be sought until the administration is at least six months in default in executing the judgment. Non-payment of the *astreinte* will give rise in turn to the remedies set out above for default in satisfying a money judgment.

In practice, the Conseil d'Etat has been circumspect in ordering an *astreinte*, but, all the same, in 1990 some eighty-four were granted. The first decision in which an *astreinte* was ordered was MENNERET (CE 17 May 1985). Mme Menneret's father, M. Saumon, mayor of Maissonais-sur-Tardoire, had been executed in 1944 by the Resistance (unfortunately by mistake, it seems). The municipal council voted on 10 July 1971 to include M. Saumon's name on the commune's war memorial as 'mort pour la France'. The mayor improperly caused the council to reconsider the matter on 17 September 1971 and it decided to proceed no further with the inscription. Mme Menneret obtained the annulment of this second decision by a judgment of the Tribunal Administratif of Limoges on 1 February 1977. The council refused to carry out its decision of 10 July 1971 and Mme Menneret referred the matter to the Report Commission (as it then was) of the Conseil d'Etat. It failed to obtain satisfaction and in June 1983 the Conseil d'Etat was requested to order an *astreinte* of 200 F a day, which it duly did.

The function of the *astreinte* is to encourage performance. It is typically ordered at a provisional level as in MENNERET (the *astreinte provisoire*) rather than as an absolute amount (the *astreinte définitive*). If the administration then complies in reasonable time, the Conseil may not in the end order the administration to pay any money to the plaintiff as a penalty when it comes to 'liquidate' the *astreinte* (e.g. CE 6 May 1988, LEROUX). Of course, the plaintiff is free to sue separately for consequential losses resulting from the failure to implement the administrative decision.

The *astreinte* is bound to have limited application in the *recours pour excès de pouvoir*. Unless there has been an administrative decision in the plaintiff's favour, there is nothing the administration can be required to do. At best, it can be enjoined from continuing to implement an illegal decision. The judgment creditor may well have to request the administration to do something positive to give practical effect to the court's decision, and it is only at the end of this second process that an *astreinte* may be appropriate. Thus, where a public employee is dismissed illegally and obtains a court order annulling the decision to dismiss him, he must then request reinstatement from the administration. If it refuses, the refusal could provide the basis for a further *recours* against the administration, leading to a second judgment, this time for reinstatement. This second judgment could be enforced by *astreinte*, whereas no *astreinte* will be ordered against the administration for failure to comply with the court's initial judgment annulling the dismissal (CE 5 May 1986, DAVID).

It should also be noted that where there has been a failure by a public authority to pay a money judgment against it or where an *astreinte* has had to be ordered to overcome the authority's recalcitrance, the Law of 1980 makes personally liable to a fine any public servant who is held to be responsible for the non-execution (or delayed execution) of the original judgment. This jurisdiction to fine (which is regarded by the French as involving financial, not criminal liability) is given to the *Cour de discipline budgétaire*, a body created by statute in 1948 and given additional powers in 1971. In practice, however, these powers are seldom invoked.[34]

10 APPEAL OR CASSATION

As has been explained, appeal from a decision of a Tribunal Administratif will lie either to a Cour Administrative d'Appel or to the Conseil d'Etat, depending on the ground of the *recours*. The appeal is both on fact and law, and the lodging of the appeal does not normally suspend the operation of the decision appealed against, except in a *recours électoral*, where the validity of the election will be in issue. From the Cours Administratives d'Appel, a *recours en cassation* lies to the Conseil d'Etat.

Naturally, there is no right of appeal in the proper sense from a decision of the Conseil d'Etat, but it may be open for a dissatisfied plaintiff to take the matter further in the following circumstances:

---

[34] See G. Braibant, *Le droit administratif français* (1st edn, Paris, 1984), 529.

116   THE PROCEDURE OF THE COURTS

Exceptionally, the Conseil d'Etat may be asked to reconsider its decision (in each case within a time limit of not more than two months) in any of three cases:

(*a*) Where the judgment has been entered in default of reply, the party who was in default may ask for the matter to be reopened by a process known as 'opposition'. Similarly, where a third party who had an interest in the original proceedings was not joined in them, he may attack the decision by the process of *tierce opposition*.

(*b*) If the party raising the matter can prove there is a case for *révision*. This is rigidly confined to cases where there is a very serious defect in procedure, or where judgment has been obtained in reliance on a forged document or through the other side wilfully withholding a document essential for the appellant's case.

(*c*) Again, a party to a judgment may apply for a *recours en rectification d'erreur matérielle* where he can show that there was a material error of fact which would have had an effect on the judgment.

Each of these three cases arises very rarely in practice (see Appendix E.)

2. Of course, a decision from the Conseil d'Etat can also be annulled by the Tribunal des Conflits in the various situations described in Chapter 6. Again, this very rarely happens.

11 SPECIAL PROCEDURES

Among the procedures which may apply, two are very important.

(a) *Le référé administratif*

Used extensively in civil procedure,[35] the *référé* or interim order procedure was not permitted so widely in administrative procedure until the Decree of 23 September 1988 which grants administrative courts substantially the same powers as exist in civil law.

Interim orders are granted either by the president of a Tribunal Administratif or a Cour Administrative d'Appel or by the

---

[35] The parallel is with the powers of the president of the *Tribunal de grande instance* under Arts. 808–10 of the Code of Civil Procedure.

president of the Section du Contentieux of the Conseil d'Etat, depending on the court which is seised of the case. The procedure can be used in both non-urgent and urgent cases. In non-urgent cases, the president can order steps to be taken to preserve evidence, to appoint an expert to determine disputed facts, or to take other measures related to the investigation of the case. In addition, the president is now empowered to award interim damages (the *référé-provision*) where the existence of an obligation (in contractual or non-contractual liability) cannot be seriously disputed. Although a plaintiff may be required to provide a guarantee against repayment of the sum if the case goes against him, this has considerable advantages in many cases, since it compensates for the delay in reaching the final decision (on which more, p. 120).

In urgent cases, the president is empowered to order such steps as are necessary to put an end to a plainly unlawful situation other than by a stay of execution of the administrative decision (as to which see below) or by issuing orders to the administration (but he may issue orders, 'vue l'urgence', to private individuals).

Although these are interim orders, in no way do they prejudice the outcome of the case.

(b) **Stay of Execution**
It must be recalled that all administrative decisions are, in principle, to be implemented once they are taken (the theory of the *décision exécutoire*), a principle reaffirmed as recently as 1982 as the rule in public law (CE 2 July 1982, HUGLO). A *recours* does not automatically result in a stay of execution (*sursis à exécution*) of an administrative decision. Indeed a stay is an exceptional measure. Unlike the *référé*, a stay requires the decision, not of the president alone, but of the formation of the court (Conseil d'Etat, Cour Administrative d'Appel, or Tribunal Administratif) which has been seised of the original *recours* challenging the administrative decision as illegal.[36] The application for a stay must be justified by reference to two criteria: that the grounds on which the *recours* to annul the administrative decision is based are serious and capable of giving rise to an annulment, and that

---

[36] See Chapus, *Droit du contentieux administratif*, 755 ff.

the execution of the decision is likely to cause loss which is difficult to compensate (*risque d'entraîner des conséquences difficilement réparables*). Such a loss is presumed in cases of deporting aliens, granting planning permission, closing an industrial plant, or preventing a student for one year from attending a course. Thus, when the Ministry of Culture commissioned the erection in the Cour d'honneur next to the Conseil d'Etat of columns striped like humbugs, neighbours alleging the commissioning was illegal were able to obtain a stay of execution, even though the work had already begun (CE 12 March 1986, MINISTRE DE LA CULTURE c. MME CUSENIER).[37]

A stay can be ordered not merely of a positive decision but also of a refusal to decide in an applicant's favour where this has the effect of altering (in law or in fact) the situation of the applicant (CE 23 January 1970, AMOROS). For example, often a lawful entrant into France can no longer remain lawfully unless he obtains a *carte de séjour* within a certain period. If such a *carte de séjour* is refused, he will automatically become an unlawful entrant after the expiry of the period unless the decision to refuse the *carte* is suspended (CE 16 March 1988, ZOLA). In this way, and indirectly, the administration is forced to make a temporary decision in the applicant's favour. A *sursis* would not, however, be granted upon refusal of an application for planning permission (see above p. 87), since the situation of the landowner remains unchanged by the refusal.

The availability of *sursis* has expanded in recent years.[38] Older restrictions on the powers of the Tribunaux Administratifs in the area of public health and public order were abolished in 1983, and the more liberal case-law of the Conseil d'Etat is having impact on the decisions of other courts. Indeed the Constitutional Council has held *sursis* to be one of the main reasons for leaving issues to be decided by administrative and not civil courts (see above p. 18).

It should be noted, however, that *sursis* is a matter of judicial discretion, even if the basic criteria for awarding it are met. It

---

[37] In the end, the Ministry won the day to keep the columns—as, alas, is apparent to a visitor to the courtyard of the Palais-Royal.
[38] Special provisions on *sursis* exist in relation to environmental law, planning inquiries, and local government law which are not discussed here, but which expand the range of its availability.

might be refused, for example, where a challenge is made to the legality of a public building project which needs to be completed in the public interest, such as new courts in Versailles (CE 13 February 1976, ASSOCIATION DE SAUVEGARDE DU QUARTIER NOTRE-DAME, a decision of the Assemblée).

12 ADJOURNMENT FOR PRELIMINARY RULING

Because of the duality of jurisdiction in the French legal system, a preliminary question of law or fact may on occasion arise in the course of proceedings properly brought before the administrative courts, or the civil courts, as the case may be, which falls for decision by the other order of courts. A *question préjudicielle* is then said to arise. In such a case it is a matter for the party most concerned to request the court of trial to adjourn, so that the *question préjudicielle* can first be determined by the court of the other jurisdiction. Thus, in GODOT (CE 24 February 1950) in the course of proceedings in an administrative court to annul an order of *remembrement* (the regrouping of agricultural holdings), a farmer claimed that the order had misstated the extent of the area of his holdings. This was a *question préjudicielle* which the administrative court, on the application of the farmer, referred to the civil court to determine.

If, however, the *question préjudicielle* does not raise a substantial point of difficulty, but the answer to the point raised is clear (perhaps because it has been given on a previous occasion), it may be determined or ignored by the court of trial; this is the so-called doctrine of the *acte clair*, which has caused some controversy in connection with the law of the European Community where Article 177 of the EEC Treaty is derived from the French procedure.[39]

Under the new procedure for obtaining an *avis* from the Conseil d'Etat, a Tribunal Administratif or a Cour Administrative d'Appel may state a point of law which is necessary for the resolution of the case. Posing such a question suspends the hearing of the case in the lower court until either the Conseil replies or the permitted period for reply (three months) has elapsed.

---

[39] Compare CE 10 July 1970, SYNACOMEX, and CE 20 October 1989, NICOLO, for examples of the Conseil d'Etat referring questions to the European Court of Justice for preliminary rulings under Article 177 EEC. See further Ch. 10.

13 GENERAL OBSERVATIONS

In addition to the *contradictoire* and inquisitorial nature of the procedure, which was discussed above, the following distinctive features are worthy of comment:

1. Each decision of the Conseil d'Etat or of a local court, is a *collegiate* one; even if the decision is given by a single Sous-section or by two Sous-sections combined, this is the decision of the Conseil itself and there can be no question of an appeal to the Section or to the Assemblée, though the original formation for judgment may prefer not to decide the case but remit it for decision by a more senior grouping of the court.

2. Each matter is examined afresh, at different stages of the procedure, by members of the court who have not seen it before; thus, at the *séance de jugement* only the *rapporteur, the réviseur,* the Commissaire du gouvernement, and those Conseillers present who may happen to be members of the Sous-section d'instruction, will have seen the case before. This careful reviewing of a case by different members of the court—by juniors and seniors alike—makes for sound judgments, but also makes the procedure slow. On average, it is understood that in the Conseil d'Etat at the present time a simple case will take eighteen months to two years from receipt of the *recours* to the publication of the *arrêt*; some will take appreciably longer. In comparing this with English procedure, it should be appreciated that much of the work of the *rapporteur—l'instruction*—will in England be undertaken by the solicitors to the parties before the writ is served.

The length of procedure is not without its problems for litigants. In *H.* v. *France* (1989) 12 EHRR 74, France was held to be in breach of Article 6 of the European Convention on Human Rights because a relatively simple case before the Tribunal Administratif of Strasbourg took nearly four years to come to judgment. The European Court of Human Rights rejected the argument that a period of three years for the Conseil d'Etat to decide the appeal in the same case was also a breach of the article: only in the Tribunal Administratif was there abnormal delay.

3. Since the substantial reforms of the local courts in 1953, one of the major problems of the Conseil d'Etat has been to ensure that the decisions of the local courts are standardized throughout the country. This has been achieved by the following means:

(a) the appellate jurisdiction of the Conseil d'Etat (up to 1989) over decisions of all courts
(b) the official circulation of all decisions of the Conseil d'Etat to the local courts
(c) the system of inspection, under which senior members of the Conseil d'Etat are detailed to take a special interest in the operation of the local courts, and to visit them from time to time

In future, the main role of ensuring consistency of decisions will belong to the five Cours Administratives d'Appel, although the role of the Conseil d'Etat will not be negligible both *en cassation* and in such appeals as it will still receive directly from the Tribunaux Administratifs.

4. Finally, in marked contrast with English practice, the plaintiff cannot be mulcted heavily in costs, even if he loses his case. Since 1972 the principle of *la gratuité des actes de justice* has ended such modest court fees as previously applied; also abolished was the requirement that certain documents (such as the *requête*) had to be on sheets of stamped paper. In a *recours pour excès de pouvoir* the plaintiff may litigate in person; in a *recours en cassation* or a *recours de pleine juridiction* (e.g. for damages) he will have to employ an advocate, whose fee may amount to the equivalent of several hundred pounds. But many plaintiffs will be entitled to legal aid to meet the whole or part of their costs: a *bureau d'aide judiciaire* exists within each Tribunal Administratif and Cour Administrative d'Appel, and a single such *bureau* administers legal aid for both the Conseil d'Etat and the Tribunal des Conflits.

The Decree of 2 September 1988 now gives an administrative court the power (in order to avoid injustice) to order costs to be paid by one party to the other. Another hazard for the vexatious litigant is the penal sanction of a fine (*amende pour recours abusif*), but this penalty is seldom imposed: a recent example is CE 4 February 1991, COMMUNE DE TARASCON-SUR-ARIEGE, where the commune appealed to the Conseil d'Etat against a decision of the Tribunal Administratif of Toulouse declaring null and void fourteen (alleged) resolutions of the municipal council—resolutions which the mayor admitted had never been passed, he himself having fabricated the minutes: rejecting the appeal, the Conseil d'Etat imposed a fine of 10,000 F upon the commune for the 'caractère abusif' of the appeal.

6

The Jurisdiction of the Courts

1 INTRODUCTION

In England the Judicature Acts of 1873 and 1875 ended the competing jurisdictions of the courts of common law and the courts of Chancery. At most the English litigant has to choose between different procedures available is the same court (writ, application for judicial review, etc.).[1] In France the dual system of administrative courts on the one hand and civil and criminal courts on the other still presents the legal adviser with a critical problem at the very threshold of his task of obtaining redress for his client, namely, to which of the two sets of courts must the case in hand be brought.

The answer to this question involves the exploration of some of the subtlest distinctions in French law.[2] Small wonder, therefore, that litigants (and their advisers) sometimes go astray and bring their case before the wrong jurisdiction, achieving nothing for themselves except another decision to swell the large body of case-law on 'conflicts of jurisdiction' (see e.g. CUVILLIER, p. 151 below).

In this chapter we will outline the general principles which govern such demarcation disputes between the French courts and describe the special court by which such disputes may be resolved. Then, in the following chapter, we will see that where an administrative court is seised of a matter which properly falls within its competence according to these general principles,

---

[1] Cf. in Scotland, where Rule of Court 260B has made the application for judicial review the exclusive remedy: see A. W. Bradley, [1987] PL 314.
[2] A standard textbook (Vedel and Delvolvé, *Droit administratif*, 11th edn., 2 vols., Paris, 1990) devotes 134 pages (out of 1,471 pages of text) to the subject of our chapter; the definitive treatment remains that of R. Odent, *Cours de contentieux administratif*, i (Paris, 1977) 471–688. But see also R. Chapus, *Droit administratif général*, i (5th edn., Paris, 1990) 566–620.

there still remain certain conditions to be satisfied in the particular case, before the court can proceed to exercise jurisdiction.

2 THE BASIC TEXTS

'Judicial functions are distinct and will always remain separate from administrative functions.' In these terms (as we saw in Chapter 3) the Law of 16–24 August 1790 expressed the principle of the separation of powers as between judiciary and executive. The Law then went on:

It shall be a criminal offence for judges of the ordinary courts to interfere in any manner whatsoever with the operation of the administration, nor shall they call administrators to account before them in respect of the exercise of their official functions.

As Professors Vedel and Devolvé have pointed out,[3] this was a corollary that did not follow logically from the premises. For

neither history nor comparative law proves that the principle of the separation of administrative and judicial authorities is tied to the constitutional principle of the separation of powers. (ibid.)

Thus, the United States maintains the latter principle without the former; and there have been periods of French history, such as that of the Convention (1792–5) or of the Vichy regime (1940–4), when the former has prevailed without the latter. It was not logic, therefore, but the history of judicial obstruction and interference which prompted the French solution and produced the administrative courts. Recently, however, the Constitutional Council, while not going so far as to state that the principle of the separation of powers, as such, is of constitutional value, has stated that a limited separation between the administrative and the ordinary courts was a constitutional requirement.

[C]onsistently with the French conception of the separation of powers, there figures among the 'fundamental principles recognized by the laws of the Republic' one whereby, except for matters which are reserved by their nature to judicial authorities, there belongs to the competence of the administrative courts the quashing or rectification of decisions taken in the exercise of prerogatives of public power by authorities exercising

[3] Vedel and Delvolvé, *Droit administratif*, i. 99.

executive power, their agents, the local authorities of the Republic, or the bodies placed under their authority or control. (CC decision no. 86–224 DC of 23 January 1987; CC decision no. 89–261 DC of 28 July 1989)[4]

This decision thus gives a constitutional basis only to a separation between the jurisdictions on matters connected with the quashing of decisions made by public authorities, and not on the full range of matters which we shall see currently fall within the competence of the administrative courts.

The provisions of the Law of 1790 were reaffirmed five years later in the Decree of 16 fructidor, an III: 'The prohibition is renewed against the courts taking cognizance of the acts of the administration, of whatever kind they may be.' It is these two texts (still in force) which provide the only legislative guidance for the respective competences of the ordinary and administrative courts. Every judgment of the Tribunal des Conflits continues to recite them as the statutory roots from which the function of that body stems. And in PELLETIER (TC 30 July 1873, discussed on p. 177) the Tribunal des Conflits invoked them to justify its view that the ordinary courts[5] could not entertain suits for damages against officials in respect of acts within the scope of their duties.

The interpretation of these texts has not remained constant but has evolved under the influence of such factors as the rise of the Conseil d'Etat as an administrative court, the emergence of public authorities other than the state (e.g. local authorities and, more recently, public corporations with diverse functions), and the search for a single satisfactory criterion to distinguish 'acts of the administration'.

3 THE SEARCH FOR A CRITERION

In the years immediately after the Revolution the administration used the above texts to deprive the ordinary courts of jurisdiction

[4] Vedel and Delvolvé (ibid. 128) refer to 'la constitutionnalisation partielle du principe de séparation des autorités administratives et judiciaires'.

[5] The use of the epithet 'ordinary' for the civil and criminal courts is not importing an anglicism: thus Vedel and Delvolvé (*Droit administratif*, i. 23) state: 'The essential function of the Tribunal des Conflits is to decide whether suits should be brought before the *tribunaux ordinaires (tribunaux judiciaires soumis au contrôle de la Cour de Cassation)* or before the administrative courts.'

in any matter connected even remotely with administrative activities. It was not until the Conseil d'Etat emerged as a court in the early decades of the nineteenth century that any attempt was made to place the interpretation of the texts upon some rational footing. Then, in the period before BLANCO (TC 8 February 1873), a number of criteria were adopted. The first was that of the state as debtor (*l'Etat débiteur*), under which the Conseil d'Etat denied the ordinary courts competence to condemn the state to any money payment. The second was the criterion of 'the act of public authority' (*acte de la puissance publique*, or *acte d'autorité*); taken up enthusiastically by the jurist Laferrière,[6] this drew a distinction between those actions of the administration which involved its public authority and mere acts of management (*actes de gestion*) which did not: the former were outside the jurisdiction of the ordinary courts, the latter were within it. The third criterion, and the one favoured by the ordinary courts, was that of 'public administration' (*gestion publique*) as distinct from 'private administration'; in the latter the administration used the same processes as the private citizen and therefore came within the scope of the ordinary courts. On the other hand, disputes arising out of its *gestion publique* belonged to the administrative courts.

These early criteria, tentative and overlapping, were discarded in BLANCO in favour of a new principle, that of 'public service'. The child Agnès Blanco was injured by a wagon which was crossing the road between different parts of the state-owned tobacco-factory at Bordeaux. The question then arose, to which court, civil or administrative, the claim for damages should be brought. The Tribunal des Conflits, adopting the analysis proposed by Commissaire du gouvernement David, held that the injury arose out of the activities of a *service public* and that for this reason the administrative court had jurisdiction. This approach was subsequently approved by such influential doctrinal writers as Duguit (in *Les transformations du droit public*, 1913), Jéze (author of a six-volume treatise on *Droit administratif*, 1925–36), and Rolland (*Précis de droit administratif*, 10th edn., 1951). According to this last 'a public service is any activity of a public authority aimed at satisfying a public need'. This definition

---

[6] Edouard Laferrière, himself a vice-president of the Conseil d'Etat, laid the foundation of *droit administratif* in his magistral, *Traité de la juridiction administrative* (2nd edn., 2 vols, Paris 1888–96).

stresses that for a public service two elements must both be present: the activity of a public authority, and the satisfying thereby of a public need.

Both these elements have undergone extensive development in the case-law, so extensive indeed that later authors[7] of public service as a 'pseudo-criterion', devoid of real value. Nevertheless such authoritative commentators as Odent (a former president of the Section du Contentieux) and Braibant (now president of the Report Section) regard it as still the fundamental principle, and the courts continue to find in it their only guiding thread.[8]

A 'public need' is not only one defined by statute, it can simply be identified by a decision of a public authority. A leading example is TERRIER (CE 6 February 1903). A *conseil départemental* decided to rid its area of vermin and announced that it would pay a quarter of a franc for every viper killed by a member of the public. So successful was this scavenging campaign that the fund set aside for the promised payments was soon exhausted. Terrier was refused payment for his dead vipers and brought an action in the Conseil d'Etat against the *département*. It was held that the Conseil was the competent court since a public service was involved, the local authority having organised a viper-destruction service in the public interest. A similar view was taken by the Conseil d'Etat in THEROND (CE 4 March 1910), where a town engaged the complainant to catch stray dogs and collect dead ones. The current view of public need extends also to setting up a dental clinic, providing a camping ground, or organizing a firework-display on 14 July. Likewise, the activity of a charitable foundation providing student residential accommodation was classified as one of public service: CE 15 October 1982, MARDIROSSIAN; and the company operating ski-lifts under a contract with the local commune was compelled to surrender the ski-lifts, pending litigation over the commune's cancellation of the contract, as the continuity of a public service was at stake: CE 9 December 1988, SOCIETE 'LES TELEPHERIQUES DU MASSIF DU MONT BLANC'.[9]

---

[7] e.g. M. Waline, *Droit administratif* (9th edn., Paris, 1963), 72.
[8] Odent, *Cours de contentieux administratif*, 1965–6, 288 (the author maintained his approach in his last edition of 1977, i. 482); G. Braibant, *Le droit administratif français* (1st edn., Paris, 1984), 133.
[9] For subsequent proceedings before the Lyons Cour Administrative d'Appel, see CCA Lyon, 9 April 1991, *AJDA* 1991, 570.

The second element in the concept of *service public*, namely, that the activity in question must be carried on by a public authority, has been extended almost to vanishing-point in recent decades. In particular, it is necessary to distinguish between the public authority's role as creator or director of the public service and its role as provider. For a public service to exist, it is not necessary that a public body actually provides the service.[10] For example, *caisses* (or, as we would say, provident societies) are the basic local organizations administering the social-insurance system. Traditionally, these societies have always been regarded as private-law, not public-law, bodies. Without departing from this view, the Conseil d'Etat held that these private bodies were engaged in providing a public service and accordingly fell within the competence of the administrative courts (CE 13 May 1938, CAISSE PRIMAIRE 'AIDE ET PROTECTION'). It is sufficient that the service itself is created by statute or by decision of a public authority. Thus, in MONTPEURT (CE 31 July 1942), this approach was extended to a statutory 'organizing committee' of manufacturers formed after the German occupation of 1940 to co-ordinate glass-production in view of the acute shortage of materials. Again, in BOUGUEN (CE 2 April 1943), it was held that the governing body of the medical profession—*le Conseil supérieur de l'ordre des médecins*—was participating in the functioning of a public service, although not itself a public institution. On this basis, the decisions of a wide variety of sporting bodies are subject to review in the administrative courts.[11] Statute requires that national sporting bodies obtain authorization from the minister in charge of sport, and this authorization is treated as sufficient to make their activities the performance of a public service. The Tribunal des Conflits so held in PESCHAUD c. GROUPEMENT DU FOOTBALL PROFESSIONNEL (TC 7 July 1980), where the French Football Association was held liable to be sued in the administrative courts in respect of its

---

[10] This is a point which is not raised typically in connection with privatization in the UK. In French eyes, transferring the provision of the service to the private sector does not necessarily mean that it is no longer a public service, but rather that it has become an ordinary profit-making activity.
[11] Cf. the restrictive attitude of the English High Court in R. v. *Football Association Ltd., ex parte Football League Ltd., The Times*, 22 August 1991, where Mr Justice Rose held the Football Association to be a private body and therefore not a body subject to judicial review.

suspension of a club's vice-chairman, pending a disciplinary inquiry. Again in the world of sport, decisions of the French Table Tennis Federation and the French Cycling Federation were both held subject to judicial control by the administrative judge as these federations fulfilled 'une mission de service public': FEDERATION DES INDUSTRIES FRANÇAISES D'ARTICLES DE SPORT (CE 22 November 1974) and FEDERATION FRANÇAISE DE CYCLISME (CE 26 November 1976). On like reasoning, the Conseil d'Etat (CE 23 June 1989, BUNOZ) accepted to review a decision of the national basketball federation. Basketball players, like fast bowlers, need to be tall, but Frenchmen are mostly of medium height. Accordingly, the federation sought by rule to limit to two the number of foreigners in a team of ten: the rule treated as foreigners even naturalized citizens of less than three years' standing. The federation's refusal to withdraw the rule when challenged by a team manager was quashed as in conflict with the *Code de la nationalité*.

A third element may be distinguished in the concept of *service public*, in addition to the meeting of a public need and the participation of a public authority. The authority must have recourse to methods and prerogatives which would be excluded in relations between private parties, that is, *des prérogatives exorbitantes du droit commun*.[12] For example, it may operate the service concerned as a monopoly or may finance it by compulsory contributions from those it benefits.

Even where the activity has the appearance of a *service public*, it may not come under the supervision of the administrative courts because the special regime of administrative law is excluded. Such exclusion may be expressed by statute, or implied because the interests involved are ones traditionally within the protection of civil courts, or because the public authority decides to function under the same conditions as private operators: see, for example, SOCIETE COMMERCIALE DE L'OUEST AFRICAIN ('LE BAC D'ELOKA') (TC 22 January 1921) discussed below.

After the First World War the state began increasingly to engage in commercial and industrial activities, as various parts of the economy were nationalized or brought under some degree

---

[12] Cf. the terminology of the European Court of Justice in *Foster* v. *British Gas* (Case No. C-188/89).

of public control. Hence the recognition of a new genre of public service—the *service public à caractère industriel et commercial*, towards which the Tribunal des Conflits adopted the view that, operating as they did under conditions comparable to private enterprise and with similar objectives, they fell within the competence of the ordinary courts. An unusual illustration is the 'BAC D'ELOKA' (TC 22 January 1921). The government of the French colony of the Ivory Coast operated a ferry boat of this name across a coastal lagoon for the convenience of the public. One night it foundered, and the owner of a motor car which it was carrying claimed damages from the colony in the local civil court; the colony challenged the competence of the court. The Tribunal des Conflits held that the civil court had jurisdiction since the colony was operating a transport service on the same conditions as an ordinary businessman.

An extension of this principle occurred in NALIATO (TC 22 January 1955). A boy of this name was injured while playing games at a holiday camp run by the government for the benefit of workpeople in certain state-owned factories and their families. The Tribunal des Conflits classified this activity in the particular circumstances as a *service public social*, no different from similar welfare-services provided by private firms and therefore subject to the ordinary courts. But doubts have since been expressed about the basis of this decision[13] in view of contrary decisions on similar facts by both the Cour de Cassation and the Conseil d'Etat (see, for example, CE 17 April 1964, COMMUNE D'ARCUEIL).

The approach exemplified in 'BAC D'ELOKA' represented a restriction of the intervention of the state, in fields normally reserved for private enterprise,[14] by compelling the state in these fields to observe the rules of private law. Decisions since the 1950s, however, have shown a shift towards a broader idea of 'public service'. In EFFIMIEF (TC 28 March 1955), a syndicate of private firms employed by the administration to undertake schemes of redevelopment in slum areas was held to be subject to the jurisdiction of the administrative courts. Similarly, a question of the liability of a private firm towards a user of the motorway,

---

[13] Vedel and Delvolvé, *Droit administratif*, i. 174.
[14] F. Bénoit, *Le droit administratif français* (Paris, 1968), paras. 1462 ff.

where the firm had been granted the concession of constructing the motorway and then of collecting the tolls payable by motorists, was held to fall within the same jurisdiction (TC 28 June 1965, RUBAN c. SOCIETE DE L'AUTOROUTE ESTEREL-CÔTE D'AZUR), and the same approach is to be seen in BARBIER (p. 149 below). This change of thought and the flexibility of the meaning of the concept of 'public service' have made the line between the two jurisdictions very difficult to draw with any precision.

Nevertheless, however unsatisfactory it may be in borderline cases, the doctrine of public service has provided the courts in the past with a single criterion to apply to very different situations. Through this flexible concept the courts have been able to decide, for example, that a contract is administrative because the person employed is associated thereby with a public service—CE 20 April 1956, BERTIN (see p. 137 below); that land or buildings fall within the public domain because they have been specially adapted to public needs—CE 19 October 1956, SOCIETE LE BETON; that an enterprise aimed at fulfilling such needs falls therefore under the special regime of public works—CE 20 April 1956, MINISTRE DE L'AGRICULTURE c. CONSORTS GRIMOUARD; and that proceedings against all public bodies, whether local authorities or agencies of the central government, should be subject to common principles (a simplification still not wholly achieved in English law despite the Crown Proceedings Act 1947). But this very flexibility has at times lent the doctrine a somewhat chimerical quality.

4 THE GUIDING PRINCIPLE

In the face of these developments some recent authors have abandoned the quest for a logical criterion to determine the respective competence of each order of courts. Some even allege that the Conseil d'Etat first decides if it wishes to intervene and then classifies accordingly; according to this view there is no such thing, in essence, as a public service: public services simply exist as and when they are revealed empirically by the administrative judge.

This existentialist view (see Chenot, 'L'existentialisme et le droit', in *Revue française de science politique*, 1953, p. 57) is one of despair. A more optimistic approach is that of Professors Vedel and Delvolvé, who suggest that it is chimerical to look for a single valid criterion; rather, one needs a general idea or directive, and this they find not in the concept of public service but in that of public authority (*puissance publique*) as the test for the competence of the administrative judge: Vedel and Delvolvé,[15] citing CE 21 May 1976, G. I. E. BROUSE-CARDELL and CE 1 February 1980, CAISSE D'EPARGNE DE COUTANCES and referring to the overall study of the problem in the preface to the seventh edition of Vedel, *Droit administratif* (1981); this remains of great value. Having adopted this as the norm, one is then able to examine the exceptions. We will follow this approach and proceed to enumerate the abnormal categories of case in which the administrative judge has to give way to the ordinary courts.

5 THE PRINCIPAL EXCEPTIONS

The basic principles for separating the functions of the administrative and the ordinary courts as set out in the previous sections would lead to giving jurisdiction to the ordinary courts only when the activity of a public body was private in character. However, these principles are subject to a number of exceptions based on convenience more than principle. As the Constitutional Council stated in its decision no. 86–224 DC of 23 January 1987:

> In implementing th[e] principle [of the separation of ordinary and administrative courts], when the application of specific legislation or regulations could give rise to various kinds of litigation which would be spread, according to the usual rules of competence, between the administrative and civil courts, it is permissible for the legislator, in the interests of the good administration of justice, to unify the rules of competence within the system of courts principally concerned.

This justifies the allocation of a number of issues to the civil courts, such as, in that case, the control of the regulatory body on competition. On the other hand, some administrative decisions are not within the jurisdiction of the administrative courts because they constitute acts of state (*actes de gouvernement*). An early example of this category was the refusal by the Conseil d'Etat to adjudicate upon the removal of Napoleon III's cousin from the list of army officers by the government of the Third Republic (CE 19 February 1875, PRINCE NAPOLEON: further discussed in Chapter 7, p. 155)

---

[15] *Droit administratif*, i. 126.

(a) The Administration of Justice

The administration of justice in the civil and criminal courts is pre-eminently an area in which public authority (*la puissance publique*) is being exercised. It is also a public service. Nevertheless, the doctrine of the separation of powers, while protecting the administration from interference by the ordinary courts, at the same time removes all matters touching the functioning of these latter from the competence of the administrative courts: CE 5 November 1976, HENIN is a modern affirmation of this principle.

Thus, there can be no complaint to the administrative judge on such a matter as the decision not to prosecute (e.g. CE 10 January 1979, KALKOWSKI) or the grant (or refusal) of a pardon (e.g. CE 28 March 1947, GOMBERT). Nor can the activities of the French police be challenged before the administrative courts, at least where they fall into the category of *police judiciaire*. As admirers of Maigret will know, a fundamental distinction is made between *police administrative* and *police judiciaire* (or PJ). The former describes the preventive function of the police in regulating traffic, controlling demonstrations, and keeping order generally. The PJ are concerned essentially with criminal investigation and the apprehension of offenders.[16] Although we distinguish, in practice, in England between the 'uniformed' branch of the force and the plainclothes men of the CID, the French distinction goes much deeper. In either country the same officers may be employed upon both kinds of duties, but in France the PJ only are under the control and discipline of the ordinary courts. The *police administrative* are subject instead to the jurisdiction of the administrative courts, but these, by a strange paradox, have asserted a much more strict and effective control over police conduct than has ever been attempted by the ordinary courts. Instances are DARAMY and LECOMTE (CE 24 June 1949) discussed at p. 186; also SOCIETE FRAMPAR (CE 24 June 1960). In the latter case the Conseil d'Etat held itself competent to receive a complaint about a prefect's seizure of a newspaper as this was an act of administrative police to prevent disorder, although ostensibly based upon powers of criminal investigation conferred upon the prefect by the *Code d'instruction criminelle*.

---

[16] Likewise, impounding an illegally parked vehicle is classified as PJ activity (CE 14 May 1982, ODDOS).

On the other hand, the organization of the judiciary, considered as a branch of the public service, falls within the competence of the administrative courts. Thus, in FALCO and VIDAILLAC (CE 17 April 1953), the Conseil d'Etat accepted jurisdiction to receive a complaint against the manner in which elections of representatives of the judiciary on the Conseil supérieur de la Magistrature had been conducted. Likewise, in RATZEL (CE 22 January 1954), the Conseil d'Etat proceeded to inquire into measures affecting the career prospects of a judge of a civil court — including now the favourable or unfavourable annotating of his official file: DUJARDIN (CE 16 January 1976). Where, however, the Conseil supérieur de la Magistrature acts as a disciplinary tribunal, it falls within the category of special administrative jurisdictions subject to review on cassation by the Conseil d'Etat (CE 12 July 1969, L'ETANG; CE 14 March 1975, ROUSSEAU).

In matter of extradition, a distinction is made between a decree of the administration to extradite an individual from France[17] and a request by the French government to a foreign government for a person to be extradited to France. In the former case, the decree is reviewable by the administrative judge under the *recours pour excès de pouvoir* (e.g. CE 1 April 1988 BERECIARTUA-ECHARRI). In the latter case, the administrative judge cannot review the request because it appertains to the administration of justice: CE 21 July 1972, LEGROS, where the Conseil declined to classify the request as an *acte de gouvernement* (see p. 155).[18]

(b) Parliamentary Proceedings

Proceedings in parliament and the organization and functioning of the two legislative assemblies cannot be the subject of judicial review, in either the ordinary or the administrative courts. This again is a consequence of the doctrine of the separation of powers, but it also reflects the reluctance of the Conseil d'Etat to come into conflict with parliament. The notion was taken so

---

[17] Of course, prior to any administrative decree of extradition, there must have been a judicial hearing in the criminal courts to decide whether the requesting state had made out a case for extradition.

[18] But the status of refugee under French law, if contested, becomes a matter for a special administrative tribunal, *la Commission de recours des réfugiés*, whose decisions are subject to review *en cassation* by the Conseil d'Etat: see Ch. 3, p. 57, and App. D.

far that it required special legislation (in the form of a *loi organique* of 17 November 1958) before the state could be held liable in the administrative courts for accidents caused by parliamentary officials (e.g. when driving on duty: see CE 15 December 1952, COMPAGNIE GENERALE D'ASSURANCES).

The same prohibition on judicial review extends to the relations of the President of the Republic or the government with parliament or with each other. Thus, no action can be brought before the Conseil d'État upon the decision of the government to introduce a bill into parliament (see CE 9 May 1951, MUTUELLE NATIONALE DES ETUDIANTS DE FRANCE), or of the Prime Minister to reshuffle his cabinet, or of the President to submit a bill to a referendum or to exercise the exceptional powers conferred upon him by Article 16 of the Constitution in the face of a national emergency (CE 2 March 1962, RUBIN DE SERVENS), or to dissolve parliament (CE 20 February 1989, ALLAIN).

(c) International Relations

The international relations of the French state are outside the competence of the administrative courts. Thus, the Conseil d'Etat has consistently refused to adjudicate upon the validity or interpretation of treaties. So far as concerns the application of a treaty, the Conseil d'Etat will intervene where the question is one of French domestic law. Thus, in the case of the extradition from France of an American citizen pursuant to the Franco-American Treaty of 1909, the Conseil d'Etat accepted a *recours pour excès de pouvoir*, directed against the act of extradition, on the basis that the Constitution of 1946 gave to treaties the force of French statutes (CE 30 May 1952, KIRKWOOD). But in carrying out its obligations as a member state under the European Community treaties or executing decisions of the European Court of Justice, the French state is regarded as always subject to the jurisdiction of the administrative courts: CE 23 March 1984, SOCIETE ALIVAR. The interaction of French law and Community law is returned to in Chapter 10, p. 267.

The Conseil d'Etat used to justify its lack of competence in the field of international relations (as in that of the relations between government and parliament) by resort to the doctrine of 'act of state' (*acte de gouvernement*). This recognized that in certain limited fields the government had an unfettered prerogative of action which could not be called in question in any court. The modern trend of the case-law, however, is to curtail this doctrine or, where it persists, to disguise its presence by referring simply to the administrative judge's not having jurisdiction in the particular category of matter in issue. More will be said of these *actes de gouvernement* in Chapter 7, at p. 155.

Exceptionally, a measure taken by the administration pursuant to a treaty may fall within the doctrine of *acte détachable*. Here the administrative courts will accept jurisdiction if the act can be regarded as not directly involving international relations: the Conseil d'Etat seems now easier to satisfy on this point (eg CE 19 February 1988, SOCIETE ROBATEL). By like reasoning, the administrative judge reviewed the legality of a building-permit granted in respect of a foreign embassy (CE 22 December 1978, VO THANH NGHIA).

(d) The Doctrine of Flagrant Irregularity (*voies de fait*)
This doctrine has been described as one of the most subtle notions of French administrative law.[19] It poses also a difficulty of translation: *voies de fait*, literally 'ways of fact', carries the connotation of self-help (as distinct from recourse to law—*voies de droit*) and also of violence (in criminal law the term is used to denote a common assault). In the administrative context it indicates some irregularity on the part of the administration which is so flagrant and gross that it cannot be regarded as an administrative act at all but is treated as if it were the act of a private body, thereby losing the privilege of being adjudicated upon only by the administrative court and falling within the cognizance of the ordinary courts.

A good illustration is CARLIER (CE 18 November 1949). Carlier was a kind of French Betjeman who had persistently criticized the Administration des Beaux-Arts for having neglected or destroyed France's national monuments. One day, while taking photographs of the exterior of Chartres cathedral, he was arrested, was taken to the police station, and had his pictures confiscated. Shortly afterwards, when he joined a queue of tourists about to visit the belfry of the cathedral, the guide,

[19] Vedel and Delvolvé, *Droit administratif*, i. 140.

acting on superior orders, refused him admission. He brought an action for damages before the Conseil d'Etat, which held itself incompetent to adjudicate upon the arrest of Carlier and the seizure of his property; for these were *voies de fait*, being acts 'manifestly incapable of being connected with the exercise of a power belonging to the administration'. Carlier had, therefore, to seek his remedy for these flagrant irregularities in the ordinary courts. On the other hand, the Conseil d'Etat decided that it could entertain that part of the claim which related to his being refused entry to the belfry; this refusal was undoubtedly irregular in the circumstances but did not lose its quality of an administrative act in respect of which the court could award damages.

The case just cited illustrates another requirement of the doctrine, namely, that the irregularity in question must have infringed some fundamental right of the individual, such as liberty of the person, sanctity of property, or inviolability of the home. Likewise, withdrawal of a passport falls within the doctrine if based simply on tax irregularities (TC 9 June 1986, EUCAT), but not if there has been a criminal conviction for fiscal fraud carrying a sentence of imprisonment in respect of the tax owed (TC 12 January 1987, GRIZIVATZ).[20]

The notion of *voie de fait* overlaps, in some degree, that of *inexistence*, discussed in Chapter 9, p. 224. As we shall see, a purported administrative decision may be simply disregarded by the administrative judge as *inexistant* where it is so gross and flagrant as to amount to the administration's acting completely outside its sphere: the judge has no need to annul the decision but merely declares its non-existence. The doctrine of *voie de fait* may then come into play before the ordinary courts if the execution of this non-existent act has infringed a person's rights, (see, e.g., TC 27 June 1966, GUIGON, p. 225 below).

(e) Administrative Activities of a Private Character

This heading constitutes an important exception to the general principle that the acts of public authorities are normally subject to the *droit administratif* and justiciable in the administrative courts. For it has long been accepted that a public authority may confer a private character upon some of its acts or activities. We

---

[20] See also TC 9 June 1986, BRUNO.

have already seen above how any public authority engaged in commercial or industrial enterprise is normally subject to the civil law (TC 22 January 1921, 'BAC D'ELOKA'). Similarly, a public authority may own what is regarded as private property (*domaine privé*) as distinct from its *domaine public*, and legal questions touching this private property will be governed by the *droit civil* and decided in the ordinary courts. Property is regarded as private where it is managed and exploited by the public authority in the manner of a private owner: the state forests provide an example (TC 25 June 1973; CE 28 November 1975, OFFICE NATIONAL DES FORETS). But the forest firefighting service was held to remain a *service public*, subject to the jurisdiction of the administrative courts, in GIUDICELLI (CE 3 November 1950).

Each category of the administrative activities referred to in the last paragraph falls *en bloc* into the jurisdiction of the ordinary courts. A more difficult problem of jurisdiction is posed by the contracts of public authorities. For the mere fact that the administration is a party to a contract does not make it necessarily an administrative contract. Rather, as a general rule, it depends on the particular characteristics of each individual contract whether it constitutes a *contrat administratif*, justiciable by the administrative judge, or a *contrat privé* subject to the ordinary rules of the Civil Code and justiciable in the ordinary courts.

Three guiding principles, however, have been worked out by the Conseil d'Etat in conjunction with the Tribunal des Conflits. In the first place, it is obvious that there can be no administrative contract unless at least one of the parties is a public authority, or some intermediary acting on behalf of the authority (TC 8 July 1963 SOCIETE ENTREPRISE PEYROT), be this a *collectivité locale* (such as *commune* or *département*), a public corporation (such as the Gaz de France), or a government department or other organ of the administration.

In the second place, contracts which closely associate a private party in meeting the public need in question are administrative. Thus, contracts with an administrative agency for the supply of goods will always be regarded as administrative in character if the supplier himself carries out the service for which the goods are needed (e.g. CE 20 April 1956, BERTIN, where the contractor undertook the provisioning and catering at a camp for French citizens being repatriated). The last case provides the more modern analysis of the criterion of public service in such matters; but it has always been clear that contracts for public works and contracts granting concessions for public services fall into the administrative category. The former are expressly declared so by legislation dating from the Revolution (Law of 28 pluviôse, an VIII). The latter, the contract of concession, is a very common feature of public administration: we have already seen (pp. 126) unusual instances in TERRIER and THEROND; more typical are concessions in respect of public utilities (eg, see CE 12 May 1933, COMPAGNIE GENERALE DES EAUX, p. 196 below) or public transport (e.g. tramways, CE 21 March 1910, COMPAGNIE GENERALE FRANÇAISE DES TRAMWAYS, p. 197 below, or, in modern times, ski-lifts, CE 9 December 1988, SOCIETE 'LES TELEPHERIQUES DU MASSIF DU MONT BLANC', p. 126 above).

If such a close association does not exist, as where a contractor supplies a government department with furniture, it will depend whether or not the contract contains *clauses exorbitantes*.

This expression provides the third and most frequently used criterion for distinguishing administrative from private contracts. It involves an examination of the terms of the particular contract rather than its object. Clauses in a contract are *exorbitant* if 'they are different in their nature from those which could be included in a similar contract under the civil law' (the formula approved by the Tribunal des Conflits in TC 19 June 1952, SOCIETE DES COMBUSTIBLES ET CARBURANTS NATIONAUX), or 'where their object is to confer rights or impose obligations upon the parties quite unlike in their nature those which anyone would freely agree to in the context of civil or commercial law' (CE 20 October 1950, STEIN). Examples are clauses imposing a penalty upon the contractor, or giving the administration (but not the contractor) an option to rescind, or allowing the administration to vary the terms of the contract in the course of its performance: many of these are standard form in contracts for public works, but their presence in other contracts, not covered by the second principle above, will convert these too into *contrats administratifs*. An actual illustration[21] is the following clause from the standard contract used by the Administration des Ponts et Chaussées:

[21] Cited by Waline, *Droit administratif*, 86.

The contractor shall comply with such alterations as are notified to him in the course of the work; he must observe the orders of the Service of Engineers even if this involve a modification in the specification of the work which he has completed.

In effect, the doctrine of *clauses exorbitantes* leaves it open to the administration to choose, by the way it frames the terms of the contract, whether it wishes its contractual obligations to be subject to the *droit administratif* or the *droit civil*. How far the two systems of law diverge in matters of contract is discussed in Chapter 8 (p. 192).

(f) Matters Traditionally Reserved to the Civil Courts

By a long tradition, based partly on legislative provisions and partly on case-law, certain categories of matters are regarded as within the exclusive competence of the civil courts. The most notable instances are questions of personal status (*état civil*), of the liberty of the individual, of the citizen's right to vote, and of title to immovable property.[22]

The traditional monopoly of the ordinary courts in protecting the liberty of the individual has been confirmed by parliament both in Article 112 of the old *Code d'instruction criminelle* and in Article 136 of the new *Code de procédure pénale*. Under the present law the Tribunal des Conflits has held that any action for damages respecting the liberty of the person falls within the jurisdiction of the ordinary courts, whether it be directed against the official personally or against the service to which he belongs, and no 'conflict' may be raised by the administration in such instances (for this term, see p. 146); however, Article 136 of the *Code de procédure pénale* does not give the ordinary courts the right to declare administrative acts illegal or to interpret individual administrative decisions (TC 16 November 1964, CLEMENT, and compare TC 27 March 1952, MURETTE, where under the previous version of the Code actions against the service in such cases were held to remain within the competence of the administrative courts). Moreover, in these cases of personal liberty there is always the possible application of the doctrine of *voies*

---

[22] But decisions to award nationality or to recognize a person as a refugee or stateless are questions for the administrative courts, e.g. CE 9 October 1981, SUBRAMANIAN.

de fait (discussed at p. 135) which may extend even further the jurisdiction of the ordinary courts.

The civil judge in France has always been looked upon as the guardian of the rights of property-owners. Thus, the civil courts have been charged since 1810 with the assessment of compensation upon expropriation (or, as we inaccurately term it, compulsory purchase). This jurisdiction has been extended to any act of dispossession, even if only temporary, on the part of the administration. A seizure of immovables in this way is known as an *emprise*, and the dispossessed property-owner can look for compensation to the civil court whenever the *emprise* is irregular. The civil court cannot, however, pronounce upon the irregularity or otherwise of the *emprise*: this is a *question préjudicielle* for prior determination by the administrative court. The function of the civil judge is to assess the damages (see TC 17 March 1949, SOCIETE 'HÔTEL DU VIEUX-BEFFROI', where it was held that the damages for the irregular requisitioning of a hotel should extend to the whole of the owner's loss, including damage to furnishings and movables); but the civil judge has no power to order the eviction of the administration from the premises (SOCIETE 'RIVOLI-SEBASTOPOL', decided at the same time as the case last cited).

(g) **Special Statutory Exceptions**

In a wide variety of circumstances the legislature has intervened to create exceptions to the general principle that questions arising out of the exercise of public authority (*la puissance publique*) are justiciable in the administrative courts. Six important instances will be mentioned here, although other examples will be found elsewhere in this book.

1. *Indirect taxation.* Taxation to French eyes is peculiarly an administrative activity, and one conducted through the special processes and prerogatives of the fiscal authorities, for, as Waline sadly observes, 'money is taken from the taxpayer by the exercise of authority and without consulting his personal convenience'.[23] It follows that *le contentieux fiscal* properly belongs to the administrative judge. Nevertheless, statute provides that indirect taxa-

---

[23] *Droit administratif*, para. 118.

tion shall fall within the competence of the civil courts. This expression includes customs and excise duties, stamp duties, sales tax, VAT and other *contributions indirectes*, as opposed to income tax and other forms of direct taxation. The latter form a substantial part of the case-load of the Tribunaux Administratifs (who combine in this respect the functions of the General and Special Commissioners of Income Tax with those of the High Court in the English tax system).

2. *Employers' liability*. Under legislation dating from 1898, accidents at work, whether in private or in state employment, may be the subject of claims for compensation in the civil courts. There was felt to be no reason in logic for not having the same rules administered in the same system of courts for either category of victim.

3. *Running-down cases*. The same logic has now prevailed in respect of accidents caused by motor vehicles. The Law of 31 December 1957 confers jurisdiction upon the civil courts to try actions for damages in respect of accidents caused by vehicles belonging to the administration. Such a reform was long overdue, the previous duality of jurisdiction having led to divergent legal rules, depending on whether the plaintiff was run over by a private or an administrative vehicle: and if in collision with both, his legal fate might be desperate indeed (as we shall when discussing the Tribunal des Conflits, p. 148 below).

4. *Nuclear accidents*. As Chernobyl demonstrates, nuclear accidents will be no respecters of nice jurisdictional distinctions. By analogy to the legislation on traffic accidents, the Law of 20 October 1968 reserves exclusively to the civil courts all claims arising from such accidents.

5. *Competition*. The Ordinance of 1 December 1986 set up a *Conseil de la concurrence* as a restrictive trade-practices council and gave exclusive jurisdiction over its decisions to a civil court, the Paris Cour d'appel (for the important decision of the Constitutional Council on this legislation, see p. 18).

6. *Postal service*. The Law of 2 July 1990 transformed Posts and Telecommunications ('PTT') from an administrative public service into an industrial and commercial public service. As a result, by Article 25 of the Law, actions by customers against the Post and France Télécom are private law-suits brought before the ordinary courts and not, as in the past, before the

administrative courts. In fact, because of the absence of the need for legal representation and differences in the burden of proof, this change makes suits against those services easier. In terms of principle, this is not really a breach of the division between public law and private law but a change of regime in relation to a service.[24]

(h) Illegality as a Defence in the Ordinary Courts

The dual system of courts may give rise to further difficulties where the defendant before the civil or criminal judge wishes to challenge the legality of the administrative act upon which the plaintiff or prosecution rest their case. This is the so-called *exception d'illégalité*, and poses a jurisdictional dilemma. On the one hand, the policy of avoiding circuity of procedure, expressed in the maxim 'le juge de l'action est juge de l'exception', favours allowing the defence to be adjudicated upon by the judge of the principal issue. On the other hand, general principle requires the legality of administrative action to be determined by the administrative courts; in technical terms, a *question préjudicielle* or preliminary point arises which has to be decided first, and by a different court, before the principal issue can be resolved by the court originally seised.

Not surprisingly, the French solution has been to add yet a further exception to the general principle, in order to permit the *exception d'illégalité* to be disposed of in the criminal court in which it is first raised. In AVRANCHES ET DESMARETS (TC 5 July 1951) two persons were prosecuted for poaching on the land of a third; they were the son and son-in-law of the tenant farming the land and argued that they were permitted to hunt over the farm by a clause in the standard agricultural tenancy for that *département*, which had been approved by the prefect; the landowner challenged the legality of the clause before the criminal court, which proceeded to declare the clause illegal and to convict the accused. The Tribunal des Conflits upheld the jurisdiction of the criminal court to determine the legality or illegality of an administrative act (such as a standard contract approved by the administration) whenever this is relevant to the case before it.

---

[24] See R.M. Chevallier, 'La mutation des postes et télécommunications', *AJDA* 1990, 667.

On the other hand, the civil (as opposed to the criminal) judge cannot himself adjudicate upon an *exception d'illégalité* but must refer it as a *question préjudicielle* to the administrative judge. But of course he can and must decide the interpretation of any administrative regulation of *general* application in the same way as he does any other provision of the general law: SEPTFONDS (TC 16 June 1923) discussed in Chapter 8, p. 170.

Historically, the *exception d'illégalité* was a valuable form of judicial control by the ordinary courts in the first half of the last century when the Conseil d'Etat was still developing its own powers. Today it preserves a value, not only because of the very short time-limit for challenging the invalid act in the administrative courts, but also as avoiding the circuity of proceedings in two courts—the court seised of the principal issue can deal also with the incidental point. The danger, of course, is that a different view may be taken in the ordinary courts from that taken in the administrative courts as to the validity of the self-same administrative act or decision. From the point of view of legal theory, the most interesting aspect of the *exception d'illégalité* is that it means the administrative courts do not have a monopoly in determining the legality of administrative action.

The same kind of collateral attack on administrative action can occur in English law even after *O'Reilly* v. *Mackman* [1983] 2 AC 237. A person prosecuted under a byelaw may allege it is illegal even before a magistrates' court: *R.* v. *Reading Crown Court, ex parte Hutchinson* [1988] QB 384. Equally, a party to a civil action may raise the illegality of an administrative decision as a defence: *Wandsworth LBC* v. *Winder* [1985] AC 461, or in reply to a defence of statutory authority raised by the administration: *Cooper* v. *Wandsworth Board of Works* (1863) 14 CB (NS) 180. Unlike in France, there is no procedure for referring such collateral issues of illegality to specialist administrative judges. Equally, where a point of private law is raised in an administrative action, judges of the Crown Office List cannot refer the matter to other judges, though the expedient of appointing another judge to try the case from the outset, e.g. from the Family Division, may be adopted.

Conversely, the administrative judge may be faced, in a case properly brought before him, by an issue of private law, upon which *question préjudicielle* he will need to have a solution from the civil courts; thus, in CE 11 July 1960, ACRE DE L'AIGLE, the question whether a person had inherited a certain property was referred to the civil court, the administrative proceedings being adjourned pending its response.

(i) Summary

By way of summarizing this long list of exceptions it may be helpful to rearrange them into the following four categories.

The first covers those matters excluded from review by any court, civil or administrative, and accordingly corresponds with the two headings

(b) 'Parliamentary Proceedings', and
(c) 'International Relations' dealt with above.

The second category consists of matters justiciable exclusively in the ordinary courts, which were dealt with above under the headings:

(a) 'The Administration of Justice';
(d) 'The Doctrine of Flagrant Irregularity';
(e) 'Administrative Activities of a Private Character', and
(f) 'Matters Traditionally Reserved to the Civil Courts'.

The third consists of those matters which are reserved to the ordinary courts by virtue of express statutory provisions—see heading (g), 'Special Statuory Exceptions', above.

The fourth is the anomalous *exception d'illégalité*—see heading (h), 'Illegality as a Defence in the Ordinary Courts', above—explicable in part by the desire to avoid circuity of proceedings.[25]

6 CONFLICTS PROCEDURE: THE TRIBUNAL DES CONFLITS

A dual system of courts, whether the dichotomy be between common law and equity or between *droit administratif* and *droit civil*, inevitably leads to conflicts of jurisdiction. In France, such conflicts have assumed even greater importance than they did in England before 1873 because the French approach is to segregate according to the nature of the factual issue rather than according to the nature of the relief sought.

---

[25] There is also the view that this competence of the ordinary courts may be justified by reference to Article 66 of the Constitution.

THE JURISDICTION OF THE COURTS 145

We have already seen in the earlier parts of this chapter how complex are the rules which determine the respective competences of the two orders of courts in France. However well-disposed either set of courts may be to observe these rules, there will clearly be a need at times for a final arbiter. This arbiter is the Tribunal des Conflits.

This body came into existence in 1848, was abolished after the *coup d'état* of 1851 and finally established in 1872. Previously conflicts had been resolved by the head of state, acting on the advice of the Conseil d'Etat—a necessarily suspect procedure in the eyes of the ordinary courts.[26] The new court, with typically French logic, is composed of an equal number of judges drawn (four from each) from the Cour de Cassation and the Conseil d'Etat as the respective supreme courts of the two orders of jurisdiction.[27] This device of a 'paired court' (*tribunal paritaire*) is found again in France in the humbler spheres of the labour courts (*conseils de prud'hommes*), composed equally of employers and employees, and the agricultural rent-tribunals (*tribunaux paritaires des baux ruraux*) composed equally of landlords and tenants. A paired tribunal risks, of course, finding itself split down the middle on issues of policy. In this event the Tribunal des Conflits is augmented by its titular president, the Minister of Justice (Garde des Sceaux), who then has the casting vote. In fact, there have been only about ten such tied decisions since 1872, the last three being GAVILLET (TC 31 March 1950), CIANELLI (TC 13 June 1960), and SAFER DE BOURGOGNE (TC 8 December 1969). It is a testimony to the close understanding and mutual respect between the two supreme courts that the Tribunal des Conflits can function without the need normally for a tie-breaking vote.

Proceedings before the Tribunal des Conflits resemble in some respects those before the Conseil d'Etat (described in Chapter 5). Thus, it is essentially a written procedure under the

---

[26] Belgium and Italy, on the other hand, entrust questions of conflict to the highest civil court in their respective countries: see below, Ch. 10.

[27] The formal process is that the Cour de Cassation and the Conseil d'Etat each designate 3 Conseillers; these 6 then choose 2 others (in practice, one drawn from each supreme court). Two further members (drawn as before) are added as *suppléants*, that is, alternates, to act as substitutes if any of the 'original' 8 is not available, thereby ensuring a bench of 8 (or 9, with the Minister of Justice). The quorum is 5 members, but usually all 8 sit.

guidance of a *rapporteur*, and one meets again the office of Commissaire du gouvernement, charged with presenting his or her independent conclusions to the court. It is a rule of practice that for each case the *rapporteur* and the Commissaire du gouvernement shall be drawn from different sides of the court. The court also sits, like the Conseil d'Etat, in the Palais-Royal (although in a special chamber—the Salle des Conflits) and its members do not robe.

Traditionally, the Tribunal des Conflits might be concerned with three types of conflict: a *positive* conflict of jurisdiction, where both the administrative and the civil courts claimed jurisdiction; a *negative* conflict, where both disclaimed jurisdiction; and a conflict, not of jurisdiction, but of *decision* (to be explained below). To these a reform of 1960 has added a fourth possibility: what we may term *incipient* conflicts. These arise when either an administrative or ordinary court (or, in some instances, only the Conseil d'Etat or the Cour de Cassation) submits a question of jurisdiction to the Tribunal des Conflits by way of a preliminary point of law.

(a) Positive Conflict of Jurisdiction

This arises whenever the administration objects to a case proceeding in one of the ordinary courts on the ground that the latter lack jurisdiction. The prefect of the *département* in which the court sits must lodge with the court a formal *déclinatoire de compétence* calling upon it to withdraw; if it refuses to do so, the prefect may proceed to 'raise the conflict' before the Tribunal des Conflits. If the Tribunal agrees with the prefect, it will declare the ordinary court incompetent in the matter, leaving the plaintiff, if he chooses, to start his action afresh in the administrative courts. If, on the other hand, the Tribunal disagrees with the prefect's view, it will quash his formal *arrêté* raising the conflict, and so leave the ordinary court free to proceed with its case.

A famous example is the Radio-Andorra case (TC 2 February 1950, RADIODIFFUSION FRANÇAISE). A commercial radio-station in the principality of Andorra put out transmissions on the wavelengths reserved by international conventions for other users; the latter complained to the French government, which ordered the state-controlled French radio to jam Radio-Andorra.

THE JURISDICTION OF THE COURTS 147

The French company which had the exclusive programme-rights on Radio-Andorra sought an injunction to stop the jamming from the Paris civil court of first instance. This was granted, the court accepting the argument that the jamming was a flagrant illegality (*voie de fait*, p. 135 above) giving rise to civil liability; and the *chambre des référés* of the Paris Cour d'appel upheld the injunction. At this point the prefect of the Seine raised a conflict before the Cour d'appel, arguing that the jamming had the character of an act of state (*acte de gouvernement*), outside the competence of the civil courts. The Tribunal des Conflits accepted this view, despite the contrary conclusions of the Commissaire du gouvernement, being persuaded, it seems, by the international character of the dispute. (For *actes de gouvernement*, see p. 155).

The raising of a positive conflict in this way is a privilege belonging to the administration alone. The procedure is one-sided in that the ordinary courts have no corresponding right to challenge a usurpation of jurisdiction by an administrative court. For example, most commentators consider that the Conseil d'Etat trespassed upon the competence of the civil court in FALCO and VIDAILLAC (for facts of these cases, both decided on 17 April 1953, see p. 133), but the Conseil d'Etat is usually scrupulous not to encroach in this way. The absence of symmetry is explicable, historically; the fear was that the ordinary courts would interfere in administrative matters, and the procedure was designed to prevent this.

(b) **Negative Conflict of Jurisdiction**

If one and the same issue has been raised before both the administrative courts and the ordinary courts, and each have declared that the case falls within the competence of the other and refused jurisdiction, the frustrated parties to the dispute are faced with a negative conflict of jurisdiction. Unlike a positive conflict, where the procedure is designed to safeguard the separation of powers by protecting the administration from suit in the ordinary courts, a negative conflict may constitute a denial of justice to the citizen. For this reason, important reforms were introduced in 1960 in order to forestall such negative conflicts, so far as possible, by seeking a preliminary ruling from the Tribunal des Conflits (discussed at p. 149 below).

There is not, of course, a negative conflict simply because an issue is declared to be non-justiciable by either order of courts (as where both the administrative and the civil judge refuse to decide upon the unconstitutionality of a statute); nor where one court declares itself incompetent in favour of another court *in the same order* (as where a Tribunal Administratif rejects a suit as properly falling within the jurisdiction not of itself but of the Conseil d'Etat).

The raising of a negative conflict (of rare occurrence since 1960[28]) is by a simple request to the Tribunal des Conflits on the part of the party or parties concerned; if a criminal court is involved, the Minister of Justice lodges the request on behalf of the prosecution.

(c) Conflict of Decisions

Jurisdiction in this third type of conflict was conferred upon the Tribunal des Conflits by statute in 1932, following the earlier stages of the litigation in ROSAY (TC 8 May 1933). In this case, the passenger in a private car was injured when it collided with an army vehicle; he sued the driver of the private car, but the civil court held this driver free from blame. The plaintiff then sued the army authorities, but the administrative judge held that the army driver was not at fault. The legislature then intervened to resolve this impasse.

The Law of 20 April 1932 provides that, where final judgements have been obtained from the administrative and the ordinary courts in 'law-suits having the same object' (*litiges portant sur le même objet*) and these are contradictory so as to lead to a denial of justice, the matter can be brought before the Tribunal des Conflits, which proceeds to decide the actual merits of the case. In this respect, a conflict of decisions differs sharply from a conflict of jurisdiction, where the Tribunal's function is limited to settling the jurisdictional issue only.

In ROSAY it was finally held by the Tribunal des Conflits that responsibility for the accident rested with the private driver. We have already seen that this particular situation could no longer

---

[28] See, e.g., TC 8 February 1965, MARTIN; a party had to raise a negative conflict after the two courts seised of the same issue had both failed to seek a preliminary ruling under the procedure introduced in 1960 (see p. 149 below); see also TC 25 January 1988, BUNELIER.

(after 1957) cause difficulty, jurisdiction being entrusted by statute exclusively to the ordinary courts even though the accident involves an administrative vehicle (p. 141 above). In general, a conflict of decisions is very rare (see, e.g., TC 12 December 1955, THOMASSON).

(d) Preliminary Rulings

The Decree of 25 July 1960 was designed to forestall conflicts of jurisdiction, especially negative conflicts. It permits a court of either order to submit a question of jurisdiction to the Tribunal des Conflits for a preliminary ruling. This 'pre-conflict' procedure of *renvoi* may be invoked in either of two situations.

First, any administrative or ordinary court *must* seek such a ruling where a case comes before it having been rejected by a court of the opposite order as being outside its jurisdiction, and the second court disagrees with this rejection. Secondly, the Conseil d'Etat or the Cour de Cassation *may* request a preliminary ruling whenever any case before it raises a jurisdictional issue of 'serious difficulty' touching the respective competences of the two orders of courts.

Whereas the first procedure is intended to anticipate, and so avoid, conflicts of jurisdiction in the interest of more speedy justice for the citizen, the second has the more academic object of encouraging the early submission of an incipient conflict to the Tribunal des Conflits, as the court best suited to resolve it, rather than (as in the past) leaving each supreme court to work out its own independent (and often divergent) solutions.

ROLLAND (TC 12 June 1961) illustrates the second procedure, the *renvoi* in this case being by the Cour de Cassation, which sought a ruling whether a regulation governing pension rights for employees of the SNCF (the French national railways) was to be classified as an administrative or a private act; the Tribunal des Conflits decided it was a matter not of administrative but of private law. Nevertheless, in COMPAGNIE AIR FRANCE c. EPOUX BARBIER (TC 15 January 1968) under a similar *renvoi*, the Tribunal des Conflits was of the opinion that, although Air France was a private company operating under private law, a regulation requiring air hostesses to resign on marriage had 'an administrative character regulating the organization of what was essentially a public service'; therefore the validity of the regulation was a matter for the administrative courts.[29]

7 CONCLUSION

If the subject of jurisdiction has been treated at some length in this chapter, it is because we feel that in the general admiration in which many common lawyers hold the Conseil d'Etat and the *droit administratif* there is a tendency to overlook this, the other side of the picture. A dual system of courts produces its own peculiar advantages and disadvantages. Conflicts of jurisdiction fall into the latter category and are the price which the French pay for their separate system of *droit administratif*. On the other hand, it would be wrong to exaggerate the number of cases in which any kind of conflict occurs. The rules which define the respective competences are now reasonably well settled. Only occasionally is the Tribunal des Conflits called upon to trace exactly where the frontier runs on the jurisdictional map. How occasional this is can be seen from the fact that in the calendar year 1990 only thirty-five cases came before the Tribunal des Conflits for decision (see Judicial Statistics, Appendix G). With this figure we may compare the 7,634 and 58,302 cases decided by the Conseil d'Etat and the Tribunaux Administratifs respectively in the judicial year ending on 31 December 1990.

Nevertheless, the 1980s witnessed a slight increase in cases, towards fifty or so a year as against some thirty a year in the 1960s. The Tribunal des Conflits needs to sit only four or five times a year to dispose of its case-load.

Finally, it should not be overlooked that the Tribunal has played, and continues to play, an important role in the evolution of the substantive *droit administratif*: see, e.g., TC 8 February 1873, BLANCO, on the principles of administrative liability (p. 174 below); TC 2 February 1950, RADIODIFFUSION FRANÇAISE, on the doctrine of *actes de gouvernement* (p. 146 above); TC 9 June 1986, BRUNO, on the doctrine of *voies de fait* (p. 136 above).

---

[29] For an example of a *renvoi* by the Conseil d'Etat, see ADAM ET AUTRES (CE 16 December 1960 and TC 12 June 1961), which was the first use of the new procedure by the Conseil d'Etat; also GUIGON (TC 27 June 1966) and PESCHAUD (CE 7 December 1979 and TC 7 July 1980; see pp. 127–8 above).

By way of postscript to this chapter we may recount the cautionary tale of M. et MME CUVILLIER (TC 2 June 1945). This began in February 1934 during the Stavisky riots, when the two plaintiffs were mistaken for rioters and badly beaten up by the police on the Place de la Concorde in Paris. They brought an action before the Conseil d'Etat for damages against the state and the city of Paris on the ground of the faulty functioning of the Paris police force. The Conseil d'Etat declined jurisdiction as it took the view that the assault occurred in the course of the suppressing of rioting and so fell with the Law of 16 April 1914, which made actions for damages justiciable in the civil courts in these circumstances. The plaintiffs brought a second action before the Seine *tribunal civil*; this court held the Law of 1914 did not apply, because the assault upon the Cuvilliers could not be connected with the rioting, which had taken place earlier in the day and some distance from where they were assaulted. Thus, eleven years after the episode, the victims came to the Tribunal des Conflits, basing their case on a conflict of decisions, rather than a negative conflict of jurisdiction, as the former would permit the Tribunal to decide the merits of the case without further delay. But the Tribunal took the view that in the previous, abortive, proceedings neither the Conseil d'Etat nor the *tribunal civil* had reached any decision on the merits, so that there was simply a conflict of jurisdiction; this it resolved in favour of the competence of the *tribunal civil*, as it considered that there was a direct connection between the riots and the unusual behaviour of the police. The plaintiffs were therefore left to recommence proceedings before the *tribunal civil*.

7

The Conditions Precedent for Judicial Review

In this chapter we are concerned with what the French term *conditions de recevabilité*, that is, the conditions which must all be present before the administrative court can be seised of a case and examine its merits. In a sense this was also the subject of the last chapter, since the paramount condition governing litigation in the administrative courts is that the matter must fall within their special jurisdiction. Assuming, however, that this *sine qua non* is satisfied and the matter before the administrative judge is truly 'administrative', what other conditions must be met before the judge can proceed to his decision?

The conditions to be examined are:

1. The nature of the act under review (*l'acte administratif*)
2. The role of the 'prior decision' (*la décision préalable*)
3. The *locus standi* (*la qualité*) of the plaintiff
4. The absence of parallel relief (*le recours parallèle*)
5. The time limits (*les délais*) for commencing proceedings

Finally, something will be said about clauses seeking to exclude judicial review.

1 THE NATURE OF THE ACT UNDER REVIEW

We have already seen in the previous chapter that the administrative courts' powers of judicial review require the existence of an *acte administratif* emanating from some organ of the administration. In this chapter we are not concerned so much, as we were in the previous chapter, with the administrative character of the act under review but rather with the notion of

CONDITIONS PRECEDENT FOR JUDICIAL REVIEW 153

acte itself.[1] There must always be some act or omission of juridical effect; as was said by the Conseil d'Etat, one cannot ask for the annulment of a vehicle accident (CE 16 October 1970, PIERRE C. EDF); if a plaintiff is asking for damages he must first obtain a refusal from the administration which then becomes an *acte administratif* open to challenge.

It is clear law that the notion covers not merely measures of general application (*actes réglementaires*), such as governmental decrees, ministerial regulations, or local byelaws, but also any decisions which apply to a single individual (*actes individuels*) or to a group of individuals (*actes collectifs*): instances of the latter would be the decision of the examiners in an open competitive examination.[2] The *actes administratifs* which have been mentioned are classified as 'unilateral' in that they commonly impose upon the citizen a change in his legal situation without his consent. In this respect they are contrasted with the administrative contract, which is also an *acte administratif* but 'bilateral' in that it requires the consent of the other contracting party (as to administrative contracts and the remedies *de pleine juridiction* to which they give rise:, see Chapter 8, p. 192).

The case-law has been very ready to extend the notion of *acte*. Thus, as we shall see below, a positive act is not essential: silence on the part of the administration in the face of the plaintiff's demand constitutes, after four months, an implied rejection, which is then justiciable as an *acte administratif*.

A decision of the Médiateur (p. 30 above) that he could not usefully pursue a complaint lodged with his office is not an administrative decision open to judicial review, although the Médiateur has the character of an administrative authority: CE 10 July 1981, RETAIL.

Any decision emanating from what the French classify as an administrative jurisdiction (as distinct from an administrative

---

[1] French lawyers talk of *actes administratifs* in this context to designate formal administrative decisions such as making decrees or contracts, granting nationality, refusing compensation, etc. A *décision exécutoire* merely implements this *acte*. Actions of the administration (*activités*) have a different character. Given the ambiguity of the word 'act' in English in relation to these two ideas, we shall distinguish between 'decisions' and 'actions'.

[2] CE 18 March 1949, CHALVON-DEMERSY, or in a degree examination (CE 27 May 1987, LOMARDI-SAUVAN), or the admissions policy adopted by a university (CE 27 July 1990, UNIVERSITE PARIS-DAUPHINE).

authority) is not an *acte administratif* but an *acte juridictionnel*; as such, as we have seen, it is reviewable *en cassation* before the Conseil d'Etat as the supreme administrative court. For examples of such specialized administrative jurisdictions, see Chapter 3, p. 55, and Appendix D, p. 290.

The French draw a distinction between *actes administratifs* and those measures which may be regarded as no more than the administration putting its own house in order. Such domestic 'house-rules' are described as measures of purely internal organization within the administration (*mesures d'ordre intérieur administratives*) and are not normally open to judicial review at all. Thus, one cannot challenge a departmental circular giving advice (as distinct from directives) on the interpretation of a statute;[3] nor an instruction about school uniform or curricula (see e.g. CE 20 October 1954, CHAPOU, where the Conseil d'Etat refused to entertain proceedings to quash a headmistress's rule forbidding the wearing of ski-trousers.[4] A similar attitude is taken towards disciplinary measures in the armed forces: see GUILLET (CE 10 October 1990), where a soldier vainly sought review of a sentence of 40 days imprisonment, of which 20 were in solitary confinement. Such an attitude is not unknown in English law in relation to managerial discretion within the administration, e.g. *ex parte Fry* [1954] 2 All ER 118 on minor discipline in the fire service. Yet in both legal systems, recent developments in administrative law demonstrate a greater willingness on the part of the courts to intervene in disciplinary decisions: CE 26 July 1985, GANDOSSI ET JOLY, and CE 10 October 1990, HYVER;[5] cf *Leech* v. *Deputy Governor of Parkhurst Prison* [1988] AC 583, *R.* v. *Civil Service Appeal Board, ex parte Bruce* [1989] ICR 171. *Mesures d'ordre intérieur*, as they are termed, are regarded, in general, as matters within the complete

---

[3] The distinction is a very difficult one to draw in practice: CE 19 January 1954, NOTRE DAME DU KREISKER; cf. CE 15 May 1987, ORDRE DES AVOCATS A LA COUR DE PARIS.

[4] On similar facts, *Spiers* v. *Warrington Corporation* [1954] 1 QB 61.

[5] Thus, a disciplinary measure of a prison authority, if it curtails an actual *right* of the prisoner (to receive books and periodicals) becomes reviewable: HYVER, above, when the Minister of Justice had stopped a burglar receiving periodicals intended to assist his burglarious profession on release, a decision upheld on the merits. For a discussion of the issues in relation to the European Convention of Human Rights see R. Errera [1991] PL 300.

discretion of the administration, or where, as Rivero expressed it, 'the administration has the benefit of a large measure of autonomy'. But above a certain point the administrative measure may become, because of its gravity, an *acte administratif* within the cognizance of the administrative judge (see CE 26 January 1966, DAVIN, where the peremptory expulsion of a pupil from a state school was adjudged so grave a sanction as to fall outside the category of such 'measures', and to be an *acte administratif* open to judicial review).

For much the same reason the Conseil d'Etat also refuses jurisdiction in regard to *actes de gouvernement*. At first sight, this may seem paradoxical since such 'governmental' acts, by their very name, must be acts of the administration. The paradox is perhaps resolved, if one adopts the translation 'acts of state'; certainly, this concept in English constitutional law has affinities with the French doctrine.

A landmark in the evolution of the doctrine was PRINCE NAPOLEON (CE 19 February 1875). After the Prussian defeat of Napoleon III and the collapse of the Second Empire in 1870, the annual Army List published in 1873 omitted the name of the former Emperor's cousin, Prince Napoleon, who had been appointed general by the Emperor in 1853. This omission was upon the instructions of the Minister of War. The Conseil d'Etat, to whom the Prince appealed, rejected the Minister's argument (based upon a previous *jurisprudence constante*) that the plea of *acte de gouvernement* could be raised in any case of a political complexion, such as this undoubtedly was; but it went on to decide that such family appointments were governed by special statutory provisions under the terms of which they could not appear, as of right, in the Army List.

This case is of importance as establishing the right of the Conseil d'Etat to determine for itself what matters fall within the doctrine of *actes de gouvernement*. In subsequent decisions the doctrine has been progressively cut down, so that at the present day it extends only to the relations of the government, on the one hand, with parliament, and on the other, with foreign states or international organizations (see, e.g., TC 2 February 1950, RADIODIFFUSION FRANÇAISE, discussed at p. 146). As we have seen in the last chapter, both international relations and parliamentary proceedings are now recognized as areas of peculiar sensitivity not suitable for judicial review in the administrative courts. This shrinkage of the ambit of *actes de gouvernement* has led one commentator to deny the continued existence of the doctrine in the modern law.[6] But this goes too far; for, as M. Weil remarks, 'in a given political society there are acts which are so important from the point of view of the preservation and defence of the society that they should not be limited by legal considerations'.[7]

The exclusion of judicial review in the situations mentioned above is usually explained by the fact that the action criticized is closely linked with bodies over which the Conseil d'Etat has no control, namely, the French parliament or a foreign state. Even here the doctrine of *actes de gouvernement* can sometimes be whittled down further by the theory of *acte détachable*: the administrative judge regains competence to review if he can separate off some *acte administratif* from the *acte de gouvernement*. For example, where a French consul expelled a French citizen from a foreign country as being undesirable, his decision to do so was detachable from his other diplomatic activity and could be reviewed (CE 4 December 1925, COLRAT). And in the matter of international relations the doctrine has been to some extent outflanked by extending the principle of liability without fault to the consequences of an international treaty (CE 30 March 1966, COMPAGNIE GENERALE D'ENERGIE RADIOELECTRIQUE, discussed further in Chapter 8, p. 190.[8]

On the other hand, by recognizing the existence of a state of war the Conseil has decided in some cases that it was impossible to hold the administration liable for its acts and that compensation could only be got by legislating specifically for such war-damage. Thus, when an Italian ship was stopped on the high seas by a French warship during the Algerian rebellion, the court ruled that the situation was equivalent to a state of war and so prevented compensation, but (significantly) it refused to consider the action as an *acte de gouvernement* involving international

---

[6] See M. Virally, 'L'introuvable acte de gouvernement', RDP 1952, 317.

[7] 'The Strength and Weakness of French Administrative Law' [1965] CLJ 252.

[8] Also, as in England, the French administrative courts have recently shown themselves willing to review such erstwhile 'acts of state' as the issue (or withdrawal) of a passport: see TC 12 January 1987, GRIZIVATZ, and TC 9 June 1986, EUCAT, discussed in Ch. 6, p. 136. (cf. *R. v. Foreign Secretary ex parte Everett* [1989] QB 811).

relations between France and a foreign power. It follows that if the same incident had occurred without the excuse of the Algerian situation, the administration would not have escaped liability (CE 30 March 1966, SOCIETE IGNAZIO MESSINA; and contrast the English rule in *Buron* v. *Denman* (1848) 2 Exch. 167).

2 THE RULE OF THE 'PRIOR DECISION'

La règle de la décision préalable is an established principle in *droit administratif* and indicates that any proceedings before the administrative courts must, in effect, be directed against a decision which has already been taken by the administration. One's action is, as it were, *in decisionem*. The rule presents no problem in proceedings to annul an administrative act; for *ex hypothesi* annulment is then being sought of a decision, whether this be formulated as a decree or regulation of general application, or as an *arrêté* affecting only the person bringing the proceedings, or as an express or implied rejection of a request.

It is in proceedings for damages (*en indemnité*) that the rule assumes significance. A right to bring an action for damages against the administration does not accrue simply because some event (e.g. the explosion of a munitions dump) has happened; there must first be a decision (such as to pay the victim no damages or damages which he deems inadequate), and it is then against this decision that the victim seeks redress from the administrative judge.

This rule does not, however, apply where the damage arises (as is common) from a contract for public works: the victim of an *opération de travaux publics* may bring an action immediately, without awaiting the negative decision from the administration.

The rule is a survival from the days of the theory of the 'minister-judge' (discussed in Chapter 3, p. 44), but is retained as ensuring at least an opportunity for the parties to settle their difference amicably before resort to litigation.[9] It cannot, however, be used as a device on the part of the administration to deny the victim justice; thus, the silence of the administration when faced with a request for compensation is, by special statutory provision, treated as an implied rejection of the request after the lapse of four months.

[9] Cf. the stricter German procedure of 'remonstrance' (*Widerspruch*) referred to in Ch. 10, p. 260.

158 CONDITIONS PRECEDENT FOR JUDICIAL REVIEW

There is no exactly comparable doctrine in English law: the rule that an applicant for judicial review seeking an order of *certiorari* or other prerogative relief, must have exhausted his statutory remedies has more affinity with the requirement (to be discussed later, p. 160) concerning the absence of parallel relief.

3 THE *LOCUS STANDI* OF THE PLAINTIFF[10]

It is a general principle of French procedure, whether in the civil or administrative courts, that the plaintiff must have some personal interest in the proceedings: 'pas d'intérêt, pas d'action'. Before the administrative judge, this means that the plaintiff must be able to show that the decision attacked is one injurious (*faisant grief*) to his interests. This requirement creates no difficulty in proceedings against the administration for damages. It is rather in proceedings to annul an administrative act that the rules governing the plaintiff's *locus standi* (or *qualité*) have been worked out in considerable detail.

Although anxious to expose administrative illegalities to attack, the Conseil d'Etat has been reluctant to admit an *actio popularis* available to every citizen and therefore open to abuse and likely to hamper effective public administration. To this end the case-law steers a middle course which does not always coincide with the path of logic.

Various interests of the plaintiff have been held sufficient to justify an action by him. For instance, a financial interest was accepted in COOK ET FILS (CE 5 May 1899), where a firm of travel agents successfully challenged a municipal byelaw subjecting excursion charabancs to the same stringent regulations as taxi-cabs. In SYNDICAT DES PROPRIETAIRES DU QUARTIER CROIX-DE-SEGUEY-TIVOLI (CE 21 December 1906) the withdrawal of a tramway service to one district of Bordeaux was challenged by an association of 'users' organized by Léon Duguit, then Dean of the Law Faculty, and the Conseil d'Etat accepted that any user of a public service had sufficient 'intérêt pour agir'. This also illustrates that a 'collective' interest will suffice; other examples are ASSOCIATION DES ANCIENS ELEVES DE L'ECOLE POLYTECHNIQUE (CE 13 July 1948), where an old

---

[10] See C. Harding, 'Locus standi in French Administrative Law' [1978] PL 144.

boys' association took proceedings out of concern for the reputation of their former school. But an association that casts the objects too widely risks being non-suited as becoming a vehicle for an *actio popularis*: e.g. UNION REGIONALE POUR LA DEFENSE DE L'ENVIRONNEMENT, DE LA NATURE, DE LA VIE ET DE LA QUALITE DE LA VIE (CE 26 July 1985); ASSOCIATION 'LES AMIS DU SOCIALISME' (CE 26 May 1986); cf. ASSOCIATION 'LES AMIS DE LA TERRE' (CE 8 March 1985). This is to be contrasted with the narrow approach adopted by the English High Court in *R.* v. *Rose Theatre Trust Ltd.* [1990] 1 All ER 954 and by the Court of Session in *Scottish Old People's Welfare* [1987] SLT 179, where associations did not have standing, despite the clear connection between their objects and the subject-matter of the litigation. Also, a French trade-union or professional association lacks standing to attack an individual decision (as distinct from a regulatory measure) that affects one or more of its members: the members must bring their own proceedings (CE 30 November 1979, MARTIN). A professional interest is accepted, as in LOT (CE 11 December 1903), where a graduate whose degree was the required professional qualification for archivists objected to the appointment of a non-graduate by the National Archives. And the interest may be not 'material' at all but 'moral', as in ABBE DELIARD (CE 8 February 1908), where a regular worshipper was held to be entitled to object to the closing of a church. Nor did the immorality of their profession prevent some prostitutes (coyly described as 'filles galantes') from obtaining review of certain administrative decisions restricting their activities, as an infringement of personal liberty: DOL ET LAURENT (CE 28 February 1919). In the field of planning, a building-permit may be challenged by an immediately adjoining owner (CE 30 September 1988, SOCIETE NATIONALE DE TELEVISION EN COULEURS 'ANTENNE 2') or by a wider area of owners affected if the project is a major development, such as a holiday village (CE 15 April 1983, COMMUNE DE MENET); but this falls far short of vesting in every citizen a sufficient environmental interest, just as the plaintiff (in CE 11 February 1949, FAVERET) was non-suited who claimed, as citizen, an interest in upholding the rule of law.

The interest of the tax-payer in attacking fiscal measures has posed difficult problems. The Conseil d'Etat has drawn a distinction between local and national taxation. The incidence of the latter is so widespread that to give every tax-payer a right of action would be tantamount to admitting an *actio popularis*. This objection does not apply to local taxation; the Conseil d'Etat admits the interest of the local tax-payer to challenge any measure which will have repercussions on the finances of the local authority. Thus, in CASANOVA (CE 29 March 1901) a tax-payer in a Corsican town was held to be entitled to attack, by means of a *recours pour excès de pouvoir*, a decision of the prefect, who had refused to annul a resolution of the municipal council to establish a clinic at public expense. Since the reform of judicial review in England in 1977 and the decision of the House of Lords in *IRC* v. *National Federation of the Self-Employed* [1981] 2 All ER 93, standing has become a subject of limited importance in England.[11] Courts may even proceed to deal with the substance of the case and only, as it were as an afterthought, address the issue of standing, e.g. *R.* v. *Boundary Commission, ex parte Foot* [1983] QB 600. This is typically the case in France. Thus, in CE 21 December 1990, ASSOCIATION POUR L'OBJECTION DE CONSCIENCE A TOUTE PARTICIPATION A L'AVORTEMENT, the Conseil d'Etat considered a *recours* brought by an association of opponents of abortion against the authorization given by the Minister of Health to the sale of the 'morning after' abortion-pill on the ground that it violated the European Convention on Human Rights. No point on the standing of the association was taken.

4 THE ABSENCE OF PARALLEL RELIEF

Unlike the other conditions which we have been examining, the present requirement is only applicable to proceedings to annul. Historically, the rule is explained by the need, before the reform of 1953, to prevent everyone aggrieved by administrative action from bringing by a *recours pour excès de pouvoir* before the Conseil d'Etat matters for which some other relief was expressly provided in a different court, such as a regional Conseil de Préfecture. But for the rule, the Conseil d'Etat might have been even more overwhelmed with work.

---

[11] Even before the reform, it was taken for granted that a ratepayer had *locus standi*: *Prescott* v. *Birmingham Corporation* [1954] 3 All ER 698 (see L. Neville Brown, 'The Ratepayer's Standing to Challenge Unlawful Expenditure' (1982) 7 *Hold. Law Rev.* 161).

CONDITIONS PRECEDENT FOR JUDICIAL REVIEW 161

Since the reform of 1953, the Tribunaux Administratifs have become the normal jurisdiction of first instance. Nevertheless the rule on parallel relief continues to serve to protect all administrative courts from large amounts of litigation. In addition, the rule on parallel relief is also justified by the priority of any special procedure over general methods of redress (*specialia generalibus derogant*)—recourse to the administrative courts is a fall-back when other means of redress have failed. Thus in income-tax matters, the tax-payer is obliged to complain about an assessment first to the regional director of taxes and thereafter to a conciliation body, the *commission départementale*. Only after exhausting these remedies can he bring a *recours* before the Tribunal Administratif. This procedure currently reduces the initial number of over half a million complaints a year against tax assessments to about ten thousand *recours* before the Tribunaux Administratifs.

The attitude of the French courts is mirrored by that of the English courts in relation to judicial review. Where an alternative remedy exists, judicial review is allowed only in exceptional circumstances: *R.* v. *Chief Constable of Merseyside, ex parte Calveley* [1986] QB 424, which provides a modern illustration of the principle propounded by the House of Lords nearly a century ago in *Barraclough* v. *Brown* [1897] AC 615 and *Pasmore* v. *Oswaldtwistle UDC* [1898] AC 387.

5 THE TIME-LIMIT FOR COMMENCING PROCEEDINGS

The amplitude of judicial review of administrative action in *droit administratif* is counterbalanced by a very short period of limitation for commencing proceedings. Usually, the plaintiff has only two months in which to take action. This may be compared with the modern rule in English law whereby proceeding for judicial review under Order 53 must normally be commenced within three months.

The period of two months is known, technically, as a *délai*, and begins to run from the date when the decision complained of is published or notified to the plaintiff, or, in the absence of an express decision, it runs from the implied decision to reject the plaintiff's demand, which (as seen above) is presumed by law after four months' silence by the administration. Where the plaintiff is alleging the liability of the administration to compensate him in damages, the *délai* runs from the date of the decision rejecting his claim, but in this case (as in all *recours de pleine juridiction*) the claim has to be *expressly* rejected before the *délai* can begin to run against the plaintiff.

After the *délai* has expired, the plaintiff will not be prevented thereafter from raising the illegality at any time by way of an *exception d'illégalité*. This would arise, for instance, if he is prosecuted for non-compliance with the administrative act of which he complains (for fuller discussion of the *exception d'illégalité*, see Chapter 6, p. 142).

He can also, after that time, raise a point which shows the *acte administratif* to have been illegal, if this amounts to a *moyen d'ordre public* (Chapter 9, p. 224).

6 EXCLUSION OF JUDICIAL REVIEW

It remains to consider the circumstances in which an *acte administratif* can be excluded from review in the administrative courts. The Constitutional Council has recognized the 'right of defence' (what Americans would call 'due process') as a constitutional value based on fundamental principles recognized by the laws of the Republic. This right includes the possibility of challenging decisions of the administration in court (CC decision no. 80–119 L of 2 December 1980), and renders unconstitutional an immunity from suit granted to an administrative official (CC decision no. 89–260 DC of 28 June 1989). As a result, any statute which sought to exclude administrative decisions from review by the administrative courts would be struck down by the Constitutional Council, unless it could be justified by reference to another constitutional value. Where such an exclusion is introduced by decree, the same principles would be applied by the administrative courts.

Such constitutional principles are no novelty in French administrative law. Already in the leading case of LAMOTTE (CE 17 February 1950) the Conseil d'Etat adopted the view of its Commissaire du gouvernement, who urged that:

> The *recours pour excès de pouvoir* is available, even without legislative warrant, to challenge every administrative act, and its effect is to guarantee respect for legality in accordance with the general principles of the law.

Moreover, as the case shows, the presumption in favour of judicial review is so strong that the *recours pour excès de pouvoir* will still lie in the face of widely phrased clauses in the relevant legislation that, until the decision of the House of Lords in *Anisminic Ltd.* v. *Foreign Claims Compensation Commission*,[12] the English courts would certainly have accepted as excluding judicial review of any kind.

The case cited concerned wartime legislation passed by the Vichy government which allowed a prefect to requisition and bring into cultivation farm-land that had been abandoned or uncultivated for more that two years. The relevant statute provided expressly that the grant to the new occupant of a concession to farm land 'could not be the object of any administrative or judicial proceeding (*recours*)' Notwithstanding this clause, the Conseil d'Etat accepted boldly its competence to receive the *recours pour excès de pouvoir* which the owner of the land lodged against the requisitioning authority.

The reaction of the Conseil d'Etat in this case was no doubt influenced by the fact that it had already quashed three previous attempts to oust this owner from her land, that the exclusionary clause had not been present in the legislation under which the previous concessions had been granted, and that the behaviour of the administration was a flagrant attempt to flout the previous decisions of the Conseil itself and accordingly illegal as a *violation de la chose jugée* (see p. 204).

Nevertheless, the decision reflects the readiness of the administrative judge to proceed, if not *contra legem* at least *praeter legem*, in defence of the general principal of *droit administratif* whereby all administrative acts are subject to judicial review. This readiness is evidenced constantly in the case-law. Thus, in D'AILLIERES (CE 7 February 1947) the plaintiff asked the Conseil d'Etat to review the decision of the 'Jury of Honour' which was set up after the liberation to inquire into the record of those members of parliament who had voted in favour of surrender in 1940 and to determine their eligibility for further political office. The relevant statute stated that the findings of the Jury were not to be open to any further proceedings ('n'est susceptible d'aucun recours'). The Conseil d'Etat decided that the Jury of Honour

[12] [1969] 2 AC 147, distinguising and criticizing *Smith* v. *East Elloe RDC* [1956] AC 736; see p. 165 below.

was an administrative agency exercising a judicial function, so that its decisions were open to review by the Conseil *en cassation*; and that the clause cited above was not effective to exclude such review.

We have already seen (in Chapter 2, p. 12) that the French legislature is sovereign and there can be no judicial review of the constitutionality of a statute once it has been promulgated. It follows therefore, as a matter of constitutional theory, that sufficiently categorical words of exclusion in a statute might oust the jurisdiction of the administrative courts. It is a striking fact, however, that there is no recorded instance of this having actually occurred. Judicial review of administrative action has become so much part and parcel of the basic republican tradition which underlies all constitutions since 1875 that it is inconceivable in the present temper of French politics that any parliament would be willing, or any government would venture, to break with that tradition.

DREYFUS-SCHMIDT (CE 8 June 1951) is sometimes cited as an example of successful exclusion. But that case concerned a parliamentary election, and this would have fallen outside the competence of the administrative courts altogether (under the principles discussed in Chapter 6) had there not been a special and limited jurisdiction conferred expressly by statute upon the local Conseils de Préfecture. The statute in question provided that the decision of the Conseil de Préfecture in such a matter was to be 'sans appel'. The Conseil d'Etat decided that it could not review 'en cassation' the decision of the local court for the reason that any statutory addition (which this was) to the normal jurisdiction of the administrative courts could only be as large as the express words of the statute allowed.

The comparison here with English administrative law is illuminating. English law starts from the same fundamental principle that there is a presumption in favour of judicial review of administrative action, but takes a different attitude to statutory provisions excluding such review. True, an English judge with a mind to it can, as Lord Denning has proved, find his way round such expressions in a statute as that some decision or other shall be 'final': see, e.g., *Re Gilmore's Application* [1957] 1 All ER 796, and *Pyx Granite Co.* v. *Minister of Housing and Local Government* [1959] 3 All ER 1. On the other hand, the English judiciary, imbued with an instinctive respect for the sovereignty of parliament, may feel compelled to accept the fact of their impotence when faced with such a phrase as that, for example, in the Acquisition of Land (Authorisation Procedure) Act 1946, 1st Schedule, para. 16: 'the order shall not . . . be questioned in any legal proceedings whatsoever'. The House of Lords in *Smith* v. *East Elloe RDC* [1956] 1 All ER 855 was unable to evade the legislative intention of this phrase, although in the later case of *Anisminic Ltd.* v. *Foreign Claims Compensation Commission* [1969] 2 AC 147, by a difficult piece of reasoning, they were able to say that a 'determination' given without jurisdiction was no determination at all, and so it could be declared invalid. The Conseil d'Etat, on the other hand, in such cases as LAMOTTE (see p. 163) did not have to resort to such an elaborate ratiocination, but could apply the unwritten 'general principles of the law' as a kind of basic legal framework into which the statute must somehow be fitted (see Chapter 9, p. 205).[13] One is conscious in these cases of the radical difference in the French and English canons of statutory interpretation, although British courts now increasingly adopt a French approach when called upon to interpret provisions of Community law.[14]

[13] Compare also the approach of the European Court of Justice in *Johnston* v. *Chief Constable of the Royal Ulster Constabulary*, Case 222/84, [1986] 3 All ER 135, which adopts an approach more akin to the French on this issue in its use of 'general principles of law which underly the constitutional traditions common to member states': see Y. Galmot, 'Réfléxions sur le recours au droit comparé par la Cour de justice des Communautés européennes', *RFDA* 1990, 255.
[14] What Lord Denning called the 'European way' in *Bulmer* v. *Bollinger* [1974] Ch. 401. See also section 3(1) of the European Communities Act 1972.

8

The Substantive Law: The Principle of Administrative Liability

1 INTRODUCTION

In this and the following chapter we consider the body of substantive law which the administrative courts, in particular the Conseil d'Etat, has evolved in the exercise of their jurisdiction as described in the previous chapters.

This *corpus iuris* bears, to Continental eyes, the unusual feature of consisting predominantly of case-law, at least as regards the general principles of administrative law. These principles have to be extracted by a process of induction from *la jurisprudence* of the Conseil d'Etat. In a sense, it is no more than a statement of the judicial practice for the time being of that court—for the time being, because, as we shall see, the Conseil d'Etat accepts no doctrine of rigid precedent. Of course, much of the substantive law regulating the administration is contained in legislation, often consolidated into 'codes' such as the *Code de l'urbanisme*, the *Code des marchés publics*, and so on. But such codes are very different in character from the systematic, comprehensive, and principled codes which form the basis for French civil and criminal law.

Common lawyers are familiar enough with the notion of a system of law constructed out of cases. In this very familiarity, however, lies a danger for them when beginning a study of the substantive *droit administratif*. For they may fail to see that this case-law differs in important respect from their own. In the first place, as the late Professor Hamson pointed out, 'the law of the Conseil d'Etat is much more purely and exclusively case-law than is the common law today in England'. In other words, the common lawyer is faced with the phenomenon of a system of law which traditionally has been allowed to develop without the intrusion of statute, though increasingly laws and decrees are playing an important part in the shaping of general principles of administrative law. Secondly, the administrative courts, including the Conseil d'Etat, have avoided the inflexibility which the English doctrine of precedent engenders.[1] The Conseil d'Etat has always felt itself free to depart from its own previous decisions, thereby reserving to itself room for manœuvre and opportunity to reform and refine its law as it goes along. In this respect, a French lawyer would still see a contrast with the practice of the House of Lords even after the Practice Statement of 1966. Thirdly, as we saw in Chapter 3 when examining the procedure of the Section du Contentieux, the succinct, almost Delphic, form of its judgments presents novel problems to the common lawyer schooled in different case-law techniques.

It is a fundamental feature of any system of administrative law that not only does it regulate relations between various organs of government and public bodies, it also regulates relations between unequal parties, an organ of government on the one hand and the citizen on the other. Again, it is essential that the administration of a state should function according to the general public interest, and the French have always considered that in forwarding this public interest the administration should not be restrained by a rigid set of rules; such limitations upon the freedom of movement of the administration as may be imposed in the interests of the individual must be liable to continual adjustment. Thus, the law applied by the administrative courts in France has come into being as the result of a delicate balance between those guarantees for the rights of the individuals that are demanded at any one time by public opinion, and the ever-changing necessities of public administration. However, a private citizen cannot be expected to see the situation created by his litigation in such a light; flexibility must be tempered with a due respect for abstract justice, and this can be achieved (say the French) only if the judge has some measure of freedom in his decision. Therefore the administrative judge in France has had only

---

[1] But, in practice, there is a general acceptance of established case-law, and it is understood that such *jurisprudence* can be reversed only by a higher formation of the Conseil d'Etat *statuant au contentieux*.

a simple pattern of remedies to apply. Such flexibility has long obtained also in Scotland, but in England it used to be necessary for the litigant to choose the necessary form of action to obtain an appropriate remedy. Only with the 1977 reforms of Order 53 of the Rules of the Supreme Court has English administrative litigation begun to offer a simple and flexible system of remedies.

2 CATEGORIES OF LITIGATION BEFORE THE COURTS

The categories of litigation coming before the administrative courts may be classified in various ways. For our purpose, that propounded originally by Laferrière remains the most helpful.[2] This makes a fourfold distinction.[3]

Le contentieux de l'annulation

Here the complainant seeks the annulment of some administrative act or decision[4] on the ground of its illegality. As we saw in Chapter 3, the actual form of the proceedings to annul will then differ according to whether or not the decision complained of emanates from an administrative body having judicial status (i.e. which is classified as *une juridiction*). If it does, a *recours en cassation* will lie to the Conseil d'Etat; if it does not, as will more often be the case, a *recours pour excès de pouvoir* will come before the appropriate administrative court, which may well be a Tribunal Administratif rather than the Conseil d'Etat.

Le contentieux de pleine juridiction

Here the complaint goes beyond illegality and concerns matters of fact and law. The function of the court is not supervisory, but to determine a person's rights or entitlement, which may well go beyond quashing an administrative decision and involve revising it. Hence the 'fullness' of this jurisdiction, the administrative court exercising its full or complete powers, just as the civil courts habitually do in compensating for torts or breaches of contract. Such litigation arises in a number of areas: local govern-

---

[2] E. Laferrière, *Traité de la juridiction administrative et des recours contentieux* (Paris, 1887).
[3] But the third and fourth categories are much less important than the first and second.
[4] On *actes administratifs* see Ch. 7, n. 1, p. 153.

ment and European elections (*le contentieux électoral*), tax cases (*le contentieux fiscal*), and actions for damages arising from contractual or non-contractual liability (*le contentieux de la responsabilité*). The 'full jurisdiction' relates both to the *subject-matter* of the claim (usually compensation in money) and to the *powers* of the court (to revise, if necessary, the original decision).

Le contentieux de l'interpretation

Here the administrative court is called upon to interpret some administrative decision in the sense of explaining its legal meaning or significance. The field of this *recours* is narrow, because the administrative judge is always loath to act as a consultant, believing that if the difficulty in understanding the decision is sufficiently serious the interested party will be prepared either to ask for it to be annulled or to claim damages in compensation. However, in some limited situations, the administrative courts are willing to give interpretations in order to prevent further litigation, provided at least there is in existence an actual *lis inter partes* linked to the question for interpretation. They may be asked to do so directly by the interested party but nearly always act on a request from the civil court before which the meaning of the decision in question has arisen incidentally in the course of the civil suit.

Waline gives this hypothetical example:[5]

The Gaz de France sues in the civil court to recover the cost of gas supplied to a consumer. The consumer alleges that he is entitled to a specially reduced tariff by virtue of a clause in the contract of concession between the local municipality and the Gaz de France whereby the latter undertook to supply the town with gas. However, this is classified as an administrative contract and as such is not open to interpretation by the civil court. The court accordingly adjourns the hearing of the civil action to enable the administrative court to pronounce upon the relevant clause. In technical terms, the consumer is said to have raised a *question préjudicielle*, that is, a question which has to be adjudicated first and by a different court before the principal issue can be decided by the original court. (see p. 119 above).

[5] *Droit administratif* (9th edn., Paris, 1963), 69.

There is not, however, such a *question préjudicielle* every time that an administrative decision has to be interpreted or its legality is called in question. Thus, the *criminal* courts may always interpret or decide the legality or illegality of an administrative act (TC 5 July 1951, AVRANCHES ET DESMARETS). Indeed, the Cour de Cassation has held that they are bound to do so, in the case of alleged breaches of regulations, by the terms of the Penal Code which refers to 'regulations *lawfully* made by the administrative authority'. Again, the *civil* courts may pronounce upon the interpretation (but not the validity) of administrative regulations of general, as distinct from individual, application; for such general regulations are, in effect, akin to statutes, which any court has power to interpret. Thus, in SEPT-FONDS (TC 16 June 1923) S lost some bags of sugar consigned by train during the 1914–18 war and claimed damages from the railway company. The company had been taken over by the government under wartime regulations which prescribed various time-limits for making claims for lost consignments. The question arose in the civil court as to the meaning of these regulations, and the Tribunal des Conflits held the civil court competent to interpret them.

As we have also seen, the 1987 reforms enable the Tribunaux Administratifs and the Cours Administratives d'Appel to refer issues of law for an *avis* of the Conseil d'Etat, and this too may be a matter of interpretation before this latter court.

Since the accession of the United Kingdom to the European Community, the procedural device of the *recours préjudiciel* has become a familiar one to common lawyers in the guise of the reference for a preliminary ruling under Article 177 of the EEC Treaty.

Le contentieux de la repression

Here the administrative court acts as a criminal court. It is an unusual jurisdiction since it is the only category of case where the administrative courts act at the request of the administration and against the individual citizen. It arises from the fact that one class of criminal offence is given to the administrative courts for trial instead of to the normal criminal courts. These are minor crimes relating to *la grande voirie*. This term encompasses, broadly speaking, those parts of the public property over which the public have a right of user (e.g. *routes nationales, routes départementales*, the streets of Paris, ports and harbours, navigable inland waterways, railways, and telegraph lines). Any damage to, or encroachment upon, these constitutes an offence triable before the administrative courts. This jurisdiction is regarded as criminal in that the court may impose a fine (but not imprisonment). All the same, both the Constitutional Council and the Conseil d'Etat consider this to be a special category of criminal jurisdiction. It does not look so much at subjective guilt as at the objective aspect of interference with public property.[6] As a result, the payment required by the court can be higher than the level of a fine for an ordinary *contravention de police* since it will include an additional amount which takes account of the damage done to public property.

The scope of these proceedings in the administrative courts was much reduced in 1926 by the transfer to the ordinary criminal courts of those *contraventions de grande voirie* which related to highways on land. Administrative courts are thus left only with those relating to waterways, ports, harbours, etc. One example is JUSSEY (CE 16 January 1930), where the corpse of a dead rabbit had been thrown into the River Saône, and a more recent example is that of ROGNANT (CE 22 June 1987), where sand was extracted by machines from the beach without a permit.

In addition to the above offences in the nature (to English eyes) of public nuisances, there are also included under the 'repressive' jurisdiction of the administrative courts a number of other miscellaneous offences, such as abuse of the process of the court (*recours abusif*), in respect of which the Conseil d'Etat, Cour Administrative d'Appel, or local Tribunal Administratif is expressly empowered to impose a fine (but seldom does), or conduct constituting contempt in the face of the court (*délits d'audience*).

3 CRITICISM OF THE TRADITIONAL CLASSIFICATION

This traditional fourfold classification is based upon what the administrative court does. For instance, does it merely interpret or declare as to the validity of an administrative decision

---

[6] See CC decision no. 87–161 L of 23 September 1987 and CE 22 June 1987, ROGNANT, *AJDA* 1988, 60.

172 SUBSTANTIVE LAW: ADMINISTRATIVE LIABILITY

(*contentieux de l'interprétation*)? Or does it punish a wrongful action of the individual in relation to the administration (*contentieux de la répression*)? Or does it quash (*contentieux de l'annulation*)? Or does it compensate the aggrieved citizen (*contentieux de pleine juridiction*)?

The classification has been criticized as unscientific. Thus, Duguit argued (in 1911) that one should classify rather according to the nature of the question in issue. The distinction then becomes twofold: (*a*) is the question in issue the breach of some rule or law of general application which the administration should observe in its dealings with every citizen; or (*b*) is the question in issue the violation of some right peculiar to the complainant? Category (*a*) is termed a *contentieux objectif*, and the best example is the proceedings to quash; the second category (*b*) is a *contentieux subjectif* of which the proceedings to indemnify are typical. For the breach of an administrative contract only infringes the subjective right of the other contracting party; or the commission of a tortious action by the administration only wrongs the victim.

Although the traditional classification is simpler and is the one commonly adopted by the courts, the objective–subjective division is valuable because it explains two differences between the *recours en annulation* and the *recours de pleine juridiction*, namely: (i) that the *locus standi* required for the former is much wider and more liberal, whereas it is only the party to the contract or only the victim of the tort who can sue for an indemnity; and (ii) the decision of the court in proceedings to quash is valid *erga omnes*, whereas in proceedings for indemnity the decision takes effect only as between the parties.

4 ADMINISTRATIVE LEGALITY AND ADMINISTRATIVE LIABILITY

Cases concerned with punishment and interpretation are only a small part of the business coming before the administrative courts. Much more important are the first two heads—proceedings to quash and to determine rights or entitlement to damages. It is around these two categories of litigation that the Conseil d'Etat has built up its special body of law to secure justice for the individual in his dealings with the administration.

Two central principles have been evolved by the Conseil d'Etat as the foundation of this law. The first is that of *légalité*. This means that the administration must act in accordance with the law; otherwise its decisions run the risk of being quashed by an administrative court. In Dicey's language this principle ensures that the administration obeys the Rule of Law. It is the justification for French constitutional lawyers describing France as an *État de droit*, a state subject to law, as compared with a state above the law or *Etat de police*.

The second principle is that of *responsabilité*. This indicates that the administration will be liable (*responsable*) to compensate the citizen who is harmed through the decisions or activities of the administration, which need not be unlawful in all cases. As we have seen above, this principle is invoked by a different form of proceedings from the first, since it involves recourse to the 'full competence' of the court. Often, however, such a *recours de pleine juridiction* will include not only a claim for compensation, but also one to have the offending administrative decision quashed.

Administrative liability, the subject of the rest of this chapter, and administrative legality, which is to be the subject of the next, may be regarded as those obligations which *droit administratif* places upon the administration to counterbalance the two great privileges which the administration enjoys. The first is the privilege of the *décision exécutoire*, which Vedel explains as the power of the administration to create unilateral rights and obligations which will bind third parties without their consent: obvious examples are decrees, regulations, and byelaws; the second is the *privilège de juridiction*, that is, the right to have administrative matters adjudicated upon by the administration's own system of courts and 'to raise a conflict' if the ordinary courts seek to meddle in affairs of the administration in defiance of the doctrine of separation of powers.

We will now examine the principle of administrative liability. This term is adopted in preference to 'governmental' or 'state' liability since, as we shall see, the principle extends in France to all public authorities.

5 THE LIABILITY OF THE ADMINISTRATION

The principle of administrative legality and the consequent right of the citizen to request that an unlawful administrative decision be quashed may be of little or no value where the citizen is faced

174 SUBSTANTIVE LAW: ADMINISTRATIVE LIABILITY

with a *fait accompli*. For instance, a mayor fails to take precautions against a riot, with the result that the citizen's shop is looted; the post office loses or misdelivers the citizen's parcel of diamonds;[7] a munitions dump blows up and destroys his house. In such cases, annulment is of little or no value, either because there is no act to annul (the dump simply blew up), or because the decision taken (that of the mayor to do nothing) has spent its force and done its damage to the citizen. The only mode of redress then open is to obtain compensation for the damage.

In English law, since the Crown Proceedings Act of 1947, the liability of the Crown and other public authorities is generally accepted, so that the citizen is able to sue them for damages in tort or contract. Moreover, as a general rule, it is the same law of tort or contract which is applied to public authorities as to private individuals.[8]

In French law, on the other hand, while the principle of administrative liability is accepted, in order to do justice in the sort of situation envisaged above, the rules governing this liability differ in important respects from those found in the *droit civil* and applied by the civil courts in suits against private individuals. In a very real sense, therefore, there co-exist in France *two* laws of tort, *two* laws of contract, the one private and the other public or administrative. This principle is expressed in the judgment of the Tribunal des Conflits in BLANCO (TC 8 February 1873) as follows:

Considering that the liability which may fall upon the state for damage caused to individuals by the act of persons which it employs in the public service cannot be governed by the principles which are laid down in the Civil Code for relations between one individual and another: that this liability is neither general nor absolute: that it has its own special rules which vary according to the needs of the service and the necessity to reconcile the rights of the state with private rights. . . .

6 ADMINISTRATIVE TORTS

The term *responsabilité* without qualifying epithet is taken in French law to refer to delictual or quasi-delictual liability: the

---

[7] See CE 6 December 1855, ROTSCHILD C. L'ADMINISTRATION DES POSTES and compare *Treifus & Co.* v. *Postmaster General* [1957] 2 All ER 387.

[8] See J. Bell and A. W. Bradley, *Government Liability: A Comparative Study* (United Kingdom National Committee of Comparative Law, London, 1991), ch. 2.

distinction between delict and quasi-delict is between deliberate harm and unintended harm arising from negligence or risk. Together they correspond broadly with the area covered by the English law of tort. In France, however, the administration may be liable in some situations even without a wrong being done, where a harm is caused to an individual for the public good. Indeed, it is argued by recent authors that, unlike private law, administrative liability is not based on the notion of causing harm (which is found in Article 4 of the Declaration of the Rights of Man of 1789), but on the principle of 'equality before public burdens' as stated in Article 13 of that Declaration.[9]

The heading for this section refers to administrative *torts* (in the plural) as a warning that the legislature has intervened from time to time to create separate sets of rules for particular categories of administrative 'wrongs'—rules which may concern either the substantive principles of liability or the question of which courts shall have jurisdiction (or both). Examples are riots, war damage, school accidents, accidents arising from public works, and compensation for criminal injuries. Nevertheless, these special cases apart, there do exist certain general principles applicable to the torts of the administration.

However, the use of the word 'torts' should not be understood as implying that French law is restricted to nominate heads of liability, as in English law. Fault, risk, and equality before public burdens are simply principles which have generated rules about liability in particular situations, but are capable of flexible expansion to meet new problems. Thus where a fifteen-year-old boy suffered paraplegia as a result of a new surgical technique in a public hospital, he was able to obtain damages even without proof of medical negligence, simply because of the risk involved in a new technique whose side-effects were not fully known (CAA Lyon, 21 December 1990, GOMEZ).

Until the Revolution the principle of 'le Roi ne peut mal faire' prevailed in France, as it still does in English legal theory. Legislation from the revolutionary period onwards established the liability in certain instances of public authorities, but it was not until BLANCO that the general principle of the liability of the

---

[9] See id., ch. 9, esp. 225–7; also R. Errera, 'The Scope and Meaning of No-fault Liability in French Administrative Law' [1986] *Current Legal Problems*, 171–2; C. Harlow, *Compensation and Government Torts* (London, 1982), 102 ff.

state was established beyond question, although the need for this principle had been gradually recognized as the nineteenth century witnessed an ever-widening range of state activities.

When Agnes Blanco was run over by a wagon crossing the street between different parts of the state-owned tobacco factory in Bordeaux, the Tribunal des Conflits in 1873 took the occasion to establish three principles. The first was that of the liability of the state for the fault of its servants; the second was, as we have seen, the principle that this administrative liability should be subject to rules which were separate and distinct from those of the *droit civil*; and the third was the principle that such questions should fall within the jurisdiction of the administrative courts.[10]

In FEUTRY (TC 29 February 1908) it was established that the general liability which BLANCO attributed to the state applied no less to all public authorities. In this case a lunatic escaped from an asylum maintained by the *département* and set fire to the plaintiff's hayrick; the *département* was held liable to indemnify.

Although BLANCO established the distinctive character of the legal principles governing administrative torts, the Conseil d'Etat has developed its rules by analogy in some measure to the *droit civil*. Thus, Article 1382 of the Civil Code makes a person liable to compensate for damage resulting from his *faute*. Likewise the liability of the administration to indemnify is based, first and foremost, on fault, this consisting of some defect or failure in the operation of the public service in question.

In FEUTRY there was held to be a *faute de service* although it was not possible to attach fault to any particular official. The more usual situation, however is that the malfunctioning of the public service can be imputed to the fault of some officer of that service. An English lawyer might well ask why then should not the victim sue the officer at fault in his personal capacity in the ordinary courts, quite apart from any possible redress against the administration. What prevented the victim doing this from 1799 until 1870 was the existence of a constitutional guarantee pro-

---

[10] Professor Waline shows how the first two principles represented, historically, a designedly cautions advance away from the old doctrine of state immunity; for the Tribunal des Conflits was saying, in effect: 'The state shall be liable, but shall not be subject to such strict rules of liability as private individuals' (M. Waline, *Traité élémentaire de droit administratif* (7th edn., Paris, 1957), para. 14).

tecting *fonctionnaires* from personal suit, save by express leave of the Conseil d'Etat (seldom accorded unless the *fonctionnaire* had acted wholly outside the scope of his duties). This guarantee disappeared in 1870 under a wave of liberal opinion which owed much to the English example. Very soon afterwards, in PELLETIER (TC 30 July 1873), the Tribunal des Conflits had occasion to re-examine the problem. This case, in which the publisher of a newspaper sued for damages a prefect and a general who had seized his paper, laid down what has become the classic distinction between *faute de service* and *faute personelle*.

There is said to be *faute personnelle* where there is some personal fault on the part of the official, that is, a fault 'which is not linked to the public service but reveals the man with his weaknesses, his passions, his imprudence' (in the famous phrase of Laferrière). Where such personal fault is present, the official can be sued personally in the ordinary courts. On the other hand, where there is simply a *faute de service* (one which is linked with the service), the official preserves his immunity by reason of the principle of separation of powers which prohibits the ordinary courts receiving actions against the administration or its officials. Instead, the injured party must sue the administration before the administrative courts.

As Professor Schwartz pointed out,[11] this difficult distinction resembles (but is not identical with) the English distinction between a servant's acts in the course of his employment and those not in the course of employment. Further complexity has been added by the desire of the Conseil d'Etat to come to the help of the victim of a *faute personelle* left perhaps with a worthless right of action against a penniless official.

In ANGUET (CE 3 February 1911) a visitor to a post office was assaulted and had his leg broken by two members of the post-office staff because he left by the staff entrance, the entrance for the public having been prematurely closed. The Conseil d'Etat found that there were two distinct *fautes*: there was the premature closing, which was a *faute de service*, and there was the unwarranted violence of the officials—a *faute personnelle*. The court went on to recognize that it was this combination (*cumul*)

---

[11] *French Administrative Law and the Common Law World* (New York, 1954), 260.

of faults which occasioned the complainant's injury, but that as it would not have occurred but for the *faute de service* the state could be held liable for the whole damages claimed.

This notion of *cumul* was taken a step further in LEMONNIER (CE 26 July 1918). A country commune was holding its annual fête, at which one of the attractions was a shooting competition, with targets floating on the local river; the complainant was struck by a bullet while walking on the opposite bank and claimed damages from the *conseil municipal*. It was proved that earlier the same afternoon other persons had complained to the mayor of being narrowly missed by bullets. The complainant commenced two actions against the *conseil municipal*, one in the ordinary civil court, the other in the Conseil d'Etat. By the time the matter came before the Conseil, the Court of Appeal of Toulouse had held the mayor liable in damages for his personal fault in permitting the shooting to continue. Nevertheless, the Conseil d'Etat held itself competent to proceed with the case and found that the mayor's negligence constituted at one and the same time a personal fault and a service fault. As the Commissaire du gouvernement (Léon Blum, later to be Premier of France) expressed it in his oft-quoted conclusions:

If the personal fault has been committed in the public service, or on the occasion of the service, if the means and instruments of the fault have been placed at the disposal of the party at fault by the service, if in short the service has provided the conditions for the commission of the fault, the administrative judge will and must then say: the fault may be severable from the service—that is for the ordinary courts to decide, but the service is not severable from the fault. . . .

How far the principle in LEMONNIER extends can be seen in the trio of cases (MIMEUR, DEFAUX, and BESTHELSEMER) all decided on the same day (CE 18 November 1949). All three concerned motor vehicles belonging to the administration but being used by officials on private and unauthorized journeys. In MIMEUR, a soldier driving a petrol tanker made a detour to his home village and crashed into the wall of a house. The houseowner was able to recover since the tanker had been entrusted to the driver to perform a public service. As Antoine Bernard remarked in his conclusions to another case:

the least that can be said is that the developments of the whole of the case-law on *cumul* in the event of the personal fault of a public employee demonstrates a clear desire of the administrative judge to extend as far as possible, out of considerations of justice, the guarantee of compensation which the finding of liability on the part of the public authority alone can provide for the victim. And that has been done at the price of such an expansive application of the notion of service fault and of performance of the service that one finds oneself in some cases having to make do with a pure fiction.

Thus in one case (CE 26 October 1973, SADOUDI), a flatmate's family was able to recover for his death, caused by an off-duty policeman when a gun went off accidentally. Since the policeman was obliged to keep his weapon at home, this created a danger for third parties such as to make the accident not unconnected with the service. But a line must be drawn somewhere, and was drawn by the Conseil d'Etat in LITZLER (CE 23 June 1954), where a customs officer in uniform but off duty used his service revolver to commit a murder.[12]

It is important to understand that the *cumul* doctrine does not permit the victim to obtain damages twice over. Rather it gives him a right to obtain judgment both in the ordinary courts (often the criminal courts in cases of personal injury) against the official for his personal fault and in the administrative courts against the administration for its service fault. Whichever of the two judgment debtors then pays the damages awarded has a right of action against the other for a contribution or, as we shall see, even a complete indemnity.

The right of contribution or indemnity between the 'joint tortfeasors' (as an English lawyer would regard the official and the administration in this situation) has been gradually evolved by the Conseil d'Etat. In LEMONNIER, the Conseil indicated that, once the commune had paid the damages to the complainant, it could require to be subrogated to the latter's rights against the mayor for his personal fault. But this was recognized to be an imperfect and clumsy device. In the result, as most injured parties very sensibly decided to sue only the administration in cases of *cumul*, having regard to its deeper pocket, the negligent official usually escaped liability.

---

[12] See also CE 18 November 1988, MINISTRE DE LA DEFENSE C. RASZEWSKI.

Fears were then expressed lest this immunity, in practice, of officials should encourage irresponsible and reckless behaviour in their dealings with the public. Prompted by such policy considerations, the Conseil d'Etat decided in the cases of DELVILLE and LARUELLE (CE 28 July 1951) that the administration should have a direct action for contribution or indemnity against the official. This action would lie in the administrative court, and it would be for this court to apportion the ultimate share of responsibility between the official and the administration in cases of *cumul*.

In DELVILLE, a government lorry was involved in an accident partly because of the drunken state of the driver, Delville, and partly because of the defective state of its brakes. Delville was sued in the ordinary courts by the other party to the accident and held liable for all the damage caused. He then claimed a 50 per cent contribution from the administration on the ground that the accident was due in part to the faulty brakes of the vehicle. The Conseil d'Etat upheld his claim. In LARUELLE, decided at the same time as the last case, a soldier, Laruelle, took away an army vehicle without authority on a private journey and knocked down a pedestrian. The latter sued the administration for damages. The Conseil d'Etat found (in an earlier decision of 12 March 1948) that there was a service fault by reason of the inadequate supervision of the garage where the vehicle was kept and that accordingly the administration was liable. The administration paid the damages awarded but proceeded to require the soldier to reimburse the whole amount. The soldier challenged this demand before the Conseil d'Etat. The Conseil upheld the demand. In its view the administration was entitled to claim a complete indemnity from Laruelle, for, although there had been a *cumul* of personal fault and service fault, it did not lie in Laruelle's mouth to allege the service fault, since this had been caused by the deception which he had practised upon the guard in charge of the garage where the vehicle was kept.

By these twin cases the Conseil d'Etat has completed its construction of a logically satisfying system of administrative liability. Where there is personal fault combined with service fault, the victim can sue either the official or the administration in the appropriate court. The ultimate division of responsibility rests, however, with the administrative courts. In this way the official who is at fault can be made to contribute appropriately towards the damage done, where, as is usual, only the administration has been sued by the victim.

The difficult question what *is* an appropriate contribution is determined according to the official's duties and responsibilities in the particular service in which he is employed, rather than by reference simply to the actual part played by him in the cause of the damage. Thus, if several soldiers are travelling together in an army vehicle which is involved in an accident through being driven negligently, the driver will not necessarily bear a heavier share of the consequences: each soldier's responsibility will be weighed by reference to his duties, and the sergeant may be held liable to a greater extent than the private (CE 22 March 1957, JEANNIER).[13]

The doctrine of *faute de service* has undergone considerable evolution. Originally, it was confined to misfeasance. Then it was progressively expanded to cover not only non-feasance but also excessive slowness, or 'late feasance', as Schwartz neatly terms it.[14] The notion of 'fault' applied by the administrative courts is wider than common-law concepts of trespass, negligence, or misfeasance in public office. As in French private law, mere illegality is in itself a fault capable of giving rise to liability without more. For example, in ORDONNEAU (CE 7 July 1989), the president of the competition commission was a civil servant on secondment and was retired compulsorily in line with the new retirement ages applicable to his grade in the civil service. This was held to be illegal in that it failed to take account of the implicit special exemption given to the president of the commission from civil service rules by the legislation creating it. No allegation was necessary that this constituted a negligent misinterpretation of the law in question, yet M. Ordonneau obtained damages for lost salary. Going back on assurances given may equally be fault, even if the assurances are more of a promise as to the future than a representation of past or present fact. For instance, in CLOUET (CE 12 February 1990), a post

---

[13] But the reform of 1957 (p. 141 above) would now confer jurisdiction for such a vehicle-accident upon the ordinary courts. For a general discussion of the law on contribution, see F. Roques, 'L'Action récursoire dans le droit administratif de la responsabilité', *AJDA* 1991, 75.

[14] *French Administrative Law and the Common Law World*, 282.

office employee in Calvados was assured by his superiors that, if he accepted a post in Caen, the postal authorities would take care of his accommodation costs. When the assurance was not met, the employee was able to obtain damages for the costs he had incurred.

But not every illegal act nor every misdemeanour committed by an official will necessarily result in a successful action for a service fault. Thus, even a minor irregularity in form affecting the dismissal of a civil servant makes the decision illegal and is undoubtedly a service fault, but the court may decide that the fault did not occasion any loss to the complainant: for example, if the court finds the dismissal to have been in fact justified, it may grant nominal damages or even refuse compensation altogether (CE 14 June 1946, VILLE DE MARSEILLE, and CE 28 July 1951, MOUILHAUD).

Another essential feature of the case-law governing administrative torts is that it distinguishes between two degrees of fault, simple and gross; it then depends on the problems posed by the particular public service which of these degrees of fault will suffice to render the administration liable. Gross fault (*faute lourde*) is held to be necessary when the task of the public service is particularly difficult or sensitive. Two obvious examples are hospital and police services. A patient who suffers injury in a public hospital must either show that there was simple fault in the organization of the hospital itself (e.g. in not providing sterile equipment: CE 8 December 1988, COHEN), or that there was gross fault in the conduct of the treatment (e.g. a failed abortion: CE 27 September 1989, MME K).

Similarly, police activities are inherently difficult or sensitive (or both). In VILLE DE PARIS c. MARABOUT (CE 20 October 1972) access to a house in a narrow cul-de-sac in a Paris suburb was constantly impeded by parked cars, although parking was prohibited by the city ordinances; the householder sued the City of Paris for damages. The Assemblée du Contentieux upheld the decision of the Paris Tribunal Administratif that the City was liable: the court recognized the special difficulties facing the traffic police in Paris, but they were guilty in this instance of *faute lourde* by their persistent neglect to enforce the law.

The blanket protection given to many public services in the past by the restriction of their liability to gross fault has come in for reconsideration in recent decisions. Notably in BOURGEOIS (CE 27 July 1990), the Conseil d'Etat held that only those parts of the revenue services which had special difficulties could be subject to the limited protection afforded by the requirement of gross fault. In this case, the revenue made a computer transcription-error in assessing a tax-payer's income and then sought to recover the inflated assessment by direct debit from his bank account. Such routine matters were held not to be covered by the limitation of the liability of the revenue's activities for gross fault, and the tax-payer recovered (subject to a reduction for contributory fault in not calling the revenue's attention to the overcharging quickly enough).

7 LIABILITY WITHOUT FAULT: THEORY OF RISK

All the cases so far discussed have involved liability for fault. At an early date, however, it was accepted that in certain special circumstances the administration might find itself liable even where guilty of no fault. Thus, a Law of 1799 imposed on the administration a duty to compensate anyone injured as a consequence of the carrying out of public works (compare the English rules about work on the highway, e.g. *Holliday* v. *National Telephone Company* [1899] 2 QB 392). Again, special legislation going back to a Decree of l'an IV (1795) has imposed upon the public authorities liability to indemnify in respect of riots.

In the course of the twentieth century and especially since 1944, the Conseil d'Etat has built a general principle of liability without fault based upon the theory of risk. The way for this, however, had been prepared by some of the 'fault' cases discussed above. BLANCO was again a landmark, by affirming the separateness of the rules of administrative liability from those enshrined in the Civil Code in which fault was the keynote. Moreover, the working-out of the *cumul* doctrine after LEMONNIER resulted in the courts finding on the part of the administration a fault of the most artificial kind. Another stepping-stone between fault and risk was the readiness of the courts to presume fault, especially in cases of injury done by administrative vehicles. Thus in SOCIETE D'ASSURANCES MUTUELLES 'LES TRAVAILLEURS FRANÇAIS' (CE 22 December 1924) the Conseil d'Etat laid down that there was a presumption of fault against the driver of such a vehicle because of 'difficulties of present-day traffic'.[15]

The jurisprudential basis for liability without fault has often been said to be that of the 'risk theory'. The activities of the state, even when conducted without fault, may in certain circumstances constitute the creation of a risk; if the risk materializes and an individual is occasioned injury or loss, it is only just that the state should indemnify him. An alternative and perhaps more profound rationale is to connect liability without fault to the fundamental principle of the equality of all citizens in bearing public burdens. This principle of 'égalité devant les charges publiques', which is founded in Article 13 of the Declaration of the Rights of Man, has been vividly expressed by Duguit in his *Traité de droit constitutionnel*:[16]

> the activity of the state is carried on in the interest of the entire community; the burdens that it entails should not weigh more heavily on some than on others. If then state action results in individual damage to particular citizens, the state should make redress, whether or not there be a fault committed by the public officers concerned. The state is, in some ways, an insurer of what is often called social risk ('risque social') . . .

Here the French are basing liability on the principle that what is done in the general interest, even if done lawfully, may still give rise to a right to compensation when the burden falls on one particular person; and as Thompson observed: 'This shouldering by the community of a burden which in English law lies where it falls (because even if a tort had been committed there would be lawful authority) is one of the more striking and laudable aspects of French *droit administratif*.'[17]

The emergence of the risk theory can be illustrated by examining four typical categories of case in which the Conseil d'Etat has imposed liability without fault upon the administration.

---

[15] This, of course, was very much in line with developments in the civil law under Art. 1384, para. 1, where a presumption of fault was being established in motor-vehicle cases: see Cass. civ. 29 July 1924, BESSIERES, in A. T. Von Mehren and J. Gordley, *The Civil Law System* (2nd edn., Boston, Mass., 1977), 621.
[16] 3rd edn., Paris, 1927–9, 469.
[17] (1968) JSPTL 470.

Risks of Assisting in the Public Service

France introduced a comprehensive system of workmen's compensation in 1898 (a year after the English Workmen's Compensation Act). But three years before, in CAMES (CE 21 June 1895), the Conseil d'Etat had anticipated this legislation. A workman in a state arsenal had his hand shattered in an accident at work for which no blame could be attached to anyone. The Conseil allowed his claim for damages against the state, taking the view that the state owed an obligation to indemnify against the risks of employment (*le risque professionnel*) to those engaged in a public service.[18] Even without statute, this principle has been used to give no-fault compensation to servicemen and women and their families for injuries on active service or in training: CE 27 July 1990, BRIDET ET AL. Subsequently the Conseil d'Etat extended this right to compensation to those assisting in the public service even in a voluntary capacity. Thus, in COMMUNE DE SAINT-PRIEST-LA-PLAINE (CE 22 November 1946), the victim was helping to set off fireworks at a village carnival when he was injured by the premature explosion of a firework; the Conseil d'Etat allowed him to recover damages without proving fault on the part of the commune. Indeed this principle may be invoked even where there is no actual public service, but where a private initiative provides what should be a service. Thus in COMMUNE DE BATZ-SUR-MER c. TESSON (CE 29 September 1970), there was no lifeguard on a stretch of beach, but the local commune did have the public duty to prevent accidents. This was held a sufficient basis to make the commune liable to the widow of a citizen who was drowned trying to rescue a child swept out to sea. Again, when an elderly lady fell into a ditch on property belonging to the local commune, her cries for help attracted the gallant M. Gaillard to the spot; he in turn fell into the ditch and so suffered injury. He was able to recover damages from the commune because, in attempting to rescue the old lady, he was undertaking a 'service public communal' (CE 9 October 1970, GAILLARD).

---

[18] The civil courts soon followed suit in establishing a presumption of fault for work-accidents based on Art. 1384, para. 1 of the Civil Code: Cass. civ. 30 March 1897, GRANGE C. COMPAGNIE GENERALE TRANSATLANTIQUE, Von Mehren and Gordley, *The Civil Law System*, 617.

Risks Arising from Dangerous Operations

Where a public authority creates what is called 'an abnormal risk for the neighbourhood' (*risque anormal du voisinage*), then it must be prepared to pay damages if the risk materializes.

A leading case is REGNAULT-DESROZIERS (CE 28 March 1919). During the First World War the military authorities installed a large dump for grenades in a residential neighbourhood on the outskirts of Paris; it blew up in 1915 with considerable loss of life and damage to property. Before the Conseil d'Etat the Commissaire du gouvernement suggested in his conclusions that the state was liable because of the faulty methods of operating the dump. The Conseil d'Etat took the different line of holding the state liable on the ground of risk: in its view there was an abnormal risk introduced into the residential neighbourhood by installing and operating a munitions dump, and it was therefore unnecessary for the victims to establish any fault.[19]

The principle has been extended beyond dangerous premises to include dangerous operations. A striking instance is provided by the twin cases of LECOMTE and DARAMY (CE 24 June 1949). Both concerned the same kind of incident: the accidental injuring of a bystander through the use of firearms by the police. In DARAMY the complainant's wife was shot dead when a policeman was pursuing the assailant of a taxi-driver; in LECOMTE the proprietor of a bar, whilst sitting at the street door, was killed by shots fired by the police at a motorist apparently evading a road block. In both cases the administration was held liable. As we have seen (p. 182), because of the peculiar difficulty and social importance of the function of the police, police action must normally constitute a gross fault (*faute lourde*) before any liability can arise—ordinary or simple fault (*faute simple*) does not suffice. Nevertheless, as in the instant cases, the use of police firearms amounted to an abnormal risk giving a right of indemnity without proof of any fault—at least for the citizen not concerned in the police operation. This principle equally covers the case of

---

[19] It is instructive to compare the breadth with which the Conseil d'Etat formulated its approach in this case with the narrow construction which the House of Lords placed on the rule in *Rylands* v. *Fletcher* when faced by not dissimilar facts in *Read* v. *Lyons* [1947] AC 156.

injury caused to a bystander by the person pursued by the police (Cass. 1re civ., 10 June 1986, POURCEL).[20]

The principle of liability for risk has been used extensively in the area of liability for the escape of prisoners and for 'liberal' penal measures. A striking example is GARDE DES SCEAUX c. BANQUE POPULAIRE DE LA REGION ECONOMIQUE DE STRASBOURG (CE 9 April 1987). Here three criminals committed a robbery at a bank. One had failed to return to prison after he had been given home leave. The second was entitled to a special regime of 'semi-liberty' whereby he worked outside prison during the day and spent the night in prison. The third had been paroled by the Minister of Justice. The bank was able to recover damages from the Minister on the ground that each of these different measures created a risk for third parties. However, there are limits in that there must be a causal link between the risk and the damage. So in HENRY (CE 27 March 1985), a prisoner had failed to return to prison from home leave, but the interval of six months between this and the murder he committed broke the chain of causation, and the Minister of Justice was not liable for the death.[21] Such decisions clearly go beyond what has been recognized by English courts in cases such as *Dorset Yacht Co. Ltd.* v. *Home Office* [1970] AC 1004.

Recent decisions on liability for other risk-creating activities also go beyond what has been accepted in England. For example, in INGREMEAU (CE 19 October 1990) a local authority was held liable for the actions of a child in care of foster-parents who injured a friend while playing with a bow and arrow.[22] Again, in ELECTRICITE DE FRANCE c. SPIRE (CE 2 October 1987) a house in the Loire valley was only 300 metres from a nuclear reactor. The owner successfully complained about the permanent noise from the reactor's operation, but failed as

---

[20] This was a case brought against the judicial police (PJ) and thus was within the competence of the civil courts, though the rules of liability are here the same as in public law.

[21] For a comparison between French decisions on borstals and English cases see C. J. Hamson, [1969] CLJ 272.

[22] The decision is not all that surprising because, in private law, parents would be liable to the victim for the acts of their child: see C. cass. Ass. plén., 9 May 1984, FULLENWARTH C. FELTEN (parents liable for a child injuring a friend with an arrow); and see also P. Bon, 'La responsabilité du fait des personnes dont on a la garde', *RFDA* 1991, 991.

regards the effect of the reactor on the landscape and the environment. This is more generous to the victim than *Allen* v. *Gulf Oil Refining Ltd.* [1981] AC 1001, but has similarities to *Merlin* v. *British Nuclear Fuels plc* [1990] 3 All ER 711.

Again, a pregnant schoolteacher runs a special risk when attending to her duties if any of the pupils are suffering from German measles; consequently, when such a teacher contracted the disease and her baby sustained pre-natal injuries, the latter was able to recover damages (CE 6 November 1968, SAULZE).

Abnormal Burdens Suffered in the Public Interest

In COUITEAS (CE 30 November 1923) the Conseil d'Etat was faced with a novel question. Couitéas was the owner of extensive tracts of land in southern Tunisia which were occupied by nomadic tribesmen. His own efforts to evict the natives having failed, he sought the help of the local police, who said it was a matter for the military authorities. The military commandant feared that action by him would provoke a rebellion in the region and referred the matter to the Resident-General, who in turn referred the matter to the government in Paris. The government decided no military assistance could be given to Couitéas because of the risk of serious disorder. Couitéas then sought damages against the administration before the Conseil d'Etat.

The Conseil d'Etat was not called upon to decide whether or not the government was guilty of fault. Instead, it invoked the principle of equality in bearing public burdens; any citizen who obtains a judgment (and Couitéas had obtained a declaration of his rights as owner from the Tunisian civil courts) was entitled to expect the assistance of the public authorities in executing his judgment; if then, perhaps for good reason, this assistance is refused, the citizen is called upon to bear an abnormal burden in the public interest; and to protect the principle of equality he must be compensated in damages for his special sacrifice. Hence Couitéas succeeded in his claim for damages.[23]

The extent of this liability is well illustrated by two cases decided on the same day and concerning the same incident. In

---

[23] Modern cases that apply the principle in COUTEAS include CAILLAT (CE 16 December 1966), MARTINOD (CE 26 June 1968), MERGUI (CE 19 March 1971), and TOURAINE AIR TRANSPORT (CE 6 November 1985, see n. 24).

August 1980, French fisherman illegally blocked the port of Calais. The port authorities decided not to use force to remove the obstruction, but as a result ships were unable to move in and out of the port. In SEALINK UK LTD (CE 22 June 1984), a British company was unable to ferry passengers over the very busy August bank holiday weekend and so lost an important part of its summer trade. The specific character of this trade and the timing of the blockade made the loss suffered special and beyond the normal inconvenience suffered by all port users. As a result it was able to recover for the loss suffered. By contrast in SOCIETE 'JOKELSON ET HANDSTAEM', a local forwarding, storage, and handling firm had been unable to process the cargo of two ships, and its adjacent office had been closed for periods of a few days. None of these losses was sufficiently serious in relation to its overall business to constitute special damage creating an abnormal burden for the firm.[24]

State Liability Arising out of Legislation

Another remarkable decision is LA FLEURETTE (CE 14 January 1938). Just one hundred years earlier (in CE 11 January 1838, DUCHATELET) the Conseil d'Etat had refused a claim for damages brought against the state by a manufacturer of artificial tobacco after the manufacture or sale of such tobacco had been prohibited by statute in the interest of the state's tobacco monopoly; the Conseil held that a statute, as an act of sovereignty, could not give rise (except expressly) to a right of compensation against the state.

In LA FLEURETTE, a dairy company of that name claimed damages against the state in respect of the losses which it sustained when it had to discontinue marketing its brand of artificial cream. The manufacture and sale of any article under the name 'cream' had been banned by statute unless the article consisted of real cream. The statute made no provision for compensation. Contrary to its previous case-law and to the conclusions of the Commissaire du gouvernement, the Conseil d'Etat decided that it was not the legislature's intention to impose an unequal sacrifice upon the complainant company and granted its claim.

---

[24] See the contrasting cases of CE 6 November 1985, TOURAINE AIR TRANSPORT, and CE 17 January 1986, VILLE DE PARIS C. DUVINAGE.

Again the principle of equality in bearing public burdens was invoked to enable the company to recover. The liability of the state in similar circumstances has even been extended to the consequences of an international treaty, which under the French Constitution has statutory force (CE 30 March 1966, COMPAGNIE GENERALE D'ENERGIE RADIOELECTRIQUE).

A further striking example of this principle of equality was MINISTRE DES AFFAIRES ETRANGERES c. CONSORTS BURGAT (CE 29 October 1976). The landlords let a flat to a young woman. She subsequently married a UNESCO official from Honduras, who joined her in the flat. The landlords endeavoured to increase the rent in accordance with the rent-control legislation then in force and, in default of payment of the increase, to expel the couple, but they were met with the plea of diplomatic immunity. They claimed compensation from the Ministry of Foreign Affairs as the state had enacted the immunity legislation which had deprived them of their normal rights as landlords, and they were successful before the Conseil d'Etat.

These decisions, however, should not be taken too far. The right to an indemnity, even where the statute is silent, only arises if the legislative history of the statute in question shows that it was not intended to make the victims of the measure bear the loss. It must also be shown that the victims were not engaged in any activity injurious to health, public order, or good morals. And the victims must show that their loss would be especially serious; thus, in LA FLEURETTE the company had to go out of business, and it was the only company so affected by the statute.

In practice, then, this doctrine is rarely applied. Indeed, in the event of pre-enactment review by the Constitutional Council, a statute which failed to provide for adequate compensation for loss caused to individuals would be likely to be struck down.

From the above account, it will be seen that the administrative courts have developed an extensive non-contractual liability of the administration, combining ideas of fault, risk, and equality before public burdens. It is certainly more extensive than English or Scots law, where the principles of no-fault liability are less fully developed.[25] Indeed, the principles are more

---

[25] For a comparison, see Bell and Bradley *Government Liability*, chs. 2, 3, 9. See also Jolonicz, [1985] CLJ 370.

extensive than those of French civil law, where vicarious liability is stricter and no-fault liability under Article 1384, para. 1, requires the intervention of a *thing* (*une chose*) in causing the damage.

Damages in Tort

In the past, the administrative courts have been less generous than the ordinary courts in the categories of loss recognized and in the measure of damages. Thus, until LETISSERAND (CE 24 November 1961) the Conseil d'Etat refused to indemnify for mental anguish (*douleur morale*). In that case the Conseil departed from its previous case-law and awarded the father a *pretium doloris* of 1,000 francs when his son was killed in a collision with an administrative vehicle which was being negligently driven. The sum awarded, however, was still niggardly compared with what a civil court might have allowed under this head. More recent decisions have been more generous. Thus in MME K, the mother recovered 50,000 francs for the mental anguish resulting from a failed abortion and the problems of bringing up a severely handicapped child, while the child recovered over a million francs to cover both pain and suffering, physical deformity, and the cost of continuing hospital treatment arising out of the operation. Such a measure of damages was very much in line with comparable cases in the civil courts.

Again, the Conseil d'Etat has recognized the civil-law idea of loss of a chance (*la perte d'une chance*). Thus in LEGOFF (CE 27 May 1987), a student was failed by the examinations committee at the University of Paris V on the basis of his final marks. The decision was quashed because the committee had failed to take account of his marks on course work throughout the year, as it was bound to do. As a result of this mistake, he received his diploma only in March 1982, not December 1978. In the meantime, he had lost the serious chance of substantial earnings to which the diploma in company management would have led. In consequence, he was awarded 100,000 francs. Even where the loss is difficult to quantify, some damages will be awarded. Thus in GIRAUD (CE 27 January 1988), 1,000 francs was awarded where a school had failed to provide compulsory courses in the national curriculum.

Exclusion of Liability

Given the importance of *la responsabilité* as a feature of administrative law, the administrative courts have been reluctant to see it excluded and have engaged in ingenious interpretation of statutes. Thus the Code of Posts and Telecommunications formerly excluded liability of the administration for the loss of ordinary mail. But the Conseil d'Etat interpreted this exclusion as inapplicable to cases of gross fault (*faute lourde*): SECRETAIRE D'ETAT AUX POSTES c. MME DOUBLET (CE 24 April 1981). Thus in GRELLIER (CE 22 January 1986), the Post misdirected a letter marked 'urgent examination', with the result that a woman missed the opportunity to sit an examination. The Post had been requested to redirect her parents' mail but not hers. This was held to be gross fault and so she could recover damages on the basis of loss of a chance. As a result, the Code of the Posts and Telecommunications was amended to include liability for gross fault. In future, such problems should not arise, as the Constitutional Council has ruled that a complete immunity of a public official from liability is unconstitutional (CC decision no. 89–260 DC of 28 July 1989, COMMISSION DES OPERATIONS DE BOURSE).

8 ADMINISTRATIVE CONTRACTS

We have seen in Chapter 6 that the French administration can enter into both administrative contracts (*contrats administratifs*) and private-law contracts (*contrats de droit privé*). The two basic criteria of the former are that the contract relates to a public service and that the contract reserves exceptional powers to the administration (the *clauses exorbitantes du droit commun*). Either criterion may suffice to make a contract 'administrative' in character. Simple to state, these criteria have given rise to a very complex case-law, which it is not necessary to recount here.

The French regard an administrative contract as essentially an arrangement between unequal parties.[26] As a consequence, the

---

[26] 'It is characteristic of the whole of administrative law that the administration has the privilege of *exécution d'office*. It can take what steps are necessary to enforce or supervise the contract, without invoking the assistance of the administrative courts. The administration is never the plaintiff.' B. Nicholas, *French Law of Contract* (London, 1982), 26.

courts are careful to examine how far an arrangement is a truly bilateral act and therefore a contract, or how far it is really a unilateral administrative decision to which the other party has merely given his assent. The extension of the idea of 'regulation by contract' in areas such as fixing prices or remuneration to doctors for medical treatment has led the administrative courts to classify some so-called 'contracts' as 'unilateral administrative acts'. This is especially the case where a collective agreement effectively has regulatory effect. As has been noted ' "administration by collaboration" ' does not seem to lead to an extension of the domain of contract'.[27] On the whole, the attitude of English law to this issue remains unclear.[28] In France, especially in the era of decentralization, 'contracts' between public bodies, including those between central and local government, are an important aspect of political strategy.[29]

The rules on administrative contracts have a number of peculiarities compared with private-law contracts. These concern the formation, the content, and the performance of contracts. These special principles stem mainly from the underlying need to recognize the predominance of the public interest, an interest which must prevail, even to the extent of overruling the express terms of the contract.

In addition, of course, administrative contracts fall outwith the jurisdiction of the civil courts and within that of the administrative courts; and being bilateral acts, they are not in general reviewable by the *recours pour excès de pouvoir*,[30] but only give rise to a remedy for damages under the *pleine juridiction* of the administrative courts.

Formation

French lawyers draw a distinction between a number of different kinds of administrative contracts. Two main categories of

---

[27] D. Harris and D. Tallon, *Contract Law Today* (Oxford, 1988), 150 and 138–44; see also R. Römi, 'La requalification par le juge des actes négociés en actes unilatéraux, *AJDA* 1989, 9.
[28] T. Daintith, [1979] *Current Legal Problems* 41.
[29] See *AJDA* 1990 no. 3.
[30] But see CE 26 July 1991, COMMUNE DE SAINTE-MARIE for an example of public-works contracts being annulled *pour excès de pouvoir* by the Tribúnal Administratif of Réunion, after being referred to the tribunal by the prefect under the Law of 2 March 1982. (On this *déféré préfectoral*, see above, Ch. 2.)

contract exist, *la concession de service public* (sometimes called *l'affermage*) and *le marché public* (whether for *travaux publics*, *fournitures*, or *services*). The concession grants the operation of a public service to an individual (often a private person). That person is responsible for running the service within the framework settled by the public authority granting the concession, and the concessionaire is paid for the service, typically out of revenue received from the users of the service. Thus, local water services in France are typically run by private companies who have concessions from local communes to provide water in their area. Water users pay these companies for the service. Although both operator and consumer are private persons and have private law contracts between them, the concession itself is an administrative contract between the local commune and the water company.

The public procurement contract (*le marché public*) is concerned with the provision of a particular object or activity for the public service. This may take the form of public works (such as constructing a road or a school), supplies (such as stationery or vehicles), or services (such as cleaning or catering). In such contracts what matters most is the quality of the product or service provided, rather than the person who makes the provision. By contrast, the concession is traditionally seen as a contract *intuitu personae* where the concessionaire is chosen for his particular characteristics by the public authority, and the concessionaire himself is providing the public service.

This divergence in the kind of administrative contract gives rise to a difference in the procedure of formation of contract. In crude terms, a private person can make a contract with whomsoever he wills and on whatever terms he wishes. The dominant position and immense contracting power of public authorities is redressed by subjecting them to certain procedural requirements. Failure to comply with these renders the decision to enter the contract (a decision separable from the contract itself) liable to be quashed for *excès de pouvoir*. In both concession and public procurement, requirements govern the publicity which must be given by the public authority to its intention to make a contract so as to invite tenders from interested parties. But the rules on the awarding of contracts are different. In concessions, the public authority has a free choice of partner. Thus in the famous case of Channel 5, the government simply negotiated with a company headed by some of its supporters before awarding it the franchise to run a newly permitted television channel in competition with the existing publicly run channels. But this was not held unlawful (CE 16 April 1986, COMPAGNIE LUXEMBOURGEOISE DE TELEDIFFUSION). By contrast, in public procurement contracts, the contract has usually to be put out to competition under the rules laid down in the *Code des marchés publics*.

Until the Local Government Act 1988, English law did not have much in the way of formal national legislation or case-law on the awarding of government contracts. Government practice, reinforced by audit, regulated the field, though discretion in the awarding of contracts was not unlimited in law.[31]

Increasingly, both English and French law on this subject are being harmonized. European Community Directive 71/305, now as amended by Directive 89/440, sets out stringent conditions for public works contracts, giving priority to competitive open tendering. Directive 77/62, as amended by Directive 88/295 on public procurement of supplies and equipment, and the (forthcoming) Directive on provision of services do the same in other spheres of public contracts. The purpose of these Directives is to establish as open a market as possible within the European Community for such contracts. The separate status of concession contracts has not been abolished, but the provisions governing both concessions and public procurement, even below the financial thresholds set out in the Directives, have been brought more into line in the recent French Law of 3 January 1991 on public procurement.

Terms of the Contract

A further difference between concessions and public procurement contracts also arises from the way the *Code des marchés publics* applies only to the latter. That Code sets out a number of standard terms and conditions (*cahiers des clauses et conditions* and *cahiers des charges*) which bind the different administrative authorities in making contracts. But in all administrative contracts, the administrative courts still retain the possibility of influencing the contents of the contract through their

---

[31] See *R. v. Lewisham LBC, ex parte Shell plc* [1988] 1 All ER 938.

interpretation of the intentions of the parties. In this interpretation, they will insist that the most reasonable terms (from the point of view of the public interest) are secured.

Contractual Equilibrium and Modification

The apparent one-sided character of administrative contracts requires special protection for the contractor. Particularly in concessions, public authorities have powers to redefine the character of the service to be performed or the work to be done in order to meet the changing needs of the public interest. This may involve additional costs, as where a transport service must provide buses to a new housing-estate. Such powers are not typically reserved to the public authority by the contract, but arise from general principles of administrative law. To protect the contractor, the law gives him the right for the financial balance of the contract to be preserved (what Blum in 1910 called 'le principe d'équation financière'. Where this is changed by the exercise of public powers, then he is entitled to an indemnity to restore the equilibrium of the contract. Payment of such an indemnity will be enforceable, if necessary, at the suit of the contractor before the administrative courts, and it may be that in such proceedings the court will decide that a particular variation or modification (in respect of which an indemnity was being claimed) should not have been made. The late Professor Mitchell pointed out that this doctrine involves considerable flexibility in the contract;[32] moreover, it may be invoked by the private contractor as well as by the administration.

Thus, in several cases concerning the supply of gas, electricity, and water, authorities in the communes have been allowed by the Conseil d'Etat to vary the terms of contracts under which monopolies had been granted to concessionaires where circumstances changed after the date of the contract. In COMPAGNIE GENERALE DES EAUX (CE 12 May 1933) a water company had had a monopoly contract since 1882 for the supply of water in La Seyne. By 1930 the population of the town had more than doubled and the quantity of water provided for in the contract was insufficient. The commune was authorized by the Conseil d'Etat to invite the company to provide adequate water at a price

---

[32] J.D.B. Mitchell, *Contracts of Public Authorities* (London, 1954), 192.

pro rata to that stipulated for in the contract, with the proviso that, if the company failed to agree, the commune would be free to enter into negotiations with another concessionnaire for the supply of the extra water. See also COMPAGNIE NOUVELLE DU GAZ DE DEVILLE-LES-ROUEN (CE 10 January 1902), where a contract for street-lighting was modified to permit the gas company to use electricity instead of gas, but, if it did not do so, the municipality should be allowed to engage a third party to provide electric lighting.

It is also clear that contractual provisions cannot exclude the power of the administration to use legislative powers to control matters covered by an administrative contract. Thus, when a contract between a government department and a tramway company empowered the department to approve the timetables, it was held by the Conseil d'Etat that this could not exclude the ordinary powers of the local prefect to regulate the number of trams put into service by the company (CE 21 March 1910, COMPAGNIE GENERALE FRANÇAISE DES TRAMWAYS).

Fait du prince

The exercise of such powers connected with the public service in question can be distinguished from the effect on a contract of the exercise of other public powers. This is covered by the doctrine of *fait du prince* and is invoked when the economic basis of the contract is upset by the act of the administration itself. This is a doctrine similar to 'act of state' in the English law of tort; as one party to the *contrat administratif* is an administrative agency, it may carry out some governmental act (unconnected with its rights under the contract) which may affect the other party to the contract. In English law it is recognized that an administrative agency cannot by contract fetter its right—or duty—to exercise an administrative discretion vested in it by statute (see e.g. *Stringer* v. *Minister of Housing and Local Government* [1971] 1 All ER 65). Nor can a public authority by its conduct estop itself from exercising such a discretion.[33] In such circumstances the other party to a contract may well have no remedy in damages. In France, however, unless the governmental act was some general legislation affecting all citizens equally

---

[33] *Western Fish Products Ltd.* v. *Penwith DC* [1981] 2 All ER 204.

(see CE 14 February 1936, VILLE DE PARIS), the contractor will be entitled to a monetary indemnity or to increase the charge to the consumer. As Mitchell said,[34] where an administrative agency which is a party to the contract takes some measure by virtue of *la puissance publique* that affects the substance of the contract, the contractor will normally be entitled to an indemnity in accordance with the *fait du prince* doctrine.

Imprévision

Under this principle, if supervening circumstances (unconnected with the administration) have arisen after the formation of the contract for which no (or inadequate) provision has been made in the express terms of the contract, making it uneconomical for the 'private' party to the contract to perform his part, he will not be allowed to resile from the contract, but may be compelled to perform the contract and will then be entitled to an indemnity from the administration against his extra expenses. This will be insisted upon by the administrative courts in circumstances where the public interest demands that the contract should be performed. In the case of a concession, the indemnity may take the form of a right to charge the consumer more than is provided under the contract. Thus, in the leading case of COMPAGNIE GENERALE D'ECLAIRAGE DE BORDEAUX (CE 30 March 1916) a commune had entered into a contract with a gas company for a supply of gas for the lighting of the streets in the town. Because the coalfields in Northern France had been overrun by the Germans in the early part of the First World War, the price of coal increased in fifteen months from 35 francs per ton at the time when the contract was made to 117 francs per ton; an increase which, unless the company were allowed to increase the contract charges for the gas supplied, would have compelled the gas company to go into liquidation. The Conseil d'Etat took the view that it was not in the public interest that this should happen, as, if the company became defunct, the streets of Bordeaux would not be lit. Therefore the court ordered the company to continue to perform the contract, but substantially increased the price of gas charged to the commune under the contract.

---

[34] Mitchell, *Contracts of Public Authorities*, 196.

This doctrine, it should be noted, goes much further than the English law of frustration; in the first place, it is not necessary that the contract should have become physically or legally incapable of performance, although the party claiming an indemnity must show there was a *bouleversement* (upsetting) of the economic substance of the contract. In the second place, *imprévision* does not operate to determine the contract; performance thereof continues. Indeed, where the contractor is unwilling to continue his performance on any terms (on any realistic terms) the administration may take over the performance of his part of the contract and provide the service, etc., themselves (*en régie*).

The principle of *imprévision* is invoked much less now that it has become common for contracts to provide expressly for such matters as inflation or monetary depreciation: what was once abnormal has become the norm for which the prudent make provision in their contracts, eg by a price-revision clause (CE 2 April 1960, PEDARD).

A modern illustration of the doctrine is provided by SOCIETE PROPETROL (CE 5 November 1982). A company had contracted to supply fuel-oil for municipal housing in Strasbourg. After the steep rise in oil prices in 1973, the company told the city it should look for a new supplier. When the city treated the contract as repudiated and sued for substantial damages, the company pleaded *force majeure*. The Conseil d'Etat agreed with the Tribunal Administratif of Strasbourg that *force majeure* was not established, as it was not impossible for the contract to be performed (see below); the company should not simply have repudiated the contract; it should have invoked the doctrine of *imprévision* and, while continuing to perform its contract, sought compensation from the city for the *bouleversement de l'économie du contrat*; the company was therefore liable in damages for having repudiated the contract.

Force majeure

In some circumstances, even the doctrine of *imprévision* cannot adequately compensate the contractor; in COMPAGNIE DES TRAMWAYS DE CHERBOURG (CE 9 December 1932) the concessionnaire was on the verge of bankruptcy, but if the tram-fares had been increased any more, the company would have lost their customers. The Conseil d'Etat considered that this was an example of *force majeure*, beyond the powers of either of the contracting parties to overcome; the object of the contract (to run a tramway service at a reasonable price) had been destroyed, and so the contractor was entitled to be released from his obligations under the contract. On the other hand, a strike by the concessionaire's employees did not necessarily constitute a case of *force majeure* so as to exonerate him from his contractual obligation: COMPAGNIE DES MESSAGERIES MARITIMES (CE 29 January 1909).

Other Rights of the Contractor

It is sometimes necessary for a public body entering into a contract to obtain the approval of some other agency; thus (before the reforms of 1982) a *département* might require the assent of the prefect under his power of *tutelle* (see Chapter 2). In such a case the contract is a nullity if the required assent is not obtained, but a contractor who has wholly or partially performed his part under such a contract will be entitled to an indemnity, on the basis of what in English law would be termed a *quantum meruit*; see, for example, PHILLIPPE (CE 27 June 1930) and BONNIOL (CE 16 January 1931). Alternatively the contractor can question before the administrative courts by a *recours en indemnité* (or en *annulation*) the decision of the superior authority to withhold approval. Similarly, the contractor may question an exercise of *la puissance publique* in the modification of a contract: this is particularly so where the variation is made by that agency of the administration which was a party to the contract (CE 9 March 1928, COMPAGNIE DES SCIERIES AFRICAINES).

Contracts of Employment

Civil servants (*fonctionnaires*) are not employed under contract but possess a special (and public) status closely defined by statute and regulation. But other public servants who are employed under contract do not necessarily fall within the jurisdiction of the administrative courts, unless their contracts are classified as administrative (see p. 137 above).

Where administrative law is applicable, complaints by civil servants and other public servants about their salary rights, pensions, and entitlement to promotion take up a considerable amount of time of the administrative courts. There is no dif-

SUBSTANTIVE LAW: ADMINISTRATIVE LIABILITY 201

ficulty in applying the *audi alteram partem* rule and other 'general principles of the law' in such a context, in contrast with the situation in common law countries. There the court has to find some excuse (usually in some expression in the statute) for importing into the ordinary law of contract, rules of administrative law (see, for example, *Malloch* v. *Aberdeen Corporation* [1971] 2 All ER 1278).

In the United Kingdom, civil servants may now complain to industrial tribunals over discrimination or unfair dismissal. Most of the principles governing their employment are those of private law. In France, although public service employment law is distinct from the law governing private employment law, the administrative courts frequently import principles from private law as 'general principles of law'. For example in PEYNET (CE 8 June 1973) a pregnant woman was dismissed from her post as an auxiliary nurse at a public hospital. Drawing on the principle behind the (private law) Labour Code, the Conseil d'Etat held that a general principle forbade the dismissal of a woman in such circumstances, and so the dismissal was unlawful. Since PEYNET, the Conseil d'Etat has frequently drawn upon private law to provide adequate protection for public employees: see, for example, CE 12 November 1990, MAHLER, concerning financial penalties imposed on strikers.

9 GENERAL OBSERVATIONS

The English law on public contracts is far less developed than that in France. Much of it remains in the realm of practice rather than law. Nevertheless, there are increasing pressures for the English law in this field to develop: changes are being promoted by legislation of the European Community, and our doctrinal writers and judges have not been idle.[35] French systematization in this area provides a useful guide to the problems which will be encountered as English law develops. The more developed French principles may also help inspire harmonization within the Community, thereby having a more direct impact on English law.

---

[35] See S. Fredman and G. Morris, *The State as Employer* (London, 1989) and *R.* v. *Civil Service Appeal Board, ex parte Bruce* [1989] 2 All ER 907.

9

The Substantive Law: The Principle of Administrative Legality

1 INTRODUCTION

The special nature of *droit administratif* as contrasted with the French *droit civil* and, for that matter, with the English common law, is particularly brought out when we consider the recognition by the administrative courts of the principle of *légalité*. As we have already explained, the Conseil d'Etat owes its present jurisdiction to Napoleon I, but this does not mean that either he or any later legislator established or predetermined the law it was to administer. In origin the Conseil d'Etat advised the head of state: not until after 1872 did it speak 'au nom du Peuple Français', and it was therefore all the easier for the Conseil to evolve the principles which it was prepared to apply to the administration. Apart from the principle of administrative liability (*responsabilité*), which we have discussed in the previous chapter, the other great principle is the principle of administrative legality: the idea that the administration must be compelled to observe the law. This is something much more than the English and Scottish doctrines of *ultra vires*, there being no question of a mere observance of statutory limitations, and it also goes further than observance of the principles of natural justice; but both these ideas, so familiar to the British administrative lawyer, can be detected as constituents of the principle of *légalité*.

As we saw in the last chapter, *droit administratif* provides two principal remedies, the *recours en annulation* and the *recours en indemnité*. The latter is very largely appropriate when the plaintiff invokes the principle of administrative liability; the former,

SUBSTANTIVE LAW: ADMINISTRATIVE LEGALITY

which we must now consider in detail, is always appropriate (but not exclusively so) when the plaintiff invokes the principle of legality. An explanation of the grounds on which a *recours en annulation* will be granted is thus at the same time an explanation of the contents of the principle of legality. Although we shall list the various accepted grounds for review of an administrative decision (*acte administratif*), it should be realized that the administrative courts may at any time recognize a new *cas d'ouverture*, based on the general idea of administrative morality—the idea that the state is an honest man and must behave properly towards *l'administré*. It must also be remembered (Chapter 7, p. 152) that judicial review is dependent upon the existence of an *acte administratif*.

Finally, what was said in the last chapter about the Conseil's pragmatic and flexible approach in fashioning the substantive law is especially true of the principle of *légalité*, the definition of which has undergone constant evolution and refinement in cases where an *acte administratif* has been held void for *excès de pouvoir*.

2 THE CONTENT OF THE PRINCIPLE OF LEGALITY

The basic principle of control by the courts over the administration is that the latter must be subjected to the Rule of Law (understood almost in Dicey's sense, and certainly as used in the historic Declaration of Delhi of 1959),[1] and all proceedings on a *recours en annulation* are founded ultimately on a violation of this principle.

The judge's first task when confronted with an allegation of *excès de pouvoir* is to determine exhaustively the rules applying to the *acte administratif* which is subjected to attack. This primary analysis of the legal situation is often difficult, because the sources of legality are complex.

In the first place any relevant enacted law must be examined; and the administration must conform to the hierarchy of written rules applying under French law to different situations. Highest in the hierarchy are, of course, the rules of the Constitution and international treaties (once ratified), and, in second place, statutes passed by parliament; there are then the various other

---

[1] See (1959) *Journal of the International Commission of Jurists*, **2**, no. 1. This Declaration was subscribed to by jurists from many countries meeting under the auspices of the International Commission of Jurists.

forms of legislation described in Chapter 2. Thus, a prefectoral order will be illegal if it contravenes a ministerial regulation, even as such a regulation will be illegal if it contravenes a statute.

This is, of course, in a different context, much the same principle as the English doctrine of *ultra vires*, and the specialized application of that doctrine to the byelaws and acts of local authorities or other government agencies. Thus, in *Blackpool Corporation* v. *Locker* [1948] 1 All ER 85, the town clerk's act in requisitioning a house containing furniture was in violation of the 'law' contained in the ministry circular giving him power to requisition only empty houses, or, as the English court said, his act was *ultra vires*. Moreover, once an administrative agency has laid down rules for its own conduct, it will be in violation of the law if it subsequently breaks those rules (*legem patere quam ipse fecisti*). In England a local authority has been held to be incapable of authorizing a departure from its own byelaws (*Yabbicon* v. *King* [1899] 1 QB 444); in France it was held that once an agency had decided that it would in future take certain measures only after consultation with a committee, the agency was acting illegally if subsequently it took those measures without that consultation: see SOCIÉTÉ MICHEL FAURE (CE 20 January 1950). On the other hand, a planning authority may permit derogations from its own development plan on a balance of competing interests: see VILLE DE LIMOGES (CE 12 May 1973).

But the principle of legality is not only concerned with enacted law. The administration is bound to respect individual rights created by its own previous decisions,[2] and this is so even if those decisions were illegal at the time but have since become immune from attack. Even more clear is its obligation to respect decisions of the courts. The doctrine of *la violation de la chose jugée* is reminiscent to some extent of the English plea of *res iudicata*. Thus, the authority of the decision of a French court is normally relative, binding only the parties; but where an administrative court annuls an administrative decision, then this decision (if not challenged by way of appeal) is valid against all, or as the French

---

[2] See, for example, CREDIT FONCIER, p. 216 below. In relation to circulars, this principle is now enshrined in Art. 1 of the Decree of 28 November 1983.

say, *erga omnes*, including, of course, the administration itself. Further, although a decision of the Conseil d'Etat does not create a binding precedent for future cases, it is nevertheless binding in the particular case on lower courts and the administration.[3]

In a significant development, the idea of legality has been extended by the case-law of the Conseil d'Etat to include certain unwritten *principes généraux du droit*. The importance of these is so great in the modern law that a separate section must be devoted to their consideration. After this we shall consider the several traditional grounds on which a *recours en annulation* may be founded, although (as we shall see) all are no more than different aspects of the same notion of *illégalité*, that is, a breach by the administration of the principle of legality.

3 LES PRINCIPES GÉNÉRAUX DU DROIT

The *principes généraux du droit* have been imported into the principle of legality through a broadening of the notion of *loi*, a breach of which (as we shall see below) is a ground for review. They have been an outstanding feature of the case-law of the Conseil d'Etat since 1944, and are a measure of the pervasive influence of the Conseil d'Etat over French public life.[4] It is in their development that *droit administratif* has shown its most elastic qualities and its ability to deal with any situation where, in broad terms, administrative morality has not been observed.

These principles may be compared with the English principles of natural justice (*audi alteram partem* and freedom from bias) but, unlike English law, *droit administratif* has refused to confine itself to restricted procedural rules, and it has considered itself free to recognize a new 'general principle' when it considers this desirable.

The principles, or rather cases where they have not been observed by the administration, are sometimes dealt with by

---

[3] The administration is bound to repeal any other administrative decisions which are incompatible with the instant decision, whether they have been brought before a court or not: see COMPAGNIE ALITALIA, p. 221 below.

[4] A valuable survey is M. Letourneur, 'Les principes généraux du droit dans la jurisprudence du Conseil d'Etat', *Etudes et Documents 1951*, 19. For a recent catalogue of these general principles, see G. Vedel and P. Delvolvé, *Droit administratif* (11th edn., 2 vols., Paris, 1990), i. 465–7.

French writers as specialized examples of *violation de la loi*,[5] sometimes as matters of *incompétence* and sometimes as applications of *détournement de pouvoir* (for these terms, see p. 223). In an important passage, reproduced in *Les Grands Arrêts de la jurisprudence administrative* (9th edn., at p. 454), Bouffandeau (a former president of the Section du Contentieux) described the *principes généraux du droit* as

unwritten rules of law, which have legislative validity and consequently are binding upon the administration in the exercise of *pouvoir réglementaire* or administrative discretion, in so far as they have not been overruled by some express legislative provision . . . they cannot, however, be described as forming a system of customary administrative law, as they have, for the most part, only recently be recognized by the administrative courts. These principles are really a creation of the courts inspired by ideals of justice (*équité*) in order to ensure the protection of the individual rights of the citizen.

As was stated in ARAMU (CE 26 October 1945), 'the principles are applicable even in the absence of a text'; compare the dictum of Byles J., 'although there are no positive words in a statute, requiring that the party shall be heard, yet the justice of the common law will supply the omission of the legislature' (*Cooper v. Wandsworth Board of Works* (1863) 14 CB (NS) 180).

Primarily, the principles are deduced as a matter of statutory interpretation, based on the assumption that the legislator is anxious to preserve the essential liberties of the individual. They do not have a 'constitutional' validity, but their operation can be excluded only by the clearest express statutory provision. They have been described by M. Chapus as 'serviteur des lois et censeur des décrets' (the servant of statutes but the censor of decrees); that is, they must give way to statute law but are a test of the validity of subordinate legislation (see *D*. 1960, Chron. 119).

Waline suggests[6] that the Conseil d'Etat has deduced these general principles from the 'general spirit' of legislation and from public opinion; when the Conseil d'Etat applies one of these principles it does so in the firm conviction that the principle is part of the law. A special application of this approach to a problem is to be seen in the CANAL judgment (CE 19 October 1962),

---

[5] M. Waline, *Droit administratif* (9th edn., Paris, 1963), paras. 762–70.
[6] Ibid., para 770.

where the Conseil d'Etat declared that a special military tribunal set up by a decree signed by the head of state was contrary to law, as violating a general principle, namely, that the liberty of the citizen required the decision of the court to be subject to review by the Cour de Cassation (see p. 54 above). It is interesting to contrast the attitude of the Conseil d'Etat in CANAL towards an emergency decree at a time of national crisis with that adopted by the House of Lords (Lord Atkin strongly dissenting) in the wartime case of *Liversidge* v. *Anderson* [1941] 3 All ER 338.[7]

The following can be identified as sources of these general principles:

1. Constitutional documents such as the 1789 Declaration of the Rights of Man and the Preamble to the 1946 Constitution of the Fourth Republic. Examples include: liberty, respect of the rights of citizens, equality in all its aspects (in taxation, public burdens, access to public services, etc.). As is said by Mme F. Batailler,[8] there is a 'tradition constitutionnelle républicaine'.

2. Some of the principles seem to have been drawn from private law and civil procedure. Thence come, for example, certain procedural rules applicable in the administrative courts even in the absence of specific legislative provisions; the rule against double jeopardy; the powers of appellate courts over courts of first instance. From the same source come such elementary procedural rules as are applied even to the purely administrative decision-process (e.g. the right to be heard and the non-retroactivity of administrative acts).

3. The residual source is 'natural law'—'ideas' (*sentiments*) of justice and equity. Thence the notion that the administration is expected to be fair and just in its dealings with the citizen (cf. the principles of 'fairness, openness and impartiality', urged by the British Franks Report). Thus, the right of review of all administrative action by the *recours pour excès de pouvoir* rests on no formal text but is a general principle derived from this source. Also under this head is the countervailing notion, favourable to the administration, of the necessity for the continuity of the public service.

---

[7] For new light on this case, see A. W. B. Simpson in (1991) *Quadrangle Notes* (Michigan Law School).

[8] See *Le Conseil d'Etat, Juge Constitutionnel* (Paris, 1966).

Before 1958, it seems, the Conseil d'Etat claimed that in establishing a 'general principle' it was only interpreting the intention of parliament, much in the same way as our own courts have evolved principles of statutory interpretation. The 1958 Constitution, however, gave the government very wide legislative powers under Article 37 as well as redefining the role of parliament (see Chapter 2, p. 8). As a consequence there arose a new emphasis on the *principes généraux* as independent principles of substantive law in their own right; indeed, the Conseil d'Etat has dignified this new phase of the evolution with a theory under which certain *principes généraux* appear as constitutional norms. This is to be found in the conclusions of the Commissaire du gouvernement in SYNDICAT GENERAL DES INGENIEURS-CONSEILS (CE 26 June 1959), although the case itself was concerned not with Article 37 but with the similar powers exercised by the government in colonial territories before 1958. M. Fournier there suggested that there were two categories of unwritten rules with which the administrative judge will enforce compliance. In the first category were certain rules of mere interpretation, to be applied in the absence of clear indication to the contrary in the statute or regulation; these include a number of procedural rules governing the functioning of both administrative authorities and administrative jurisdictions (for this distinction, see Chapter 3, p. 55). Fournier gave as an illustration of such rules 'les droits de la défense', but in 1989 the Constitutional Council elevated such rights to a constitutional principle: CC decision no. 88–248 DC of 17 January 1989, CONSEIL SUPERIEUR DE L'AUDIOVISUEL.

The other category is the more important and consists of the *principes généraux* properly so called. These are those fundamental human rights contained in the Declaration of the Rights of Man of 1789 and the Preamble to the 1946 Constitution or which may be deduced from them.[9]

The executive, even when acting under Article 37, cannot transgress these principles. For, although the division of powers between parliament and government is now expressly regulated by the Constitution, the administrative judge is still entitled, indeed obliged, to examine the validity of governmental action

---

[9] Both the Declaration and the 1946 Preamble are reaffirmed in the Preamble to the 1958 Constitution.

by reference to those principles which constitute the very basis of the republican regime, such as the essential liberties of the citizen (*Liberté*), equality before the law (*Egalité*), the doctrine of separation of powers, the principle of non-retroactivity and the right to judicial review. Such fundamental rights, which are mostly entrenched in the text of the United States Constitution, in French law are protected to a large extent by resort to the unwritten *principes généraux*.

Categories of General Principles

In the years since 1971, the concept of general principles has received further endorsement by decisions of the Constitutional Council, in which the Council has ruled bills to be unconstitutional by reference to 'fundamental principles recognized by laws of the Republic' or 'the political, economic, and social principles particularly necessary for our time', both phrases being enshrined in the Preamble to the 1946 Constitution and reaffirmed in the Preamble of 1958. Characteristically, the Constitutional Council prefers to derive its general principles from constitutional texts, whereas the Conseil d'Etat is prepared to create general principles in praetorian fashion through its case-law.[10]

Despite the wealth of this case-law, it is quite impossible to give with confidence a complete list of the general principles or of the unwritten rules of interpretation; moreover, individual principles often overlap, and some writers allege the existence of more principles than others. All that can be done here is to outline those of the more important principles that have already been recognized and applied in the administrative courts.

Because of the new activism of the Constitutional Council, a distinction has now to be made between those general principles of *legislative* force and those having *constitutional* value (such as the 'fundamental principles recognized by laws of the Republic'). The former bind the administration in its decisions or rule-making, but the legislature is free to derogate from them. The latter may be applied by the Constitutional Council to invalidate even a statute before it is promulgated, as well as any decree or regulation; and they are invoked equally by the administrative courts to annul administrative decisions, unless they

[10] Vedel and Delvolvé, *Droit administratif*, i. 474–81, and Ch. 2, p. 21 above.

have been expressly authorized by statute: for, in principle, the administrative judge is unable to challenge a statute as unconstitutional.[11]

Prerogatives of the Administration

The administrative courts have to recognize that the administration has a job to do, and while it will not be permitted to do by force what it has no legal power to do, the courts will accept a general power in the administration to function with regularity and certainty. This is the basis of the powers of the court to interpret the 'police' powers of the administration. The administration must always be allowed to take any necessary measures to preserve public order, even in spite of laws protecting personal liberty, the right of assembly, or the right of freedom of expression. The courts will determine the limits of such administrative action by the necessity of the situation; thus, in BENJAMIN (CE 19 May 1933), the Conseil d'Etat recognized that a mayor, by forbidding the holding of a conference, had gone further in interfering with the right of assembly than was demanded of him by the necessity to maintain order; the risk of a disturbance was not so acute that it could not have been averted by less drastic measures. Likewise, the seizure of a newspaper could be justified only on the grounds of emergency and public order: ACTION FRANÇAISE (TC 8 April 1935); and in the famous *arrêt* FILMS LUTETIA (CE 18 December 1959), the Conseil d'Etat refused to be a judge of morality in considering whether the mayor of Nice was justified in banning the showing of certain films in his commune, but decided the case in the mayor's favour on the basis that there must be (and these were here shown to have existed) 'local circumstances' justifying the ban.

By a similar process of reasoning, in DEHAENE (CE 7 July 1950) it was held that the administration was entitled to limit by regulation the right of civil servants to strike, although the Preamble to the Constitution recognized in general terms that individual workers were entitled to strike. So too SYNDICAT NATIONAL DE RADIODIFFUSION ET DE TELEVISION (CE 20 January 1975), where the Conseil d'Etat upheld the right of the

[11] Proposals in 1989 that Conseil d'Etat and the Cour de Cassation might refer the constitutionality of a statute to the Constitutional Council proved abortive, but were revived by President Mitterrand in 1991.

SUBSTANTIVE LAW: ADMINISTRATIVE LEGALITY

minister (acting under the law governing the French radio and television network) to curtail the strikers' rights to the extent necessary to maintain a basic radio and television service. This curtailment of workers' rights has now been given constitutional recognition by the Constitutional Council in its decision of 25 July 1979 on strikes in radio and television (CC decision no. 79–105 DC of 25 July 1979). Again, the doctrine of the separation of powers is really nothing more than one of the most ancient *principes généraux*, that which protects the administration from interference by the ordinary courts.

Liberties of the Individual

Those *principes généraux* which pave the way to efficient administration are counteracted by principles designed to protect the rights of individuals; personal liberty, the right of assembly, and freedom of expression by the press are expressly enforced through specific statutes, but those other human rights guaranteed only by the Declaration of the Rights of Man depend for their enforcement on the *principes généraux*. Thus, the courts have recognized the right to strike (DEHAENE, above), freedom of thought and opinion (CE 1 April 1949, CHAVENEAU), and freedom of movement (CE 20 January 1965, VICINI) where a prefect had prohibited camping by gypsies: see p. 243 below).

The right to a normal family life was upheld in GROUPE D'INFORMATION ET DE SOUTIEN DES TRAVAILLEURS IMMIGRES, (CE 8 December 1978), where the Conseil d'Etat annulled a decree which sought to restrict the entry into France of members of the family of an alien who was holding a residence permit, unless they undertook not to take any employment in France; the decree was held to be contrary to the right of aliens regularly residing in France 'to lead a normal family life' as enshrined in the Preamble to the Constitution of 1946, such right being adjudged to include the freedom of such aliens to be joined by their spouse and minor children.

Freedom of conscience was the issue raised in ASSOCIATION FAMILIALE DE L'EXTERNAT SAINT-JOSEPH (CE 20 July 1990). Under the *Code du travail*, private bodies employing more than twenty persons must have a set of internal rules (*un règlement intérieur*) for such matters as safety, health, and discipline. But such rules must not unduly restrict the employees' fundamental rights and freedoms. The rules are systematically reviewed by the *inspecteur du travail*, who may ask the employer to withdraw or modify a rule. This case concerned the rules of a private Catholic school, linked to the public service by a contract of association. Such schools have been recognized by the legislature as having a distinct character (*caractère propre*), and in 1977 the Constitutional Council upheld the validity of a statute requiring teachers to respect that distinct character, provided that this requirement did not limit their freedom of conscience, a basic right under the Constitution.[12] The Saint-Joseph School had a rule imposing this requirement upon *all* employees, whether teachers or not. The *inspecteur du travail* and, on appeal, the Minister of Labour directed the rule to be withdrawn, the Nice Tribunal Administratif upheld the Minister's decision, whereupon the school appealed to the Conseil d'Etat. The Conseil d'Etat rejected the appeal. The school could lawfully require *teaching* staff to respect the school's *caractère propre*, but only if the rules also confirmed that freedom of conscience was to be respected. Also, as it stood, the rule was unacceptable by being applied indiscriminately to all categories of employees—teachers, administrative and secretarial staff, etc. Distinctions needed to be introduced, which were lacking in the rules, between these different categories. Thus, the rule, as it stood, was unlawful.

In BERECIARTUA-ECHARRI (CE 1 April 1988) a decree of extradition to Spain was quashed by the Conseil d'Etat in respect of a Basque militant who had obtained refugee status in France; the Conseil based its decision not on the 1951 Geneva Convention on the Status of Refugees, but on a general principle applicable to all refugees which prevented him being returned to his country of origin.

A different general principle relating to extradition arose in SAIA (CE 29 September 1989). Saia challenged a decree for his extradition to Italy on the ground that he could not be tried or imprisoned there because he benefited from an amnesty declared by the Italian government. The French extradition statute of 1927 does bar extradition wherever criminal proceedings cannot be brought for whatever reason (for example, amnesty) in the

---

[12] CC decision no. 77–87 DC of 23 November 1977.

requesting country. Unfortunately, France had ratified the European Convention on Extradition of 1857, which makes no provision for the situation of amnesty, and the Convention prevails over the statute in the event of a conflict. The Conseil d'Etat affirmed the general principle of the law of extradition whereby extradition is prohibited to a country in which an amnesty has been declared applying to the person concerned; but in the case of Saia, the Conseil d'Etat upheld the decree of extradition because the crimes for which extradition was being sought did not fall within the amnesty declared by the Italian government.

A commentator has referred to a 'somewhat spectacular extension since 1977 of the scope of judicial review' by the Conseil d'Etat in extradition cases.[13]

Economic and Social Rights

The general principles that have gradually been formulated to protect the economic and social rights of individuals and institutions are particularly interesting, in that these provide an example of the manner in which the jurisprudence of the court adapts itself to meet changing economic and social conditions. Originally, the administration was not allowed to compete with private enterprise (see, for example, CE 29 March 1901, CASANOVA, p. 160 above, where a commune in Corsica were held to have exceeded its powers in establishing a post of public doctor for the inhabitants of the commune), but in recent years this trend has been reversed and local authorities have been allowed to establish a dental clinic, a local insurance company, a garage, an open-air theatre, and food shops, in cases where local need could be proved (see CE 30 May 1930, CHAMBRE SYNDICALE DU COMMERCE EN DETAIL DE NEVERS, and CE 24 November 1933, ZENARD, concerning a butcher's shop). Similarly, the right to strike, to join a union, or to abstain from joining a union operating a 'closed shop', are recognized by the Constitution only in its Preamble, and therefore are enforced in practice as *principes généraux*. The nearest parallel to this idea in English law is *Nagle* v. *Fielden* [1966] 1 All ER 689, where the Court of Appeal were prepared to recognize a 'right to work'.

[13] See R. Errera, [1985] PL 328, where there is a valuable review of French extradition law and practice.

Again in the field of labour law, an important case before the Conseil d'Etat (CE 23 April 1982, VILLE DE TOULOUSE) established a new general principle in the nature of a right to a minimum wage for everyone engaged in the public sector of employment. The *Code du travail* confers a statutory right to an index-linked minimum wage only upon employees working in the private, not the public, sector, although all civil servants and a few limited categories of other public employees have in practice enjoyed the same guarantee. In VILLE DE TOULOUSE three persons had been engaged as casual staff by the city to serve as leaders in its *colonies de vacances* for local children; after their appointment they sued the city for extra pay in order to bring their low wages up to the level of the statutory minimum wage. The Conseil d'Etat admitted their claim as well founded on the basis of a general principle applying to every wage-earner and underlying the particular provision in the *Code du travail*. Already in DAME PEYNET (CE 8 June 1973, discussed chapter 8, p. 201) the Conseil, invoking the doctrine of general principles, had extended to public servants the protection accorded in the *Code du travail* to private-sector employees against their dismissal during pregnancy. But the later case established a more far-reaching general principle, as French commentators have been quick to note.[14]

Protection of the Environment

The French administrative courts in recent years, to match the mood of contemporary society, have begun to develop general principles for the protection of the environment.

Thus, in SYNDICAT CFDT DES MARINS-PECHEURS DE LA RADE DE BREST (CE 25 July 1975) a group of sea-fishermen in the Brest roadsteads opposed the creation on the shore of an industrial complex which was planned to include an oil refinery. The Conseil d'Etat, adopting the approach established in VILLE NOUVELLE EST (CE 28 May 1971, p. 247 below), laid down the principle that the compulsory acquisition of land can, in law, be declared to be in the public interest (*d'utilité publique*) only if, in particular, the ecological disadvantages which it may entail are not excessive, having regard to the advantages of the proposal. In the instant case, only a small amount of agricultural land was

[14] F. Tiberghien and B. Lasserre, *AJDA* 1982, 433.

to be acquired, precautions would be taken against pollution of the sea, and the proposal did not affect a site classified as of special historic or ecological value; accordingly, the declaration *d'utilité publique* was upheld.

In the environment field the pioneering approach of the Conseil d'Etat has prompted statutory intervention. The Law of 10 July 1976 now requires an *étude d'impact* to be prepared before any governmental project of environmental importance (as specified in the statute is undertaken). This corresponds to the 'environmental impact statement' of the federal National Environmental Protection Act 1969 in the United States and the environmental directives of the EEC.

Equality before the Law

This is, perhaps, one of the best known of the general principles and one which is frequently applied in practice. Thus, in S. A. DES GRANDS MAGASINS ECONOMIQUES (CE 10 February 1937) a traffic regulation of a local mayor was quashed because it forbade the parking in certain streets of private cars and not of all vehicles. In many cases this principle has an economic application, in that the courts will insist that all citizens should have an equal opportunity of benefiting from public services. Thus, in SOCIETE DES CONCERTS DU CONSERVATOIRE (CE 9 March 1951), two members of the Society had been 'disciplined' by its committee, for playing in a concert organized by Radiodiffusion Française (a state broadcasting service) instead of playing for the Society. The administration of Radiodiffusion thereupon refused to broadcast any concerts organized by the Society. The Conseil d'Etat annulled this decision as an *excès de pouvoir*, in that one of the general principles of law had been overlooked, namely the principle of equality, which governs the operation of all public services. The famous BAREL case (28 May 1954, p. 238 below) applied this principle in a different context, namely, to ensure equal access to the civil service. Equality of the sexes was upheld in relation to employment in the public service in SYNDICAT CHRETIEN DU CORPS DES OFFICIERS DE POLICE (CE 21 April 1972), a decision of the Assemblée, which cited not only the principle of equality of the sexes contained in the *Statut général des fonctionnaires* of 1959 but also referred to the constitutional source of that principle in the Preamble to the Constitution of 1946.

The principle of equality before the law was also applied by the Conseil d'Etat in CREDIT FONCIER DE FRANCE (CE 11 December 1970) to an application for a grant from a national fund in respect of house improvements. The application in question was rejected because it did not comply with standards prescribed in a circular issued by the commission administering the fund, which circular did not have the force of law. It was held that such a commission was entitled to lay down standards, but that they must be such that all applicants for grants were treated on an equal footing.[15]

A recent illustration of the principle of equality of the sexes in public employment is provided by FEDERATION DES SYNDICATS GENERAUX DE L'EDUCATION NATIONALE (CE 26 June 1989). Here a teachers' union challenged the personnel regulations for state boarding-schools on the ground that they provided for a different composition of disciplinary boards which dealt with supervisory staff, depending whether a man or a woman was being disciplined. The regulations dated from 1937-8, and the union argued that they were outdated in view of the 1946 Preamble and of a 1975 statute affirming equality of men and women in the public service. The union asked the Minister of Education to repeal the regulations, and, when he refused, appealed to the Conseil d'Etat, which proceeded to quash his decision. The Conseil applied the general principle which it had adopted in COMPAGNIE ALITALIA (CE 3 February 1989, p. 221 below), namely, that the relevant authority is bound to repeal any unlawful regulation, when asked to do so, either if the regulation was void *ab initio* or if it had become so by reason of a subsequent change in the legal or factual circumstances. The Conseil again cited the 1946 Preamble: 'The law guarantees to women, in all areas, rights equal to those of men.' This principle applied to the public service.

This principle of equality, to which French administrative lawyers attach so much importance, has a parallel, of course, in Dicey's analysis of his 'Rule of Law' as requiring equality before the law, and, as a corollary, the denial of special privileges

---

[15] English readers may be reminded of the jurisdiction assumed by the Parliamentary Commissioner for Administration to investigate allegations of injustice arising from the so-called 'bad rule'. For this notion and the so-called 'Crossman Catalogue', see *Garner's Administrative Law* (7th edn., 1989) 93-4.

SUBSTANTIVE LAW: ADMINISTRATIVE LEGALITY 217

for the administration. Modern legislation in Britain has introduced special rules and special machinery of enforcement in relation to racial and sexual equality, especially through rules of employment law which are applied both to public and private employment. In English administrative law, there is not, however, such a highly developed case-law on the principle of equality as we find in France. If a member of the public were to complain of having been excluded from participating in a public service or a public benefit, his or her complaint would either have to fall within the special legislation against discrimination or be based generally upon some breach of the *ultra vires* principle.[16]

Impartiality

The administration must observe the principle of impartiality in the course of arriving at a decision. Therefore a decision taken in the presence of an interested party was annulled in LACAMBRE (CE 28 January 1948), and in TREBES (CE 4 March 1949) the court quashed the decision of a committee advising about personnel on the ground that the committee included several members whose careers would be directly affected by the decision taken. This principle is similar to the English 'freedom from bias', and English cases where a decision was annulled on this ground include *Dimes* v. *Grand Junction Canal* (1852) 3 HLC 759; *R.* v. *Hendon Rural District Council, ex parte Chorley* [1933] 2 KB 696, and *Steeples* v. *Derbyshire County Council* [1984] 3 All ER 486.

Audi alteram partem

The necessity to hear both sides—the right to a hearing or 'due process' before a decision is arrived at—is recognized by *droit administratif* as it is in English administrative law. Clearly all procedural requirements laid down in statutes or regulations must be observed, but also a decision will be annulled whenever the aggrieved party has not been advised in advance of the case he has to meet and given an adequate opportunity of presenting his views. This was one of the steps by which the Conseil d'Etat reached its decision in BAREL (see below). In S. A. COOPERATIVE

---

[16] See, for example, *James* v. *Eastleigh Borough Council* [1990] 2 All ER 607, HL.

D'HABITATION A BON MARCHE DE VICHY-CUSSET-BELLE-RIVE (CE 24 April 1964), an order of the minister winding up the Society was declared void because the president of the Society had not been given an opportunity of being heard before the decision was taken.

In TROMPIER-GRAVIER (CE 5 May 1944) a concession to sell newspapers from a kiosk on the Boulevard Saint-Denis in Paris was revoked by the prefect when he received a complaint to the effect that Mme Trompier-Gravier (the concessionaire) had sought to extort a substantial sum of money from the manager of the kiosk. The grant or revocation of these concessions was entirely within the discretion of the prefect, but his decision was quashed by the Conseil d'Etat because he had not first given Mme Trompier-Gravier an opportunity of answering the charge brought against her. In SOCIETE DES ETABLISSEMENTS CRUSE (CE 9 May 1980) this principle was extended to the procedure for depriving wine of its *appellation contrôlée*. This may be compared with the English case of *Ridge* v. *Baldwin* [1964] AC 40[17] and the Franks principle of 'openness', which governs generally the procedure of British administrative tribunals.

Duty to Give Reasons

The Conseil d'Etat refused to elevate to a general principle a requirement that the administration should give reasons for its decisions and communicate these to all interested parties who request them. This gap has now been filled, however, by the statutory obligation to give reasons introduced by the Law of 11 July 1979 (see p. 88). Nevertheless, the duty does not cover all administrative bodies, an omission criticized by the European Court of Justice in relation to a French committee which refused recognition of a football trainer's Belgian qualification (Case 222/86, UNECTEF [1989] 1 CMLR 401).

Proportionality

The notion of proportionality has assumed increasing importance in recent years. This requires a certain proportion or bal-

---

[17] See also *R.* v. *Wear Valley DC, ex parte Binks* [1985] 2 All ER 699, where the Council withdrew permission for Bink's mobile snack-bar at a market without warning and without giving any reasons, thereby contravening natural justice, and so the High Court quashed the decision.

SUBSTANTIVE LAW: ADMINISTRATIVE LEGALITY 219

ance between the administrative measure taken and the end to be achieved. Thus, a disciplinary penalty should not be excessive in relation to the wrongful conduct of a civil servant (CE 26 July 1978, VINOLAY). Again a police measure should not be a disproportionate restriction on basic liberties of assembly, free movement, or commercial activity. For instance, the principle was invoked to challenge the pedestrianization of certain streets as an excessive interference with motorists' freedom to drive where they will (CE 8 December 1972, VILLE DE DIEPPE).

The principle of proportionality guides the exercise of the power of the French Minister of Culture, under a decree of 1961, to ban the showing of a film to minors under 13 or 18 or (exceptionally) to ban a film totally. In PICHENE (CE 9 May 1990), a decision of the Minister to allow the showing of the notorious film *The Last Temptation of Christ* was referred to the Conseil d'Etat with a request for the annulment of that decision. But the Conseil upheld the decision: the danger to public order was not so great as to warrant the banning of the film, nor was it (as Pichene alleged) an *atteinte grave aux consciences*. In a decision of the same date (DE BENOUVILLE), M. de Benouville, a hero of the French Resistance, challenged the Minister's release of a film *How Bitter is the Truth*, which concerned the arrest in Lyons in 1943 of the Resistance leader, Jean Moulin, and which suggested that M. Hardy, also of the Resistance, was implicated in Moulin's arrest. The film also dealt with certain letters of Klaus Barbie (who was tried in Lyons in 1987 for his wartime activities as Gestapo chief in that city). De Benouville argued, first, that the film infringed his individual rights as a member of the Resistance, secondly, that the film was an attack on the Resistance in general and so a danger to public order and an *atteinte grave aux consciences*, and, thirdly, that the film could influence both the forthcoming trial of Barbie for crimes against humanity and his own libel action against the film's producer. The Conseil d'Etat rejected each of the three arguments. As to the first, the film's content did not violate the fundamental rights of individuals featured in it to such an extent that public order would be affected. As to the second, the film did indeed have a historical as well as a polemical content on matters widely debated among the general public: accordingly, in the circumstances, it was neither a danger to public order nor an *atteinte grave aux consciences* such as to justify a ban. As to the third, the influence alleged was not established on the facts, but the Conseil d'Etat indicated that, if it were established that a film would prejudice the 'serenity' of court proceedings, the Minister would be entitled to take whatever steps were necessary (such as a ban or postponement of the film's showing) to protect the fundamental rights and interests of the parties; but this was not the case here.

The principle of proportionality is linked with the doctrine of the 'balance-sheet' (*le bilan*) and the modern concept of gross error in appreciation of the facts (*erreur manifeste d'appréciation des faits*). This will be discussed in detail later in relation to the leading case of VILLE NOUVELLE EST (see p. 247). Proportionality has some resemblance to the doctrine of reasonableness in English administrative law, but the two notions are not identical, according to the House of Lords: see *R.* v. *Secretary of State for the Home Department, ex parte Brind* [1991] All ER 720, where the law lords were not prepared to import the principle of proportionality into English domestic law, however well established it might be in the law of the European Convention on Human Rights. The language of proportionality is, however, to be found in English cases involving impositions of penalties. Thus in *R.* v. *Barnsley Metropolitan Borough Council, ex parte Hook* [1976] 1 WLR 1052, the Court of Appeal quashed the Council's decision to deprive a market stallholder of his livelihood on the ground that it was excessive and out of proportion to his minor breach of market discipline (cf. the approach of the Conseil d'Etat in VINOLAY, above).[18]

Non-retroactivity

As in English law, the French administrative courts will if possible give such a construction to a legislative provision as will

---

[18] For suggestions that English law could incorporate the proportionality principle see J. Jowell and A. Lester, in J. Jowell and D. Oliver, *New Directions in Judicial Review* (London, 1988), 51–72. For a critical and contrary view, see S. Boyron, 'Proportionality in English Administrative Law: A Faulty Transplant?' (1992) *Oxford Journal of Legal Studies*. A seminal study of the French law is G. Braibant, 'Le Principe de proportionnalité', in *Mélanges Waline*, ii. 297; and see also J.-P. Costa, 'Le principe de proportionnalité dans la jurisprudence du Conseil d'Etat', *AJDA* 1988, 343.

prevent it from having retrospective effect. Further than this, however, a similar principle is applied to acts of administration, so that once a decision has been made in favour of a citizen it cannot subsequently be revoked. Thus, in CACHET (CE 3 November 1922), the plaintiff had before the First World War let a market garden to a tenant. Under wartime legislation the tenant was exonerated from payment of any rent; the plaintiff then claimed compensation under the terms of the statute. The local official granted her compensation but she did not consider this was sufficient, so she appealed to the minister. He not only refused to increase the compensation but revoked the local official's decision, holding that the property was not a market garden but an agricultural holding. The Conseil d'Etat quashed the minister's decision and restored that of the local official, holding that the administration had no legal power to rescind a decision that had vested legal rights in the plaintiff.

Similarly, JOURNAL L'AURORE (CE 28 June 1948) made it clear that an order of a minister fixing the price of electricity could do so only for the future and not retrospectively.

This principle is constantly reaffirmed in the case-law and applies particularly to fiscal provisions (e.g. CE 16 March 1956, GARRIGOU) and to regulations on the status and pay of public servants (e.g. CE 11 July 1984, UNION DES GROUPEMENTS DE CADRES SUPERIEURS DE LA FONCTION PUBLIQUE).

Duty to Abrogate Unlawful Regulations

In COMPAGNIE ALITALIA (CE 3 February 1989) the Italian airline asked the Prime Minister to abrogate certain articles of the *Code général des impôts*, adopted by decree in 1967 and 1979, which were contrary to the sixth VAT directive of the European Communities dated 17 May 1977. His refusal to do so was quashed by the Conseil d'Etat on the ground that it breached a general principle of law. The Conseil d'Etat drew upon previous case-law (especially CE 18 January 1930, DESPUJOL) to enunciate the principle that, whenever the competent authority is called upon to abrogate an unlawful decree, it is bound to do so, whether the decree was unlawful *ab initio* or because it has become unlawful by reason of subsequent legal or factual circumstances.[19] In the present case, the VAT directive was such a legal circumstance. For one commentator, the decision is welcome as a strong and unambiguous affirmation of the full consequences of the principle of *légalité*.[20] It is also, of course, a further recognition of the supremacy of Community law when in conflict with French law; to this aspect we return in the next chapter.

The Right to Judicial Review

The Conseil d'Etat considers, as a *principe général*, that the legality of an administrative decision can always be reviewed by the administrative judge, even if the administration has been expressly empowered to act 'without recourse' to any court. This has been explained in Chapter 7, p. 163; see CE 7 February 1947, D'AILLIERES, and CE 17 February 1950, LAMOTTE, there discussed.

The CANAL judgment applies the parallel principle that the decision of any court, be it administrative, civil, or criminal, is liable, even without statutory provision, to be reviewed *en cassation*. The decree in that case was declared void by the Conseil d'Etat because the special military tribunal which it created was not submitted to the supervisory jurisdiction of the Cour de Cassation, the ultimate review body for all criminal courts, and the exclusion of such supervision violated an essential guarantee of the citizen (see further Chapter 3, p. 54).

The Inalienability of the Human Body

General principles as applied by the administrative courts may be inspired by doctrines of French private and criminal law. In ASSOCIATION 'LES CIGOGNES' (CE 22 January 1988), the Conseil d'Etat was faced with a problem of bio-ethics. Alsace-Lorraine has special legislation relating to the registration of associations. The prefect may veto a registration if the association has an illicit purpose. The prefect of Strasbourg exercised his veto against 'The Storks', an association for the promotion of surrogate motherhood, on the ground that the activities directly breached Article 353 (1) of the Penal Code. This makes

---

[19] For a subsequent application of this same general principle, see CE 26 June 1989, FEDERATION DES SYNDICATS GENERAUX DE L'EDUCATION NATIONALE, discussed on p. 216 above.
[20] R. Errera, [1990] PL 650.

it an offence, *inter alia*, 'to induce future parents to sign an instrument according to which they undertake to abandon their child on birth'. In the view of the prefect, contracts negotiated by the association clearly fell into this category of instruments. The Tribunal Administratif of Strasbourg agreed, as did the Conseil d'Etat on appeal.

The Conseil d'Etat did not, however, seek to justify its decision on any general principle, such as that of the inalienability of the human body. This principle has been affirmed in similar cases before the civil courts: see e.g. ASSOCIATION 'ALMA MATER'.[21] In this advisory capacity, the Conseil d'Etat has published a wide-ranging report, *Sciences de la vie: de l'éthique au droit* (1988), which examines such issues as surrogate motherhood, pre-birth diagnosis, embryo research, and medical experimentation. No doubt, in its judicial capacity, the Conseil d'Etat will soon be called upon to provide appropriate general principles on these controversial bio-ethical issues.

4 GROUNDS FOR REVIEW

The principle of legality prescribes a line of conduct for the administration from which it cannot depart without committing an *excès de pouvoir*. Any violation of the principle can be a ground for review (the French speak of a *cas d'ouverture*) and will make the *acte administratif* void. It is mainly for historical reasons that proceedings are based on one of four grounds, namely, *incompétence*, *vice de forme*, *violation de la loi*, and *détournement de pouvoir*. These traditional grounds are by no means mutually exclusive, and it is not always easy to see why a particular case is considered under one head rather than another. Moreover, in the evolution of the case-law of the Conseil d'Etat, the distribution of cases between these headings has by no means remained constant. In particular, whilst *détournement de pouvoir* is undergoing some decline, *violation de la loi* (under which most violations of the *principes généraux* are included) has tended to expand.

---

[21] Cass 1ʳᵉ Civ., 13 December 1989, *D.* 1990, 273: see J. Bell, in P. Cane and J. Stapleton (eds.), *Essays for Patrick Atiyah* (Oxford 1991), 106–8. The results reached in this case have been qualified by subsequent decisions of the Cour de Cassation: see Ass. plén. 31 May 1991, *D.* 1991, 417, rapport Chartier, *JCP* 1991 II 21752.

Of course it is for the plaintiff, through the legal arguments or *moyens* in his *recours*, to raise one or other of these grounds for review. Exceptionally, however, the administrative judge, once seised of the case, may himself on his own initiative take notice of a *moyen d'ordre public*, which would call for annulment of the *acte administratif* complained of. This idea of *moyen d'ordre public* is not precisely defined but it may be raised by the judge (or by a party) where the illegality is so grave that he (the judge) considers it should be raised in spite of all ordinary procedural rules. Examples are cases where the administrator has acted totally outside his jurisdiction, or where a commune had failed to obtain prefectoral approval to some act where this was required by law, or where (as in CE 23 January 1951, COURAJOUX) a measure has been adopted without prior consultation of the Conseil d'Etat, where such consultation was obligatory. Again, conformity with binding rules of international law and Community law is now a *moyen d'ordre public* (CE 20 October 1989, NICOLO, discussed on p. 269 below).[22] Moreover, such a ground may even be raised outside the ordinary *délai* of two months.

The doctrine of *moyen d'ordre public* frequently arises where there has been a misunderstanding of the scope of application of the law (*la méconnaisance du champ d'application de la loi*). Even if the administration does not make the point that the plaintiff is asking for more favourable treatment than the law permits, the administrative court will answer *d'office* (if it be the case) that the plaintiff is not within the scope of the law in question and has no entitlement at all.[23]

Inexistence

French law takes the view that an administrative decision may be simply non-existent for want of some essential element. There is no need to annul it, as the court has only to declare its non-existence. Thus, in ROSAN GIRARD (CE 31 May 1957) the mayor of a commune in Guadeloupe had declared the result of

---

[22] But the Conseil d'Etat will not, of its own motion, test the validity of a decree by reference to a Community directive: CE 11 January 1991, MORGANE. The conclusions of the Commissaire du gouvernement, invoking the 6th VAT Directive, were impliedly rejected on this point: see E. Honorat and R. Schwartz, *AJDA* 1991, 111.
[23] See R. Chapus, *Droit du contentieux administratif* (2nd edn., 1990), 470.

a communal election on the basis of the votes contained in three only of four ballot-boxes. The prefect thereupon declared the election void and set the procedure in train for a new election; but under the electoral regulations the prefect should have referred such a case to the local administrative court. As this necessary step had not been taken, the Conseil d'Etat, four years later, declared that the prefect's decision purporting to avoid the original election itself was itself void and had no existence in law.

This ground of jurisdiction is not often invoked, because in most cases it is sufficient (however serious the illegality may be) to annul the administrative decision on one of the four traditional grounds discussed below. *Inexistence* assumes that the judge has gone so far as to doubt whether the administration has really taken any decision for him to review: the illegality is so gross and flagrant as to amount to the administration's acting completely outside its jurisdiction. The administration has virtually behaved as an outlaw, and as such deserves no consideration from the judge. Thus disallowed the administrative act becomes a *voie de fait* (Chapter 6, p. 135), and the ordinary courts recover jurisdiction to consider the consequences. In the case of GUIGON (TC 27 June 1966), the military authorities had locked and sealed the private lodgings of an officer against whom disciplinary proceedings were being taken, and refused him permission to recover his belongings. An official had purported to authorize this action in writing, and the civil courts therefore apparently had no jurisdiction in the matter. The Conseil d'Etat, convinced that this so-called decision did not really exist, doubted whether the administrative court could in such a case annul, and so deferred judgment until the question had been resolved by the Tribunal des Conflits. The Tribunal des Conflits then held that the Conseil d'Etat could declare the authorization to have no existence, and that the civil court could have done the same had it been seised of the case initially; but that in accordance with the doctrine of *voie de fait* the plaintiff in such a case would have to look to the civil court for any award of damages.

An even more striking illustration of *inexistence* is the case of MAURICE (CE 15 May 1981). Maurice was condemned to death by a Paris criminal court but successfully challenged the legality of the appointment of one of his judges. This judge had been one of several former judges brought out of retirement on short-term appointments to help relieve congestion in particular courts. The relevant statute limited their functions to a particular grade within the judicial hierarchy. The judge in question was appointed for three years in the relevant grade on 21 December 1979 and assigned to the court at Meaux, but only five days later he was re-assigned in a higher grade to the Paris court which sentenced Maurice. The Conseil d'Etat held that the first appointment was one of pure form (*une nomination pour ordre*) to satisfy the terms of the statute and unrelated to the needs of the court at Meaux; it was therefore *ultra vires* the enabling statute and null and void as an *acte inexistant*, as too was the appointment to the Paris court. Moreover, thanks to the doctrine of *inexistence* both acts could be challenged by Maurice even after the lapse of some twelve months.

The doctrine of *actes inexistants* has been described as one of the subtlest and most uncertain in administrative law.[24] By the very nature of this ground of attack it cannot be precisely delimited: however, the argument that an act has no existence—is void *ab initio*—can be pleaded in any administrative proceedings by any person who has an interest so to plead and (which is the real value of the plea) at any time, as the ordinary *délai* of two months does not apply to the commencement of such proceedings. English law knows the distinction between void and voidable administrative decisions, but the category discussed here concerns a more radically vitiated decision.

Incompétence

Very close to non-existence in the worst cases is the *cas d'ouverture* known as *incompétence* or want of authority. Translated into English terms, this is perhaps the nearest the French get to substantive *ultra vires*; if an official acts completely without authority, his decision will be declared void for *incompétence*. Thus, if the administration takes a decision in a domain reserved for parliament by the Constitution, the decision will be annulled on this ground. Everyday examples of this situation include cases of civil servants who are dismissed by their immediate superiors where the superior has not had delegated authority from the minister to take such action.

---

[24] Vedel and Delvolvé, *Droit administratif*, ii. 256.

Where a 'caretaker' government in France was carrying on the administration until a new government had come into office, and they purported to extend to Algeria a requisitioning statute passed in Paris, this was declared void, as such a government could only concern itself with current business: SYNDICAT REGIONAL DES QUOTIDIENS D'ALGERIE (CE 4 April 1952). Moreover, in such a case it was a matter for the administrative courts to decide what was 'current business', and it was within their jurisdiction to find, as they did, that such a matter as requisitioning did not fall within this category.

Vice de forme

This may be rendered into English either as 'procedural *ultra vires*' (as distinct from substantive *ultra vires*) or as 'procedural impropriety', the term proposed by Lord Diplock in *CCSU* v. *Minister for the Civil Service* [1985] AC 374. A leading illustration of such a procedural defect is the English case of *Vine* v. *National Dock Labour Board* [1956] 3 All ER 939. As might have been expected from *droit administratif*, the courts will not insist on a rigid observance of all procedural formalities, and they are quite prepared to accept the distinction made in some English decisions between mandatory and directory procedural requirements;[25] the French also make a distinction between essential and inessential formalities (*les formes substantielles et non substantielles*).[26] Only failure to observe an essential formality will lead to avoidance of the subsequent proceedings. Thus, a slum clearance order had been posted in all the districts most concerned, but not (as was required by law) in every district of Paris within 5 kilometres of the building; this was an 'insubstantial' error, insufficient to invalidate the proceedings: COMMUNE D'ISSY-LES-MOULINEAUX (CE 12 November 1909). However, in PETALAS (CE 18 November 1955), an order of extradition which was made without consultation with the appropriate government department was held void for want of compliance with an essential procedural requirement.

---

[25] See, for example, *London & Clydeside Estates Ltd.* v. *Aberdeen DC* [1980] 1 WLR 182.
[26] The Treaty of Rome of 1957 adopted this French distinction: Art. 173 provides that the legality of Community acts may be reviewed on grounds (*inter alia*) of 'infringement of an *essential* procedural requirement' (italics supplied).

Vice de forme also covers a failure to give a fair hearing. For example, in DAVIN (CE 26 January 1966) it was held that where the headmistress of a *lycée* had required the father of a pupil to remove her from school, she should first have stated the reasons for this requirement and given the father an opportunity of replying to those reasons. The headmistress's decision was therefore quashed for *vice de forme*. A general right to put one's case before an individual administrative decision is taken has now been established by Article 8 of the Decree of 28 November 1983. Equally, the administrative courts would also hold a decision of a minister taken on the recommendation of an advisory commission to be void, if it could be shown that some essential matter had not been investigated by the commission.[27]

The statutory obligation since 1979 for any administrative decision to be supported by reasons also means that the lack of adequate *motivation* may constitute a *vice de forme* so as to justify the quashing of the decision (see p. 239).

Violation de la loi

Historically, this was the last of the four traditional grounds to acquire separate identity, although, in the ultimate legal analysis, it is the justification for most cases of review of administrative action by the administrative courts. Vedel and Delvolvé find its origin in the Decree of 2 November 1864 which (as we have seen, p. 46) allowed the *recours pour excès de pouvoir* to be brought without the services of an *avocat*.[28]

The expression 'violation de la loi' is doubly misleading: first, because *loi* is here to be understood not simply in its usual French meaning of written law, and secondly, because such matters as *incompétence, vice de forme*, and even *détournement de pouvoir* (see below) are no less departures from legality, and therefore constitute violations of the law in a wider sense.[29] Moreover, in its written form, *loi* may be a rule of the Constitution, a statute passed by parliament or a ministerial regulation.

As a head of classification in the modern jurisprudence of the Conseil d'Etat, *violation de la loi* goes beyond a consideration of

---

[27] Another example of *vice de forme* is to be seen in MOSCONI, Appendix H.
[28] *Droit administratif*, ii. 236.
[29] For an example, see CE 20 December 1967, MINISTRE DE L'INTERIEUR c. FABRE LUCE.

the merely formal validity of an *acte administratif*: this is the concern of the heads of *incompétence* and *vice de forme*. But granted that the administrative agency is competent and has acted in due form, under the head of *violation de la loi* the administrative judge moves on to examine the actual contents of the administrative act itself, in order to decide whether it conforms with the legal conditions set upon administrative action in the particular case (including, since 1944 at least, *les principes généraux du droit*—see p. 205). An example may make this clear.

Disciplinary action cannot be taken against a civil servant unless he or she is at fault. Accordingly, if such action is taken, then quite apart from questions about the procedure adopted and the competence of the disciplining body, the court may inquire whether the sanction imposed is authorized by the relevant legislation, whether the facts alleged are made out, and if so, whether they amount to fault so as to justify disciplinary action. In TEISSIER (CE 13 March 1953) a senior official (the director of the National Centre for Scientific Research) was dismissed for having refused to dissociate himself from an open letter (which included his name in the letter-heading) attacking the government; on a review of the case the Conseil d'Etat found that he was at fault since he had exceeded the limits of freedom of opinion which were permissible to an official in his position.

Again, *violation de la loi* will be in issue where an administrative agency has what the French call a *compétence liée*, or, in English terms, is under a mandatory duty and has no discretion. For example, an applicant for a licence to open a chemist's shop is entitled as of right to a licence if the number of dispensaries in the locality is still below a certain statutory maximum: the refusal of a licence (where all other conditions are met) is then a *violation de la loi* and would be quashed. Similarly, once an association has complied with the legal formalities leading to a declaration of charitable status (*déclaration d'utilité publique*), the prefect has no choice but to accept its registration: CE 25 January 1985, ASSOCIATION 'LES AMIS DE SAINT-AUGUSTIN'. The corresponding English remedy, of course, would be a mandamus.

Détournement de pouvoir

This is an important and distinctive ground of review. Like *violation de la loi*, it enables the administrative judge to control the actual content of an *acte administratif*: hence the French regard both those grounds as concerned with the 'internal legality' (*la légalité interne*) of the decision in question, whereas *incompétence* and *vice de forme* relate to the external legality (*la légalité externe*), or, as Vedel and Delvolvé put it, 'the formal elements of the decision'. But *détournement de pouvoir* is essentially subjective, involving as it does an inquiry into the motives which inspired the administrator so to act. English students of *droit administratif* have perhaps in the past paid more attention to *détournement de pouvoir* than to the other grounds of review because it seemed so very different from anything known in English law. This attitude has not been altogether justified, because *détournement de pouvoir* is not of such great importance in modern French practice and also because in the modern English decisions there is often some provision in the relevant statute which the court can seize hold of as a justification for much the same kind of review.

The French approach is concerned less with statutory language and more with the analysis of motives. In simple terms it may be said there has been a *détournement de pouvoir*, or 'abuse of power' if an administrative power or discretion has been exercised for some object other than that for which power or discretion was conferred by the statute; and the court will, in exercising this review, be bound not only by the precise terms of the statute, but will also infer an object from what is reported to have been said in the legislature, or from any other relevant *travaux préparatoires*. Waline points out[30] that this search for the intentions of the legislator may be a difficult task, and even the *travaux préparatoires* may be too few to give any clear indication; indeed, in the case of certain forms of subordinate legislation, there may be no *travaux préparatoires* at all. In such cases the Conseil d'Etat may be prepared to invent an object for the legislature; as in TABOURET ET LAROCHE (CE 9 July 1943), where it had been provided in a wartime statute that a purchase of real estate should be invalid unless it was approved by the local prefect. No guidance was given in the statute or the *travaux préparatoires* as to why it had been passed, but the Conseil d'Etat expressed the view that the statute had clearly been passed for

---

[30] *Droit administratif*, (8th edn., 1959), para. 734.

SUBSTANTIVE LAW: ADMINISTRATIVE LEGALITY 231

the purpose of preventing speculation in land, and that any refusal of an approval based on any other ground would be void as being a *détournement de pouvoir*.

Similarly, in order to apply this ground for review the examining court must ascertain the object behind the particular exercise of discretion—the *acte administratif* complained of. Only if this object does not agree with, or fall within, the purpose of the statute (the 'intention of the legislator') can the court intervene on the basis of *détournement de pouvoir*. This second problem is sometimes more difficult to solve than the first, but in this instance the plaintiff in proceedings before the courts will be assisted by the general principle of administrative morality. If he can point to unexplained behaviour on the part of the administration which on the face of it is not within the purpose of the statute, it will be for the administration to rebut these suspicions. Thus, when it appeared that a mayor had made certain parking and traffic regulations after a meeting of commercial vehicle owners in the locality, the Conseil d'Etat quashed the regulations on the grounds that they had been made not for the purposes of traffic regulation but so as to protect commercial interests: COMPAGNIE PYRENEENNE DE TRANSPORTS PAR TAXIS (CE 10 February 1928).

A few examples of decided cases may serve to show how *détournement de pouvoir* has operated in practice:

1. Regulations made by the mayor of a commune controlling the holding of dances were quashed when it appeared that they were made so as to encourage people to patronize his own inn (CE 14 March 1934, RAULT).

2. A school was started, as it eventually appeared, in the sole interest of the man it was proposed to appoint as director; the decision to open the school was quashed (CE 5 March 1954, SOULIER).

3. A mayor made an order prohibiting dressing and undressing on the beach; the order was quashed when it appeared that the real purpose underlying the order was not public decency, but so as to compel would-be bathers to use the municipal bathing-establishment (CE 4 July 1924, BEAUGE).

4. In purported exercise of his *police* powers a mayor refused permission for a local band to parade and play in the streets at the funeral of one of its members; the refusal was quashed when

his reason was found to be, not a threat to public order, but the fact that he favoured another band, subsidized by the commune and, unlike the plaintiffs, well disposed towards his administration (CE 17 May 1907, SOCIETE PHILHARMONIQUE LIBRE DE FUMAY).

5. The opening-hours of the town hall were so decided as to prevent the local schoolmaster, because of his classes, from acting as a town-clerk; the decision was quashed (CE 2 April 1971, ZIMMERMANN).

The distinctive feature of this ground for review, from the point of view of an English lawyer, is that it is not dependent on an application of the *ultra vires* doctrine, in the sense of the excess or misuse of *statutory* powers. The powers vested in the administrative authorities concerned in the examples given above were discretionary, and not subject to statutory limitations. In an English context, however, administrative authorities invariably derive their powers from statute, and therefore if the courts are asked to investigate the validity of an administrative act, they will scrutinize the precise terms of the statute to see if these have been contravened or exceeded in any way. Thus, in *Roberts* v. *Hopwood* [1925] AC 578, the House of Lords were able to question the wages paid by a local authority, under a statutory power to pay such wages as they might think fit, on the basis that the statute was to be construed as subject to a proviso that the authority must act reasonably. Again, in *Padfield* v. *Minister of Agriculture, Fisheries and Food* [1968] AC 997, the House of Lords showed themselves prepared to search for a *purpose* within the statute, and not to be restricted by its precise terms. Sometimes procedural error may be seized upon as a ground for intervention by the English courts (as in *Webb* v. *Minister of Housing and Local Government* [1965] 2 All ER 193). Unlawful motive too can arise as a problem in England. For example, in *R.* v. *Derbyshire County Council, ex parte Times Supplements Ltd.* [1991] COD 129, the decision by a local education authority not to advertise teaching-posts in the *Times Education Supplement* was found to be motivated by a pending libel action brought by the Council leader against one of the *Times* group of newspapers. This was held to be a bad-faith decision and not a proper exercise of the duty to advertise teaching-posts under Section 38 of the Education (No. 2) Act 1986.[31]

Thus, although the English courts will not typically quash an 'acte administratif' on the grounds of bad faith alone, it is true that they often arrive, by an elaborate process of statutory construction, at a result similar to that which would have been reached in the *droit administratif* by a simpler route. Thus in the famous London 'fair fares' case, the English courts were driven to splitting hairs over the meaning of 'economic' in the relevant statute: *Bromley London Borough Council* v. *Greater London Council* [1982] 1 All ER 129 (CA and HL). The French administrative courts would not be conscious of any need to be sensitive or over-respectful towards the precise terms of any relevant statute; this is partly because French lawyers do not apply the theory of parliamentary sovereignty to the same logical extent as the English, and in France a statute would never be allowed to be the instrument of fraud. In England it has been said that 'fraud unravels everything'[32] but the courts cannot always avoid the 'plain words' of the statute. This is not to deny that the English judges, under the guise of statutory interpretation, may embody in their decisions certain values and attitudes which are not expressly derived from the statutes being interpreted.[33]

The French approach, then, is concerned less with statutory language and more with the analysis of motives. The following are examples of cases concerned with motivation where the French courts have applied the *détournement de pouvoir* principle:

1. where the act was inspired by some motive other than the public interest (RAULT, above and MOSCONI, Appendix H.);

2. or if it was inspired by the interest of some third party; where for example, a doctor had been refused a licence to practice in a certain district, in the special interest of the majority of other doctors in the neighbourhood (CE 2 February 1938, RUHLE);

3. or in the case of an act inspired by party political motives, for example, the granting of a subsidy to a free school, not in

---

[31] See also *R.* v. *Port Talbot BC, ex parte Jones* [1988] 2 All ER 207, where a council house was allocated to a person so that he might more effectively fight the ward seat at local elections.
[32] Per Denning LJ in *Lazarus Estates Ltd.* v. *Beasley* [1956] 1 All ER 341 at 345.
[33] See Professor A. W. Bradley's comment at [1968] CLJ 144.

the interests of education, but as a protest against the religious teaching being given in the local state school (CE 12 April 1935, COMMISSION DEPARTEMENTALE DU BAS-RHIN);

4. or where the act was inspired by some public interest other than that for which the power was conferred: BEAUGE (p. 231) provides a good example of this, as there was no suggestion in that case that the mayor was in any way acting to his own personal advantage.[34] In another case (CE 6 January 1967, BOUCHER), a compulsory-purchase order for the acquisition of a château, ostensibly for the provision of a local museum, was quashed when it appeared that the true purpose of the acquisition was to frustrate a proposed purchase of the property by a person coming from another part of France. Again, in FRAMPAR (CE 24 June 1960) certain newspapers had been seized by the police in Algeria for the alleged purpose of taking criminal proceedings against the editors; actually they had been seized in the interests of public order so as to prevent the newspapers being read by the public. This seizure was categorized as being a *détournement de procédure* and therefore illegal.[35]

At the present time, *détournement de pouvoir* is somewhat declining in importance as a *cas d'ouverture*. In the late nineteenth century it was the principal means whereby the Conseil d'Etat was enabled to insist on the observance of administrative morality, but with the recent expansion of *violation de la loi* it is thought of rather more as a 'last ditch' or residuary ground for review.

Thus, in MAUGRAS (CE 16 November 1900) this ground was invoked to quash a mayor's disciplinary action against a policeman who had submitted a justifiable report about the conduct of a personal friend of the mayor. At the present day, as Vedel and Delvolvé observe,[36] the court would not need to explore whether the mayor's action was motivated by loyalty to his friend but could quash on the basis of *violation de la loi*; since the policeman was in no way at fault, it could hold that the disciplinary action lacked legal foundation or was based on a ground

---

[34] Cf. *Wheeler* v. *Leicester City Council* [1985] AC 1054.

[35] We have followed Vedel and Delvolvé (*Droit administratif*, ii. 333) in treating *détournement de procédure* as one species of *détournement de pouvoir* rather than as a separate head; the former notion is also a familiar one in English law; see *Webb* v. *Minister of Housing and Local Government* (cited p. 232 above).

[36] *Droit administratif*, ii. 325.

SUBSTANTIVE LAW: ADMINISTRATIVE LEGALITY 235

which did not exist. M. Letourneur and his co-authors[37] conclude that if there is a choice between *détournement de pouvoir* and *violation de la loi* (or *de la règle de droit*), the Conseil d'Etat always chooses the latter as being objective, whereas the former is often difficult to prove.

5 CASSATION

Where the decision sought to be annulled is that of an administrative jurisdiction (see Chapter 3, p. 57), the appropriate form of proceedings is that of a *recours en cassation*, not a *recours pour excès de pouvoir*. Fortunately, three of the grounds of review are identical in both types of proceedings, namely, *incompétence, vice de forme*, and *violation de la loi*. Only the fourth ground of *détournement de pouvoir* is not available for a *recours en cassation*, presumably because the French cannot imagine persons charged with a judicial function offending in this respect.

The test of whether a particular administrative body constitutes a jurisdiction subject to cassation is a complex one, involving a combination of various criteria. These include: the function of the body in question (e.g. Is it a disciplinary body?); the organization and composition of the body (e.g. Is it collegiate?); the procedure which it follows (e.g. Does it give a hearing?); and the force of its decisions (e.g. Do they have *l'autorité de la chose jugée*? Is it a jurisdiction of last resort?). Some examples of bodies satisfying the test were mentioned in Chapter 3 and for a list of the most important see Appendix D.

The distinction between the *recours en cassation* and that *pour excès de pouvoir* is not merely formalistic but is important on four accounts. First, cassation is the exclusive business of the Conseil d'État, whereas *excès de pouvoir* is usually a matter for the local Tribunaux Administratifs (for some exceptions to the latter rule, see Chapter 3, p. 52). Secondly, there is the very practical point that legal representation is obligatory in most cassation proceedings. Thirdly, there is the theoretical point mentioned above, namely, that *détournement de pouvoir* is not available as a ground of cassation. Lastly, the degree of control exercised on cassation

---

[37] See M. Letourneur in J. Bauchet and J. Méric, *Le Conseil d'Etat et les tribunaux administratif* (Paris, 1970), 157.

under the head of *violation de la loi* used not to be so great as in a *recours pour excès de pouvoir*. Thus, the Conseil d'Etat has always been prepared to review whether there is any mistake of fact ('whether the facts contained in the decision under review are correct', says Waline),[38] but for a long period it was very reluctant to interfere with the legal characterization (*qualification*) of those facts. This rather subtle distinction is neatly illustrated by Waline: a finding that a schoolteacher taught her pupils hymns during a lesson is one of fact; but to find that she had thereby infringed her professional obligation of religious neutrality (*neutralité scolaire*) is to characterize her behaviour as one of fault, involving disciplinary consequences. A classic statement of these distinctions is to be found in MOINEAU (CE 2 February 1945) where Dr Moineau challenged the decision of the Disciplinary Committee of the National Order of the Medical Profession rejecting his application for registration as a medical practitioner. Today, however, the Conseil d'Etat in cassation proceedings is no less rigorous in reviewing the characterization of the facts than it is in the *recours pour excès de pouvoir*, e.g. CE 14 March 1975, ROUSSEAU.

The reluctance of the Conseil d'Etat to upset the view (*appréciation*) of the facts taken below is explained by Odent in these terms: 'the decisions of a jurisdiction benefit, by comparison with those taken by an administrative authority, from a much stronger presumption of legal correctness.' Nevertheless the approach of the Conseil d'Etat is the exact converse of that adopted by the Cour de Cassation, for the latter restricts its review to errors of law, including the legal characterization of the facts, but will not correct errors of fact. But here, typically, the Conseil d'Etat is concerned less with orthodox legal theory than with good administration, and in particular with assuring *l'administré* that the administrative decision is based on correct findings of fact: what view is then to be taken of the facts is a matter for the particular administrative jurisdiction rather than for the Conseil. The English approach to judicial review of tribunal decisions still seems closer to that of the Cour de Cassation unless the error of law or fact appears on the record, but Professor Purdue argues that the expanded grounds for review

---

[38] *Droit administratif* (9th edn.), para. 433.

SUBSTANTIVE LAW: ADMINISTRATIVE LEGALITY 237

logically require the courts to be prepared to engage in fact-finding.[39]

Since the reform of 1987, the Conseil d'Etat has an expanded jurisdiction in *cassation* as the reform confers on it a general power of review of the decisions of the Cours Administratives d'Appel.

6 THE EXTENT OF JUDICIAL REVIEW

It may here be convenient to draw together certain central ideas which are at the root of the control exercised by the administrative courts over the acts of the administration. Although we have followed modern French writers in classifying the several *cas d'ouverture*, the plaintiff before the administrative courts is not in practice required to choose his ground of complaint, nor does the judge feel himself compelled to fit the facts of the case into any existing pigeon-hole. The administrative courts will not interfere with administrative policy—a term which (significantly) is untranslatable into French, the nearest equivalent being, *l'opportunité* ('the merits'); but policy will not be allowed to run riot or offend against principles of administrative morality.

Most cases before the courts fall into one of three categories; the attitude adopted towards these by the court may serve to illustrate the general approach.

(a) Where the Administration has no Discretion

Here, where the administration is subject to a *compétence liée* (see p. 229) the task of the court is simple and is confined to examining whether a situation of *compétence liée* really existed and whether the administration has failed to observe the terms of the law. If so, the administration can find no refuge in *opportunité*. Waline[40] provides a classic example of this doctrine: 'the *sous-préfet* is bound to issue a hunting-licence to anyone who satisfies all the conditions required by law; he is not to concern himself whether or not it is *opportun* to issue the licence—on the pretext, for example, that this hunter is a very bad shot' (citing CE 13 November 1946, LETENDARD).

---

[39] M. Purdue, 'The Scope for Fact-finding in Judicial Review', in G. Hand and J. McBride (eds.), *Droit sans frontières: Essays in Honour of L. Neville Brown* (Birmingham, 1991), 193.
[40] *Droit administratif* (9th edn.), para. 739.

238 SUBSTANTIVE LAW: ADMINISTRATIVE LEGALITY

Conversely, which is more often the case, the doctrine of *compétence liée* means that the citizen must be refused his request, once the administration decides (correctly) that he does not meet all the conditions prescribed: the administration has no choice to do otherwise.

Some writers[41] use *compétence liée* more loosely. Thus Braibant distinguishes between *la compétence liée* (as in the example of Waline above) and *la compétence partiellement liée* (as in CE 19 May 1933, BENJAMIN, where the mayor's discretion to ban a public meeting was held to be hedged around with various restrictions). Vedel and Delvolvé follow Waline: for them, as the passage cited below, n. 42, shows, all discretion is in some way restricted.

(b) Where the Administration has 'Absolute' Discretion (*pouvoir discrétionnaire*).

Only in the quite exceptional case of the *acte de gouvernement* (Chapter 7, p. 155), does this mean that the French courts have virtually no control at all over the administration. In all other cases, although the discretion is absolute, the judge will ensure that the administration has committed no mistake of law or fact (including under the former any infringement of a *principe général du droit*) and is innocent also of any *détournement de pouvoir*. For, as M. Braibant has remarked, the administrative judge in France is in principle opposed to admitting the existence of 'zones de non recours'.[42]

In BAREL (CE 28 May 1954) the minister responsible had refused five young men permission to sit the examination for entry to the Ecole Nationale d'Administration. It was suggested in a national newspaper that the government had decided to refuse entry to the examination to any candidates who were communists, but a short time later the minister denied this categorically in a statement made in the National Assembly. The

---

[41] e.g. G. Braibant, *Le Droit administratif français* (1st edn., Paris 1984), 236.

[42] In English law, judicial activism over recent decades has led Professor Wade to comment that to argue that the administration has an absolute discretion is to commit 'a constitutional blasphemy'. See H. W. R. Wade, *Administrative Law* (7th edn., Oxford, 1977), 39. Cf. Vedel and Delvolvé, *Droit administratif*, i. 525: 'there is never an absolutely pure discretion. The idea that there are *actes discrétionnaires* which escape all control for legality disappeared in the case-law more than 50 years ago.'

young men concerned then referred the matter to the Conseil d'Etat, who (in an unusually speedy decision) quashed the minister's refusal, because he gave no reasons justifying his decision and refused to produce the files to the court; therefore it was presumed that the candidates were refused entry for the motive alleged, namely, that they were communists. Accordingly, the Conseil d'Etat quashed the minister's decision as a mistake of law (*erreur en droit*) namely, his violation of the general principles of freedom of opinion and equal access to the civil service.

In the somewhat similar case of IMBACH (CE 14 May 1948) the court quashed the decision of the minister refusing a passport to the plaintiff on the ground that he had collaborated with the Nazis during the occupation, only because they were not satisfied that the facts alleged had been established. In SOCIETE 'MAISON GENESTAL' (CE 26 January 1968) the Conseil d'Etat extended this principle to the economic sphere. The plaintiffs had applied to the appropriate ministry for a special grant towards a redevelopment project; this application was refused by the ministry because a departmental committee advised against the project. In justifying its decision before the Conseil d'Etat the ministry simply stated that the project had not appeared to have sufficient economic advantages, in the public interest, to justify a grant. The Conseil d'Etat ruled that this expression of reasons (*motivation*) was formulated in far too general terms to allow the court to exercise its control over the legality of the refusal, and accordingly referred the case back for further *instruction* to the Tribunal Administratif of Rouen, a court of first instance, so that the ministry could indicate more precisely the *raisons de fait et de droit* for its decision. Whereas in BAREL the Conseil d'Etat was content to require the ministry to produce its files, it was now demanding that the ministry give its reasons fully.

As a result of such cases, it used to be said that every administrative decision must be *motivé*. This meant that there had to be substantial reasons justifying the decision that could be adduced to, or discovered by, the court. Traditionally, those reasons did not need to be disclosed to the *administré* at the time when the decision was made; but the administrative judge could call for the reasons if litigation ensued (as, for instance, in BAREL). Now, however, the Law of 11 July 1979 (see p. 88) lays down that reasons must be stated, in writing and with sufficient precision, at the time when any administrative decision is made which directly affects the interests of an individual in a prejudicial manner:[43] the French refer to the wish to ensure *la transparence* of public administration, which we may equate with the Franks concept of 'openness'.

Reverting to our threefold analysis we may conclude that, in a case of absolute discretion, the administrative court will examine whether there is any justification for interfering with the *acte administratif*, based usually on one of the general principles. Where there is no such justification, the court will not interfere; the administrative judge is not, as the French say, the 'supérieur hiérarchique' of the administrator.

(c) Where the Administration has Limited Discretion

The third class of case is perhaps the most important of all; it falls between the extremes of no discretion and absolute discretion, and in practice is one of the most frequent causes of litigation before the administrative courts. In this class of case the administration has a discretion exercisable within statutory limits, and the judge sees that these limits are observed. But this judicial control over *vires* extends to what an English lawyer would characterise as a mistake of law or even a mistake of fact.

Here the discretion must be exercised not only in accordance with the general principles of the law, but also within the limits set by the written law. The administrative courts have—and freely exercise—an effective control over the limits of *l'opportunité*, by interpreting the statute and assessing, or *qualifiant* (classifying) as the French say, the facts of the situation in the light of the terms of the statute as so interpreted. Moreover, the decision of the administration must be *motivé*, and the *motif* as declared by the administrator (or as subsequently elucidated by the court) must be one which fits the terms of the statute.

In practice, the Conseil d'Etat is prepared to curb its desire to review the facts only in two types of case. The first concerns those cases where extremely important questions of *police* are

---

[43] See F. Wagner, 'The Statement of Reasons in French Administrative Law' (1980). 5 *Hold. L. Rev.* 54.

SUBSTANTIVE LAW: ADMINISTRATIVE LEGALITY 241

involved, such as measures concerning foreigners. The second concerns cases involving scientific or similar technicalities. Thus the court would not review the question whether a hair-lotion was poisonous, in SOCIETE TONI (CE 27 April 1951); nor whether a wine was worthy of an *appellation contrôlée* in SYNDICAT AGRICOLE DE LALANDE-DE-POMEROL (CE 14 October 1960). A similar deference to the expertise of the administration is shown in the English case of *Puhlhofer* v. *Hillingdon LBC* [1986] AC 484.

This timidity of approach in the face of technical questions is somewhat surprising, especially when they have arisen in cases involving the status or reputation of a civil servant or a doctor. But in this context the Conseil d'Etat has now shown itself willing to quash obvious mistakes, or as the French say 'manifest errors', by the administration in its appreciation of the facts— see, e.g., DENIZET (CE 13 November 1953), and COMMUNE DE MONTFERMEIL (CE 9 May 1962), where the question was whether a civil servant had been reinstated in an equivalent position to that he held previously; also MINISTRE DE L'AGRICULTURE c. BRUANT (CE 19 April 1961), concerning the equivalence of agricultural lands which were being exchanged, and SOCIETE IRANEX (CE 6 November 1963), concerning the disadvantages of adding gum arabic to ice-cream. Again, the Conseil d'Etat was prepared to review the area designated for production of wine *à appellation d'origine contrôlée* (CE 2 March 1979, INAO c. ROUSSOT) and the modification by the Ministry of Health of the rates of reimbursement upon certain medecines (CE 20 May 1988, SOCIETE LABORATOIRES DE THERAPEUTIQUE MODERNE). But the Conseil has held back from reviewing, even for *erreur manifeste*, the marks awarded in a *concours* (CE 20 March 1987, CAMBUS) or the assessment of the merits of those seeking the Legion of Honour (CE 10 December 1986, LOREDON).

Likewise, in the delicate area of questions of *police* concerning foreigners, the traditional reluctance of the administrative courts to review the facts is now tempered by the doctrine of *erreur manifeste d'appréciation des faits*. Thus, with regard to the deportation of an alien who has entered France illegally or exceeded a limited duration of stay, the prefect has a discretionary power of deportation (*la reconduite à la frontière*) under the Law of 2 August 1989, subject to a right of appeal against the order within

24 hours to the local Tribunal Administratif; the court then has 48 hours to come to a decision, and the order is meanwhile suspended. In its recent case-law, the Conseil d'Etat has ruled that, even where the conditions of the statute are met, the prefect must satisfy himself that the deportation order will not create for the alien exceptional hardship because of his personal or family circumstances, a notion akin to that of 'respect for family life' safeguarded in Article 8 of the European Convention on Human Rights. The administrative courts will review and quash the prefect's decision for *erreur manifeste d'appréciation* if it is found to be wholly unreasonable or grossly disproportionate to the facts.

Such a review took place in PREFET DE DOUBS c. OLMOS QUINTERO (CE 29 June 1990), but in the event the Conseil d'Etat declined to quash the decision of the prefect. Quintero was a Mexican woman who had entered France on a temporary visa and then overstayed for several months. Because of her overstay, the prefect refused to issue her a *carte de séjour* and ordered her deportation, despite the fact that she had lived in France for several months with a Frenchman whom she planned to marry and by whom she was pregnant.[44]

More will be said below concerning the doctrine of *erreur manifeste d'appréciation*, a major development of the case-law of the Conseil d'Etat.[45]

Apart from sensitive *police* and highly technical matters, in the great majority of cases the administrative courts feel free to exercise their fullest powers of control. In GOMEL (CE 4 April 1914) a prohibition on building was justifiable under the statute only on the ground of preserving an existing view of architectural value (*une perspective monumentale*). The Conseil d'Etat considered itself competent in such a case to consider whether such a view existed: in French parlance, there was no question of *inexactitude des faits*; as Vedel observes,[46] 'the existence of the

---

[44] For a fuller discussion of this and previous cases, see R. Errera [1990] PL 432, and J. Bell, 'The Expansion of Judicial Review over Discretionary Powers in France' [1986] PL 99.

[45] Although it is an obvious Gallicism for 'glaring mistake', the term 'manifest error' has been given respectability by its adoption in the English-language version of the reports of the European Court of Justice: e.g. Case 131/77, *Milac* v. *HZA Saarbrücken* [1978] ECR 1014; Case 109/79 *Maiseries de Beauce* v. *ONIC*, [1980] ECR 2883.

[46] Vedel and Delvolvé, *Droit administratif* (11th edn., 1990), 762.

Place Beauveau was not in dispute'; but rather a question arose of *qualification*: 'Could the Place Beauveau, being what it was, be regarded as falling within the category of *perspectives monumentales*?' The English reader is reminded of *Re Ripon (Highfield) Housing Order 1938, Application of White and Collins* [1939] 3 All ER 548, where the court held itself competent to consider whether the parcel of land being compulsorily acquired was in fact part of a 'private park', as the enabling statute expressly excluded any compulsory powers where the factual situation could be so classified. Thus, although a prefect has a wide discretion in matters of public order or public health, when in the expressed interests of public health a prefect forbade camping by gypsies in eighty-six communes of the *département* of the Alpes Maritimes, the decision was quashed as offending against the general principle of personal freedom of movement; there could not have been a genuine public-health reason for excluding gypsies from such a large area: VICINI (CE 20 January 1965). In CAMINO (CE 14 January 1916) a prefectoral order had suspended Dr Camino, the Mayor of Hendaye, from his post for an alleged mishandling of the affairs of his office. When the Conseil d'Etat inquired into the matter, it appeared that the complaint against the mayor was that at a funeral in the town he had failed to pay proper respect to the corpse of the deceased, as he had made the undertakers carry the coffin into the cemetery by a route which lay across a ditch and through a hole in the cemetery wall. The prefect's order was quashed as the reason given for it was considered to be insufficient and the decision was therefore not a proper exercise of his discretion. And in BENZ (CE 11 December 1970), when the Minister of the Interior exercised his power (under the relevant treaty) to order the expulsion of an alien from France 'for compelling reasons of national security', the Conseil d'Etat considered itself competent to examine whether such 'compelling reasons' had been made out.

Moreover, the facts must not only be capable of the correct classification; they must also be true. In VILLE DE NANTERRE (CE 20 November 1964), a decision of a municipal council to establish a municipal dental clinic under its general powers to provide facilities for the good of the commune was questioned, and the Conseil d'Etat then proceeded to examine whether such a clinic was in fact for the good of the commune and was necessary in view of the number of private dentists in the area, or whether such a venture would not justify expenditure of public funds.

The reasons given must be the correct ones; in ŒUVRE DE SAINT NICOLAS (CE 7 July 1950), a decision refusing a state subsidy to a school for technical education was quashed when it appeared that the subsidy had been withheld because of the devotional nature of the particular Roman Catholic school. In JEUNESSE INDEPENDANTE CHRETIENNE FEMININE (CE 5 March 1948), the Prefect of Paris had given permission for the holding of a special outdoor Mass in the gardens of the Palais de Chaillot. A few days before the Mass was due to be held, the permission was withdrawn, the reason being that the crowds of people who would attend the Mass would spoil the gardens. The Conseil d'Etat declared this withdrawal of permission was illegal, on the ground that there could have been no valid justification for it, as during the interval between the grant of the original permission and its withdrawal no new facts or circumstances had arisen.

Mistakes of fact also will be reviewed by the administrative court, following the same kind of reasoning. Thus, in TISSOT (CE 14 December 1934) a decision to dispense with the services of a senior civil servant had been expressed to have been taken at his own request. When it appeared that this was false, the decision was quashed. Again, in MINISTRE DES AFFAIRES SOCIALES c. SHAUKAT (CE 26 February 1988), the application of Shaukat, a Pakistani, for naturalization was considered inadmissible on the ground that he did not reside in France. For residence in France, it is not enough to prove one's presence in the country, but (according to the case-law) one must establish that France is one's 'centre of interests', as indicated, for example, by the presence of one's family and one's occupation in France. Shaukat was held not to have proved his residence in France because his family were still living in Pakistan. In fact, by the time his application was refused, his wife and one of their children had been living with him in France for some three months. The Conseil d'Etat annulled the minister's decision for lack of a correct factual basis (*faits matériellement inexacts*).

It can always, in *any* matter, be argued that the facts, as they were found by the administration, were incorrect. It is in this

SUBSTANTIVE LAW: ADMINISTRATIVE LEGALITY 245

matter of the control over the facts of the situation that the administrative courts have made the most subtle efforts to reconcile conflicting interests. They must, and in most cases they seem to do this with amazing ability, hold the balance between *l'opportunité* and the interests of the individual. A specialized example of this balancing of interests in practice is to be seen in the doctrine of 'exceptional circumstances', under which the administrative courts will allow a wider discretion to the administration, for example in times of emergency, than they would under normal conditions. A much stricter control over the use of bars by prostitutes was allowed in the wartime case of DOL ET LAURENT (CE 28 February 1919) than would normally have been permitted during peacetime.

In recent years, judicial control in this third class of case, namely, where the administration has a limited discretion, has been considerably restricted by the trend towards more precise statutory provisions, the effect of which has been to leave the administrative courts with less scope for classifying and controlling the action of the administration under the guise of statutory interpretation.

(d) The Doctrine of Manifest Error in Assessment of the Facts

The readiness of the Conseil d'Etat to restrict the domain of *l'opportunité* (or, to adopt the phrase of M. Braibant, to reduce the 'zone de non-recours') and to extend its control over administrative decisions is well illustrated by the development by the Conseil of the doctrine of *erreur manifeste d'appréciation des faits* (often referred to simply as *erreur manifeste*). This doctrine is a development of the last two decades, although its roots go back into earlier case-law (see e.g. DENIZET, CE 13 November 1953 p. 241 above). In the 1960s, however, the Conseil d'Etat felt the need, in exceptional cases, to intervene in what had hitherto been accepted as undoubtedly the domain of *l'opportunité*, such as to render this aspect of the decision not open to judicial review: the principle that *opportunité* was for the administrator, not the judge, needed, it was felt, a safety-value to enable justice to be done in extreme cases, and the doctrine of manifest error met this need.

Accordingly, where the Conseil d'Etat was convinced that the administration, although making no mistake in its finding of facts, had nevertheless committed a 'manifest error in the assessment of those facts', the Conseil was prepared to quash the decision. In English terminology, it is as if the Conseil d'Etat were to say that no reasonable administrator could have reached that view of those facts. In other words, the administrator has the right to err, to decide wrongly, but not to make a *manifestly* wrong decision. Thus, in the typical case of ROUGEMONT (CE 7 July 1967) the court categorized a decision of a local medical committee refusing to place a well-known and highly qualified surgeon on a special list which would entitle him to higher fees, as being a manifest error, and the decision was therefore quashed.[47]

The doctrine has been widely applied: e.g. control over aliens (CE 8 December 1978, MINISTRE DE L'INTERIEUR c. BENOUARET): refusal to issue a passport (CE 3 February 1975, FOUERE); assessment of professional ability (CE 10 April 1974, MINISTRE DE JUSTICE c. DELAMARCHE); fitness of a civil servant for promotion (CE 17 February 1978, SPIRE). But its most notable extension has been into the sphere of disciplinary sanctions imposed upon civil servants.

Traditionally, judicial control in such matters was confined to checking the correctness of the facts alleged against the *fonctionnaire* and to their *qualification* as falling within the category of disciplinary offences. Otherwise, in general, the administrative court did not question whether 'the penalty fitted the crime': the administrative authority had a complete discretion to fix whatever penalty it felt was appropriate (see, for example, CE 10 February 1978, DUTRIEUX). In LEBON (CE 9 June 1978), where a teacher had been charged with indecent behaviour towards very young girl pupils and dismissed by the education authority, the Conseil d'Etat rejected the appeal, after finding explicitly that the authority had not committed any manifest error and that the penalty was in no way disproportionate to the offence. On the other hand, in VINOLAY (CE 26 July 1978), where the director of a local agricultural committee was dismissed for his delay in replying to an inquiry from the Ministry of Finance, the Conseil accepted that this delay amounted to a *faute professionelle* but quashed the decision to dismiss on the ground that the

---

[47] Cf. The somewhat similar facts of CE 28 May 1971, LANGLAIS.

penalty was so disproportionate as to amount to manifest error in the assessment of the facts. See also MINISTRE DU TRAVAIL c. STEPHAN (CE 16 March 1979), and BERNETTE (CE 5 May 1976); in the latter case the Conseil d'Etat found that 'the faults alleged against M. Bernette were not sufficiently grave to justify the approval (by the Minister of Agriculture) of his dismissal by the company'. Again, in WAHNAPO (CE 27 February 1981), the Conseil d'Etat quashed a disciplinary measure taken by the government against a local mayor as excessive.

As these last cases show, manifest error may be linked with the concept of proportionality (or disproportionality), a concept familiar (as *Verhaltnismäßigkeit*) in German administrative law and now imported also into European Community law. A more distinctive French doctrine of recent years uses rather the metaphor of the balance sheet (*le bilan*) and may perhaps be regarded as a particular application of the broader doctrine of *erreur manifeste*.[48] The doctrine of *le bilan* was first given full expression in the decision of the Assemblée du Contentieux in VILLE NOUVELLE EST DE LILLE, a striking illustration of the readiness of the Conseil d'Etat to extend its control of *l'opportunité* in the area of land-use planning (CE 28 May 1971).[49]

New faculties of law and arts for the University of Lille were to be built on the outskirts of that city. As experience had shown the undesirability of segregating students from the rest of the population, it was proposed to build a whole new town adjoining the faculties in order to form a single residential and academic complex. To acquire the extensive site (involving the demolition of some hundred houses), the usual procedures of compulsory purchase were set in train. The consequent public inquiry provoked a storm of protest led by a 'defence association' of threatened property-owners and residents. The Association

---

[48] For the interconnection of these doctrines, see J. Lemasurier, 'Vers un nouvean principe général du droit? Le principe "bilan coût–avantage" ', *Mélanges Waline*, ii. 551, and the references at n. 18 above; also X. Philippe, *Le Contrôle de proportionnalité dans les jurisprudences constitutionnelle et administrative françaises* (Paris, 1990), 177–83.

[49] The full name of the case is MINISTRE DE L'EQUIPEMENT ET DU LOGEMENT C. FEDERATION DES DEFENSES DES PERSONNES CONCERNEES PAR LE PROJET ACTVELLEMENT DENOMME VILLE NOUVELLE EST. For an authoritative comment on the case see D. Labetoulle and P. Cabanes in *AJDA* 1971, 404.

challenged the subsequent declaration of the minister that the acquisition was in the public interest (*d'utilité publique*). This challenge was upheld by the Tribunal Administratif of Lille on account of certain procedural irregularities, and the *déclaration d'utilité publique* quashed. The minister appealed to the Conseil d'Etat.

The Conseil d'Etat found there had been no irregularity in the procedure leading to the *déclaration d'utilité publique*. The importance of the case, however, as marked by the convening of a full court, was the readiness of the Conseil d'Etat to extend its previous case-law so as to examine whether the minister's decision was warranted by the evidence before him. The Conseil adopted and applied the criteria proposed by Commissaire du gouvernement Braibant and held that

an operation cannot be legally declared *d'utilité publique* unless the interference with private property, the financial cost, and, where it arises, the attendant inconvenience to the public is not excessive having regard to the benefits of the operation.

On the facts of the present case the court found these criteria to be satisfied and accordingly allowed the minister's appeal.

Although the outcome was a victory for the minister, the judgment marks, as French commentators have pointed out, a remarkable extension of judicial control in cases of compulsory purchase, a realm where hitherto administrative discretion was almost sovereign, procedural defects apart. For the future, the administrative court will weigh for itself the advantages and the disadvantages of a challenged expropriation; only if the balance is in favour of the former will the court adjudge the *déclaration d'utilité publique* to be well founded. To cite M. Braibant's conclusions before the Assemblée:

There is no question that you should exercise discretions that belong to the administration; questions such as whether the new airport for Paris should be built to the north or the south of the capital, or whether the eastern motorway should pass close to Metz or close to Nancy remain matters of *opportunité*. It is only above and beyond a certain point, that is, where the cost, whether in social or financial terms, appears abnormally high, that you ought to intervene. What matters is that you should be able to review decisions which are arbitrary, unreasonable or ill-considered, and that you should compel local authorities to put before the public in the first place (and later, if need be, before the court) solid and convincing reasons for their proposals.

Since VILLE NOUVELLE EST, the doctrine of *le bilan*, whereby the court will make a legal evaluation of the decision by drawing up a balance sheet in relation to the facts showing the respective advantages and disadvantages to the public of a project, and assessing whether the one outweighs the other, has become well established in the case-law. Thus, a proposed slip-road to a new motorway was struck out of the scheme by the Conseil d'Etat when it appeared that this would make an important mental hospital virtually unusable: CE 20 October 1972, SOCIETE CIVILE SAINTE-MARIE DE L'ASSOMPTION. In CE 26 October 1973, GRASSIN, the Conseil d'Etat rejected a proposal to build a new recreational airport near Poitiers when it was shown that the cost was too high a price to pay for the limited benefits to the area of the small airport. But in ADAM ET AUTRES (CE 22 February 1974) the Conseil d'Etat refused to interfere with the decision of the administration in its choice of the route for a new motorway when there was little to choose between two routes proposed.

Finally in DREXEL-DAHLGREN (CE 27 July 1979) the Conseil d'Etat quashed a proposed compulsory acquisition by the Prefect of Paris of a town house in the Rue des Saints-Pères; the Conseil, in applying the doctrine of *le bilan*, found that the cost of acquisition and refurbishing (in the region of 20 million francs) outweighed the advantages of securing the premises as an annex to the overcrowded Ecole Nationale des Ponts et Chaussées next door, especially as the School was likely to be moving out of Paris in the course of the ensuing twelve months; the balance which the Conseil had to strike in this case was between two *public* interests; the cost to the national exchequer in the one scale, and in the other, the educational needs of the School.

As in BAREL twenty years before, the Conseil d'Etat in VILLE NOUVELLE EST and the subsequent case-law, has shown itself, in its approach to contemporary issues, to be capable of adopting and adapting established principles, such as *erreur manifeste* and *proportionnalité*, to solve novel problems. By expanding *les principes généraux du droit*, the Conseil has widened the *cas d'ouverture* of *violation de la loi* with vigour and determination. This attitude

250   SUBSTANTIVE LAW: ADMINISTRATIVE LEGALITY

compares favourably with the piecemeal reforms introduced into English administrative law in recent years, cribbed and confined as the English courts are by the doctrine of parliamentary sovereignty and the necessity to introduce any reform by beneficent statutory interpretation. The wonder is that the judiciary has been able to achieve so much through the vehicle of applications for judicial review. We return to this development in our final chapter.

7 GENERAL OBSERVATIONS

To conclude this section on the extent of judicial control in France, reference must be made to the distinction which is often expressed, both in doctrinal writing and in case-law, between 'minimum' (or 'limited') review and 'normal' (or 'full' review). As M. Errera explains

> Where full review takes place the courts review the application of statutory and regulatory requirements and the use made by the administration of its powers. Hence a full review of the legal characterization of the facts (*qualification juridique des faits*) leading to an answer to the question: given the facts, the law and the general context, was the decision taken by the administration within its powers? Limited review is more restricted. The decision will be annulled if the court finds that there was a gross error, that the administration acted unreasonably.[50]

He adds that the boundary between these two levels of review is not fixed. As we have seen, the Conseil d'Etat is ready, where appropriate, to expand judicial review. The doctrine of manifest error itself illustrates how minimum review was extended into previously unreviewable areas. Review may also be enlarged from minimum to full. Thus, in SOCIETE 'LES EDITIONS DES

---

[50] [1985] PL 510. Vedel and Delvolvé (*Droit administratif*, ii. 336–7) adopt a threefold distinction between minimum (or limited), normal, and maximum control. Minimum control allows review for errors of fact, mistake of law, and *détournement de pouvoir*, expanding (but not invariably) into manifest error. Normal control is where there is also review of *la qualification juridique des faits*. Maximum control is where the court is prepared to ask whether the measures taken go beyond what is necessary or reasonable: an example is where police measures restrict civil liberties. See further Bell, 'The Expansion of Judicial Review', 113–17.

ARCHERS' (CE 17 April 1985) the government banned a foreign publication that reprinted extracts from a wartime Nazi magazine. The Paris Tribunal Administratif quashed the ban as exceeding the powers conferred by the *décret-loi* of 6 May 1939; it did so after a full review, although previous case-law (based on S. A. LIBRAIRIE FRANÇOIS MASPERO, CE 2 November 1973) had limited review to the ground of manifest error.

The Conseil d'Etat, while upholding the decision on appeal, declined to exercise full review but accepted that there had been a manifest error. The approach of the Conseil has been criticized, full review being desirable whenever freedom of expression is in issue: the bolder approach of the Paris Tribunal may well be adopted on some future occasion by the Conseil, if past experience is any guide.

10

The Influence of droit administratif Outside France

1 INTRODUCTION

The most outstanding contribution made by France to legal science has undoubtedly been the Civil Code of 1804, but almost as important has been the separate system of administrative jurisdiction and administrative law created by the Conseil d'Etat during the nineteenth and twentieth centuries. Most European countries follow the French practice of the double jurisdiction; even Belgium, which was strongly opposed to the French example in its Constitution of 1831, has since established a separate Conseil d'Etat. The Netherlands, Luxembourg, Italy, Spain, Portugal, and Greece all have separate administrative courts administering principles of administrative law not far removed from those of the *droit administratif*. In this chapter, therefore, it is proposed briefly to outline the systems in five of France's partners in the European Communities, namely, Belgium, the Netherlands, Italy, Germany, and Greece, and then to discuss what may well in time become the most important French export of all, the procedure and law of the Court of Justice of the European Communities at Luxembourg.

Upon the United Kingdom becoming a member of the European Communities in January 1973, a British judge and a British Advocate General joined the European Court. Their presence and that of the English and Scottish practitioners appearing before the Court have already had an influence upon the development of the procedure and substantive law of the Court,[1] but

---

[1] See Case 155/79, *Australian Mining & Smelting Europe* v. *EC Commission* [1982] 2 CMLR 264, where the English concept of legal professional privilege, as presented in the Opinions of the English Advocates General (Mr J.-P. Warner and Sir Gordon Slynn) helped shape the eventual judgment of the European Court in an appeal brought by an English-based company against a demand by the Commission that the company produce certain documents passing between the company and its lawyers.

it should be appreciated that the Court (including its previous incarnation as the Court of the European Coal and Steel Community) has been in existence for nearly forty years, a period long enough for the Court to have acquired a substantial body of case-law and to have worked out and established an effective system of procedure. No doubt in turn the French background of the Court will in time influence British attitudes to their own judicial institutions. But first we must say a little about the influence of *droit administratif* on the legal systems of some of the member states.[2]

2 BELGIUM

When the Constitution of Belgium was written in 1831 the founding fathers, anxious to reject all apparent legacies of foreign despotism, refused to create a Conseil d'Etat on the French model, although, in accordance with generally accepted European constitutional doctrine, they subscribed to the theory of the strict separation of powers, and in the course of the nineteenth century the Belgian ordinary courts worked out a system of substantive *droit administratif* very similar to, and strongly influenced by, the French.[3] Later, the need was felt for a separate and specialized administrative court, and eventually, after the Liberation, a Law of 23 December 1946 established a Conseil d'Etat, constituted very much on the French pattern. At the present day, the Conseil d'Etat consists of twenty-four members, who are supported by an *auditorat* of forty-six persons of various grades. In its function as *juge de l'excès de pouvoir*, it is the court of first and last instance: there are not, as in France, administrative tribunals of first instance nor administrative appellate courts. The Conseil d'Etat exercises its judicial functions through a *section d'administration*, whereas its legislative section has primarily a consultative function. The administrative section

---

[2] For further information the reader is referred to J.-M. Auby and M. Fromont, *Les Recours contre les actes administratifs dans les pays de la Communauté Economique Européenne* (Paris, 1971).

[3] See A. Mast, *Précis de droit administratif belge* (Brussels, 1966) and 'Le Conseil d'Etat de Belgique [1971] PL 51; F. Delpérée, 'Belgique: le contentieux administratif' *RFDA* 1988, 227. Also L.-P. Suetens, in J. Bell and A. W. Bradley, *Government Liability: A Comparative Study* (United Kingdom National Committee of Comparative Law, London, 1991), ch. 8.

has six chambers, two French, two Flemish, and two bilingual, and any two chambers may unite to form an *Assemblée générale*. The primary jurisdiction of the administrative section of the Belgian Conseil d'Etat is the power to quash a decision of the administration (*contentieux d'annulation*) based on any one of several *cas d'ouverture*, which correspond to the normal grounds of French law.[4] In a few cases, Belgian law ascribes to the Conseil d'Etat jurisdiction to advise the award of compensation (*contentieux d'indemnité*) where the plaintiff has suffered some exceptional injury and no other remedy has been provided by law.[5] As a general rule, however, the ordinary courts retain the *contentieux de pleine juridiction*, that is, the jurisdiction in actions for damages against organs of the administration.

The continued superiority of the ordinary courts as compared with the Conseil d'Etat is assured by the fact that the determination of conflicts between the two jurisdictions is entrusted to the highest ordinary court, the Cour de Cassation. Moreover, the members of the Conseil d'Etat are thought of as being primarily judges, not members of the administration, and therefore, unlike their French brethren, are allowed to 'porter la robe' like *magistrats* of the ordinary courts.[6] French decisions and the views of French jurists are commonly cited in Belgian legal literature.

3 THE NETHERLANDS[7]

When the Kingdom of the Netherlands was created in 1814, French *droit administratif* was a strong influence on Dutch administrative law, and this remained so until about 1880.

In this period, the administration was its own judge of disputes over administrative decisions. As in the early *droit administratif*, from the *ministre-juge* the dissatisfied citizen could refer his case to the Crown, which reached its decision after being advised by the Council of State: since 1862, the advisory organ has been the Administrative Litigation Division or 'Section du Conten-

---

[4] See above, p. 223.

[5] The Conseil d'Etat also acts as a court of cassation from decisions of a few administrative tribunals, especially those concerned with military service.

[6] A. Mast, *Précis de droit administratif* and [1971] PL at 62 and cf. p. 107 above.

[7] We are indebted to Professors Pieter de Vos and C. J. Bax of the Erasmus University of Rotterdam for their valuable advice on Dutch law.

tieux' (*Adfeling geschillen*) of the Council. It was not until 1988, as a consequence of the decision of the European Court of Human Rights in *Benthem* v. *The Netherlands* (1985) 8 EHRR 1, that the Dutch legislature finally accepted the principle of *la justice déléguée*, whereby the Administrative Litigation Division should decide appeals in its own right.

The Constitution of 1815, as a reaction to the French doctrine of the separation of powers, declared that the ordinary courts should have exclusive jurisdiction over property and civil rights. But in the period 1822–44, any attempt by the citizen to seise the ordinary courts of his dispute with the administration was met by the administration's 'raising conflict'—even if it was concerning a civil right. This 'conflict order' ran counter to the Constitution.

By 1879, the principle of legality (or the Rule of Law) was accepted in relation to the legislative powers of the administration. From 1915, judicial review of other administrative decisions was gradually developed in the case-law, the ordinary courts accepting to review administrative acts for illegality or for breach of civil-law principles of tortious liability. Since 1988, ordinary courts also measure the behaviour of the administration by the yardstick of its compatibility with the general principles of proper administration, not simply for conformity with the strict letter of statute or regulation.

In 1976, a Judicial Division of the Council of State was created to serve as a residuary administrative court in any case where no other relief was provided for an individual. This judges a broad range of disputes between the citizen and the administration.

At the present day, the ordinary courts have surrendered to administrative courts the power to annul individual administrative decisions. But the ordinary courts retain jurisdiction over administrative contracts and where the citizen is suing the state for damages in tort. They also have jurisdiction where the government or another administrative body has issued a regulation that infringes a higher legal norm such as the Constitution or an ordinary statute. But the courts are forbidden by the Constitution to review the compatibility of a statute with the Constitution.

The Netherlands have developed a number of administrative courts, each with their own special jurisdiction. But there is still no separate system or order of general administrative courts such as one finds in France. The specialized administrative courts include, first and foremost, the Council of State: this now has two judicial divisions, the Administrative Litigation Division and the Judicial Division. Secondly, there is the Central Appeals Board for social security cases: this *Centrale Raad van Beroep* was established in 1902 and, since 1929, has also heard appeals from the Civil Service Tribunal (*Ambtenarengerecht*). Thirdly, in 1954 there was established the *College van Beroep voor het Bedrijfsleven* as an appellate court for disputes in economic law. Fourthly, on taxation matters, there is the commission on customs and excise, and the tax chamber of the Supreme Court, which hears appeals against decisions of the tax chambers of the high courts. In general, these administrative courts deal with actions for annulment of administrative decisions (*administratieve beschikkingen*). Damages for unlawful decisions can be claimed in both administrative courts and ordinary courts.

As is apparent, the Dutch system of administrative justice is one of great complexity, with control of administrative action being dispersed among a variety of different courts.[8] Not surprisingly, important reforms are currently under consideration by the Dutch parliament, and the first phase of these is expected to be implemented in 1992.[9] These reforms are part of the total restructuring of the Dutch court-system. Basically, this will entail an integration of the administrative courts into the ordinary court-system. Every court of first instance will have an administratif chamber to deal with administrative cases. These chambers will, however, operate on the basis of a separate procedural law, so that they will not use the law of civil procedure. During the first phase of the reforms, appeal from these administrative chambers is to be to the Central Appeals Board, and a new division for administrative jurisdiction of the Council of State. This division is to be the result of the fusion of the two existing divisions, and it will also deal with cases at present still decided by the Crown. How this appeal will operate will be settled in the later phases of the reforms.

---

[8] In addition, from 1982 a national Ombudsman has been established to receive complaints against the administration.

[9] See further, C. J. Bax, 'Dutch Administrative Law' (1989) 1 *European Review of Public Law*, 119; W. Konijnenbelt, 'Les "paysages-bas" des contentieux adminstratifs hollandais', *RFDA* 1988, 252; N. Verheij, 'Dutch Administrative Law after Benthem's Case' [1990] PL 23.

4 ITALY[10]

In the days of Napoleon's Cisalpine Republic (1805–14) most of Italy possessed a Conseil d'Etat on the model of that established by Napoleon in France in Year VIII, but it was not until 1865, after many political vicissitudes, that the Consiglio di Stato was established for the Kingdom of Italy, very much in the form in which it exists in the modern Republic. Following Belgian views, the Risorgimento had favoured a single system of justice whereby the administration would have been subjected to control by the ordinary courts, but this view prevailed only to the extent that conflicts between the Consiglio di Stato or the local administrative tribunals (*Tribunali amministrativi regionali*) and the ordinary courts fall to be resolved by the highest ordinary court, the Corte di Cassazione. However, the administrative courts have power to annul decisions of the administration on the grounds, familiar to French law, of *incompetenza* (*incompétence*), *eccesso di potere* (*excès de pouvoir*) or *violazione di legge* (*violation de la loi*). A remedy may be given by the administrative courts if the plaintiff's legitimate *interests* are prejudiced by some act of the administration, but if he is complaining of an infringement of a private *right* he must look for his remedy to the ordinary courts. This distinction between an 'interest' and a 'right' is peculiar to the Italian system. In practice it means that actions for damages against the administration must be brought in the ordinary courts, whereas proceedings for the annulment of an administrative act by a party having sufficient standing must be brought in the administrative courts.[11] But this is only a general rule, for in recent years, the distinction between the two systems of courts has become less and less clear-cut, as the legislature has intervened to extend the exclusive jurisdiction of the administrative courts to specific matters, e.g. land-use planning. The administrative courts, for their part, have expanded their judicial control into certain areas involving *rights*, whilst the ordinary courts will now hear cases involving *interests*, e.g.

---

[10] On this topic, see V. Pazienza, 'Italie: la Justice administrative', *RFDA* 1988, 246; and G. L. Certoma, *The Italian Legal System* (London, 1984), 251–62. S. Galeotti, *The Judicial Control of Public Administration in England and Italy* (London, 1953) remains a classic work on this topic.

[11] See generally M. Clarich, in Bell and Bradley, *Government Liability*, ch. 10, esp. 239–42.

employment disputes in the public sector. There is also a tendency for the procedure of the administrative courts to come closer to that of the ordinary courts; thus, the procedure for appeals from the local administrative courts to the *Consiglio di Stato* is largely borrowed from civil procedure. In the long term, doctrinal opinion predicts that the dual system will be replaced by unity of jurisdiction.[12]

The *Consiglio di Stato* of Italy has six sections, three being consultative and administrative and three concerned with litigation (*il contenzioso*), but the members of the two groups of sections respectively do not interchange; in its double role as a court and a consultative body it has similarities to the French Conseil d'Etat which are perhaps the closest of the higher administrative courts in Europe.

5 GERMANY[13]

Administrative law in the Federal Republic has been considerably affected by the Constitution of 1949 (*Grundgesetz*), and by the federal system itself.[14] In the Republic there are not only separate administrative courts and tribunals in each *Land* or state (*Verwaltungsgerichte* at local level, and appellate *Oberverwaltungsgerichte* at the capital of each *Land*), and at the federal level (the *Bundesverwaltungsgericht*), but also there is separate hierarchy of commercial courts, of labour courts, of tax courts, and of social security courts, as well as the ordinary courts. Moreover, none of these specialized courts may offend against the *Grundgesetz*; if a constitutional question arises, the matter has to be determined by yet a sixth court—the *Bundesverfassungsgericht*, the superior (federal) constitutional court. Questions of conflict may thus frequently arise, but they will always fall to be determined in the final resort by the *Bundesverfassungsgericht*.

[12] See E. Midena, 'Les Juges et l'administration publique en Italie: Dualisme ou unité de juridiction', *RFDA* 1990, 882.

[13] The reunification of Germany in 1990–1 means that the territory of the former German Democratic Republic is now united with the Federal Republic, and its five *Länder* share the federal system, including its pattern of administrative courts.

[14] See generally, M. P. Singh, *German Administrative Law in Common Law Perspective* (Berlin, 1985); H. Maurer, *Allgemeines Verwaltungsrecht* (7th edn., Munich, 1990); C. Autexier, 'République fédérale d'Allemagne: Juridiction administrative et contentieux administratif', *RFDA* 1988, 257.

It is difficult to generalize about the several German administrative courts, as details of the organization of the courts of the several *Länder* are a matter for the legislatures of these *Länder* to regulate, but as a general rule an administrative court will consist of five members, of whom two may be non-lawyers. The *Bundesverfassungsgericht* is organized in a manner similar to the French Conseil d'Etat, having eight separate 'Senates' (corresponding to the French Sous-sections), each with a president and five judges, but each of these Senates is concerned only with '*contentieux*' and not with advisory or consultative matters. The *Bundesverfassungsgericht*—and indeed the lower administrative courts—do not have any consultative role, either in the legislative process or by way of advising the executive. The judges at all levels in the hierarchy are lawyers with a common professional training, who have chosen to become judges and are otherwise similar to the judges in the other courts, except that they specialize in this branch of law.

The principal jurisdiction of the administrative courts is to pronounce on the validity of an administrative decision, and the Germans, with typical thoroughness, have expended a great deal of academic argument on what precisely constitutes such a decision (sometimes described as action taken by an administrative authority in relation to an individual case). The procedure to annul corresponds very closely to the French *recours pour excès de pouvoir*, except that it cannot be used in respect of subordinate legislation. In addition, the administrative courts can adjudicate on disputes between two or more organs of the administration (such as between two *Gemeinden*—local authorities), and here they can revise a disputed decision as an alternative to annulment. Proceedings for damages against an organ of the administration are not entrusted to the administrative courts,[15] but are reserved for the ordinary courts,[16] and the latter follow either the principles of the civil law, holding the administration liable if fault can be established, or if the action of the administration amounts to expropriation or 'quasi-expropriatory encroachment'

---

[15] But a claim for compensation (e.g. payment of a pension etc.) may on occasion be combined with proceedings for annulment.

[16] This is a part of the legacy of suspicion of the administration, as an aftermath of the Nazi regime.

(*Aufopferung*), principles which introduce almost a kind of strict liability for actions of public authorities, especially in the area of public works.[17] Substantive administrative law is not codified in a single text, and, subject to the *Grundgesetz*, generally follows parallel principles to French *droit administratif*. All the same, texts such as the Administrative Procedure Act of 1976 establish important general rules and principles. The Germans are even fonder of classification and abstract explanations than are the French, and the principal '*cas d'ouverture*' for judicial review, such as *incompétence, vice de forme, vice de procédure*, and *violation de la loi*, are as well known to the German administrative lawyer as they are to the French but the *principes généraux du droit* (*Rechtsgrundsätze*) are not in all respects as fully developed as in France, and such ideas as the principle of equality depend more closely on the Constitution (*Grundgesetz*) and are classified by the Germans themselves as fundamental rights of the individual.

The procedure before an administrative court is principally written, and it closely follows French practice. The Germans, however, differ in requiring a prospective plaintiff before an administrative court first to follow a 'remonstrance' procedure, or an 'appeal' (*Widerspruch*) to the administration; this takes place before a local 'remonstrance board', consisting of civil servants, lawyers, and lay members. The object of these boards seems to be to give the administration an opportunity to rectify any wrong informally, expeditiously, and cheaply, before the matter is ventilated in formal legal proceedings, and only if the board cannot find a satisfactory solution will the plaintiff be allowed to proceed.[18] The other special feature of the German system is the power given to the administrative courts to make an order on a complaint (*Verpflichtungsklage*) requiring the relevant administrative agency to perform a specific administrative act (such as to grant a licence to an applicant). They may also make a declaration as to the validity of an administrative decision on the application by a plaintiff having a sufficient interest in the matter.

---

[17] See W. Rüfner, in Bell and Bradley, *Government Liability*, ch. 11.
[18] Cf. the French procedure of conciliation committees, above p. 27.

6 GREECE[19]

The Greek system of judicial review of administrative action has been greatly influenced by that of France. The reason is historical: the Bavarian regents and advisers of King Othon, the first King of the Hellenes in 1832, were greatly inspired by the French model in setting up the Greek system.

Today, under the Constitution of 1975, the administrative courts which have jurisdiction to review administrative action are the following:

1. The ordinary administrative courts (twenty-eight courts of first instance and seven courts of appeal). Since 1985 these have general jurisdiction:
    (a) in matters of *'plein contentieux'*, that is, where damages are being sought for contractual or non-contractual liability on the part of the administration, and
    (b) in suits to annul, *'pour excès de pouvoir'*, decisions affecting individual civil servants, other than dismissal, demotion, and most disciplinary sanctions (as to which, see below).

2. The Council of State. The Council's judicial functions are exercised by four *'sections du contentieux'* and the plenary assembly. Its jurisdiction is threefold, extending to:
    (a) actions to annul, *'pour excès de pouvoir'*, administrative acts, whether individual or general: thus, the Council is the normal court to annul administrative decisions (that is, *'le juge d'annulation de droit commun'*);
    (b) actions *'en cassation'* to quash decisions of the lower administrative courts; and
    (c) cases specifically reserved to the Council by the Constitution or statute: in particular, disputes by civil servants and employees of local authorities and public corporations concerning their dismissal, demotion, or disciplinary sanctions.

3. The Court of Auditors (the 'Cour des Comptes'). This court, whose decisions are final, deals with civil servants' disputes

---

[19] We are indebted to Professor Glykeria Sioutis of the University of Athens for information on Greek law. See further K. D. Kerameus and P. J. Kozyris, *Introduction to Greek Law* (Deventer, 1988), 37–42; P. Spiliotopoulos, *Droit administratif hellénique* (Paris, 1991); id. 'Grèce, l'organisation de la justice administrative', *RFDA* 1988, 240.

concerning their pensions, as well as the audit of the accounts of government departments, local authorities, and public corporations.

The Council of State is thus very similar to the French Conseil d'Etat. It was first established in 1835, abolished in 1844 and, after many transformations, re-established in in its present form in 1929. According to the Constitution and relevant statute, the grounds to seek annulment of an administrative act closely resemble their prototypes before the Conseil d'Etat, namely:

1. want of authority ('*incompétence*') of the administrative body responsible for the act
2. breach of an essential procedural requirement
3. violation of law
4. '*détournement de pouvoir*'

In its jurisdiction '*en cassation*', the Council may quash decisions of the lower administrative courts on grounds of:

1. the lower court's lack of jurisdiction
2. its illegal composition
3. violation of an essential procedural requirement
4. wrong interpretation or incorrect application of the law
5. violation of the principle of *res iudicata*

7 THE COURT OF JUSTICE OF THE EUROPEAN COMMUNITIES

It is not surprising, in view of the strong influence of French *droit administratif* on the law and procedure of the other five founder members of the European Communities (Italy, Germany, Belgium, the Netherlands, and Luxembourg—all countries at one time part of the Napoleonic empire), that the same *droit administratif* has had a profound influence on the law and procedure of the Court of Justice of the European Communities. The extent of that influence is outlined briefly in the passages that follow.[20]

---

[20] For a full discussion of the jurisdiction of the European Court of Justice, see Brown and Jacobs, *The Court of Justice of the European Communities* (3rd edn., London, 1989).

(a) Substantive Law

The law applied by the European Court has had to be of its own making; for sources it is of course primarily concerned with the terms of the Treaties themselves,[21] the regulations, directives, and decisions made thereunder, and the various protocols and conventions. In general terms, it is the duty of the Court to ensure that the law is observed in the application and interpretation of the relevant treaty. Sometimes the Court may be required by a treaty to apply the national law of one of the member states, such as when deciding whether or not a company has legal personality entitling it to sue before the Court (see e.g., Case 18/57, *Nold* v. *The High Authority*, [1957] ECR 121); but the Court has no jurisdiction to annul the legislative or administrative acts of a member state (see, e.g., Case 6/60, *Humblet* v. *Belgium* [1960] ECR 559), although it may declare such acts to be contrary to Community law.[22] In other circumstances, in particular in cases of non-contractual liability of the European institutions for damage caused by their actions, the treaty may require the Court to apply 'the general principles common to the laws of the member states'.[23] The treaties are silent about the source of these general principles, although it is clear that a particular 'principle' need not form part of the law of all the member states; typical examples of such principles would be the procedural rules of *audi alteram partem* and freedom from bias. French law, as we have seen, applies the 'risk' doctrine in questions of liability of the administration, while Belgian, German, and Italian laws remain based on 'fault' to varying degrees. Which 'general principle' is to be applied by the European Court? The answer seems to be in such circumstances that the

---

[21] 'Community Treaties' are originally defined in section 1(2) of the European Communities Act 1972.

[22] See Case 106/77, *Simmenthal* [1978] ECR 629, where the Court declared that the Italian court should refuse to apply a subsequent law which was in conflict with a prior provision of Community law; likewise, in Cases C-213/89 and C-221/89, *Factortame*, as the Court ruled, the British Merchant Shipping Act 1988 could not be applied by the House of Lords as conflicting with Community law: see *Factortame Ltd.* v. *Secretary of State for Transport* [1991] 1 All ER 70 (ECJ and HL); [1991] 3 All ER 769 (ECJ).

[23] Art. 215(2) EEC Treaty. But the Court has invoked such principles beyond the narrow context of Art. 215(2) by treating them as an integral part of 'the law' which it shall ensure is observed (Art. 164).

Court is free to make such choice as in its opinion appears to be most in accord with the spirit of the treaties.[24]

The law evolved by the Court is concerned primarily with interpretation of the treaties and subordinate instruments. Like French administrative law, it is to be found in the decisions of the Court itself: as Judge Koopmans has observed, 'There is legislation galore, in the treaties and in the thousands of regulations and directives. Nevertheless, the main rules for judicial activities are unwritten: the treaties are practically silent on the relationship between Community law and national legislations, and criteria for assessing the legality of Community decisions are expressed in a way which, though recalling the grounds for review known to French administrative law, leaves much room for interpretation, precision and further elaboration.'[25]

However, it is in the extent of the jurisdiction of the Court and the procedure it habitually follows that the greatest resemblance to French law is to be found.

(b) Jurisdiction of the Court

This of course depends on the provision of the treaties, and can be summarized as follows:

1. Applications by individuals and enterprises for annulment of a decision of a Community institution on the ground of illegality (EEC Treaty, Article 173) or proceedings against such institutions under the Court's plenary jurisdiction (e.g. to award damages): ibid., Articles 178, 215(2), etc. From a comparative point of view, Article 173 is perhaps the most important head of jurisdiction, for it is by this means that the Court exercises effective supervision over the legality of measures taken by the Council and the Commission of the Communities and the European Parliament in a manner very similar to the control over *légalité* exercised by the French Conseil d'Etat. This similarity is emphasized by the grounds of jurisdiction specified in this article, namely, 'lack of competence, infringement of an essential procedural requirement, infringement of this Treaty or

---

[24] See, e.g., Case 14/61, *Hoogovens* v. *High Authority* [1963] CMLR 73, and G. Eike zur Hausen, in Bell and Bradley, *Government Liability*, ch. 12. For recent developments in this area see Cases C-6/90 and C-9/90, *Francovich* v. *Italian Republic, The Times*, 20 November 1991; A. Barav, (1991) 141 NLJ 1584.
[25] *Essays in European Law and Integration* (Deventer, 1982), 15.

of any rule of law relating to its application, or misuse of powers'. An allegation of *détournement de pouvoir* was made, albeit unsuccessfully, in one of the earliest cases to come before the Court, Case 1/54, *France* v. *The High Authority* [1954–56] ECR 245, and again in Case 15/57, *Compagnie des Hauts Fourneaux de Chasse* v. *The High Authority* [1957–58] ECR 199. And Case 32/62, *Alvis* v. *Council of EEC* [1963] CMLR 396, involving the dismissal of an employee of the Community, applied the rule

generally accepted by the administrative law in force in the member states of the EEC [that] the administrative departments of these states must grant their officials the opportunity of replying to incriminating facts before any disciplinary decision is taken in respect thereof.

2. Preliminary rulings on questions concerning the interpretation of the EEC Treaty, the validity and interpretation of measures taken by institutions of the Community, and the interpretation of statutes of bodies established by the Council of the Communities. References for such rulings may be made by any court of a member state in which such a question is raised, but if it is raised in a court of last resort, the court must refer the question for determination by the European Court (EEC Treaty, Article 177). The exclusive nature of this jurisdiction of the European Court was emphasized in the leading case from Italy of Case 6/64, *Costa* v. *ENEL* [1964] CMLR 425. However, the Conseil d'Etat of France successfully avoided Article 177 by holding that the 'question' of interpretation of the treaty in issue was not really a question at all, in that the answer was quite clear: SOCIETE DES PETROLES SHELL-BERRE (CE 19 June 1964) and the French Cour de Cassation, in its turn, held that the law of the EEC was not relevant to the case it was deciding: *Etat Français* v. *Nicolas* [1965] 4 CMLR 36. On the other hand, in SYNACOMEX (CE 10 July 1970) the Conseil d'Etat did refer for the first time a question of interpretation to the European Court[26] and then accepted that court's ruling by annulling a decision of a French public authority. But the *acte clair* doctrine was again invoked by the Conseil d'Etat in COHN-BENDIT (CE 22 December 1978): the Conseil held it to be clear (despite

[26] See Ch. 5, p. 119.

case-law of the European Court establishing the contrary) that the notorious German radical could not invoke an EEC directive to prevent his exclusion from France and refused to allow a preliminary reference under Article 177. But as we shall see (see below, p. 269) the Conseil d'Etat has now, in effect, overruled COHN-BENDIT.[27]

3. Disputes between the Communities and any of their employees relating to their service regulations or conditions of employment; but from 1988 this jurisdiction will only be exercised on appeal from the Court of First Instance, which was set up under the Single European Act 1986 to lighten the burden of the European Court. In such employment appeals the European Court is exercising a truly administrative jurisdiction of a continental type; for example, in France proceedings by *fonctionnaires* make up a significant slice of the business of the administrative courts.

Most of the Court's jurisdiction is thus of an administrative character; but to some extent the Court is an international tribunal, as it also has jurisdiction (much invoked by the Commission) to decide upon infringements of Community law by a member state (e.g. Articles 93 and 169, EEC Treaty) or to determine disputes between member states arising under the treaties (e.g. Articles 170 and 182). In general, however, the jurisdiction of the European Court is closer to that of a superior administrative court, such as the French Conseil d'Etat, than to that of a traditional international tribunal, such as the International Court of Justice at The Hague.

(c) Procedure and Composition of the Court[28]

Here perhaps is to be seen the greatest resemblance between the European Court and the Conseil d'Etat. The Court itself now consists of thirteen judges in plenary session, but it may sit as chambers of three or five judges for certain purposes. One of the judges will act as rapporteur for the particular case. In addition, the court is assisted by six advocates general,[29] who must be

---

[27] On this topic generally, Brown and Jacobs, *Court of Justice of European Communities*, ch. 10.

[28] See generally ibid., ch. 12.

[29] The first two Advocates General were a member of the French Conseil d'Etat (M. Lagrange) and a German civil lawyer (Herr Roemer). Subsequently, several other members of the Conseil d'Etat have served as either Advocates General or judges of the European Court.

chosen from the same kinds of persons[30] as the judges, and enjoy the same status and security of tenure (see EEC Treaty, Articles 166 and 167). The functions of the Advocate General in any particular case are closely similar to those of the Commissaire du gouvernement before the Conseil d'Etat, it being his duty, under Article 166 of the Treaty, 'acting with complete impartiality and independence, to make, in open court, reasoned submissions on cases brought before the Court of Justice, in order to assist in the performance of the task assigned to it in Article 164'.

An actual decision of the Court may be quite short, and the full reasoning for the decision may have to be sought in the conclusions of the Advocate General, published with the decision itself.

The procedure is essentially a written one, but the parties' legal representatives are given an opportunity to address the judges in open court, and this is made use of much more frequently than is the case before the French Conseil d'Etat. Decisions of the Court are reported in English in the *Common Market Law Reports* and the *European Court Reports*, and a glance at these makes obvious the resemblance to the reports of French decisions such as seen in Lebon. Yet, to quote Judge Koopmans again:[31]

Although the Court's way of formulating principles, or general propositions of law, is closely akin to methods used by the French Conseil d'Etat, its techniques of relying on previous cases, of invoking the authority of its own case-law and of determining the ratio decidendi of earlier judgements are not dissimilar to those used by the English common law courts.

8 FRENCH LAW AND COMMUNITY LAW IN CONFLICT: REACTION OF THE CONSEIL D'ETAT

Although strictly outside the title of this chapter, this important matter may conveniently be introduced here. It illustrates the readiness of the Conseil d'Etat to initiate a *revirement*, or reversal, of its previous case-law, where necessary.

---

[30] 'The Judges and Advocates-General shall be chosen from persons whose independence is beyond doubt and who possess the qualifications required for appointment to the highest judicial offices in their respective countries or who are jurisconsults of recognised competence': EEC Treaty, Art. 167.
[31] Koopmans, *Essays in European Law*, 27.

The initial reaction of the Conseil d'Etat was demonstrated in SYNDICAT GENERAL DES FABRICANTS DE SEMOULES DE FRANCE (CE 1 March 1968). In this case, there was a clear conflict between a French statute and a Community regulation. The Conseil d'Etat declared itself unable to ignore the statute: if the text was unconstitutional by reason of its conflict with Community law, that was a matter for the Constitutional Council, not the Conseil d'Etat. The Conseil d'Etat could not challenge the legality of a statute, as being in conflict with a prior treaty obligation. This view it maintained in a number of subsequent cases up to 1989.

The Cour de Cassation, however, reached the opposite conclusion in DIRECTEUR GENERAL DES DOUANES c. SOCIETE DES CAFES JACQUES VABRE [1975] 2 CMLR 336: faced with a conflict between Article 85 of the EEC Treaty and a subsequent French statute, the court found in favour of the Treaty.

Again, the grudging approach of the Conseil d'Etat to Community law may be seen in COHN-BENDIT (CE 22 December 1978), where, as we saw above, the Conseil d'Etat was unwilling to comply with the settled case-law of the European Court concerning the right of an individual to challenge in a member state an administrative decision which conflicted with an EEC directive.

A sea-change in the attitude of the Conseil d'Etat took place in 1989 with the decisions in COMPAGNIE ALITALIA (CE 3 February 1989) and NICOLO (CE 20 October 1989). The former case we have already met in the previous chapter when discussing *les principes généraux du droit* (p. 221–2).

Alitalia challenged before the Conseil d'Etat the refusal of the French Prime Minister to abrogate several articles of the *Code général des impôts*, adopted by regulation in 1967 and 1979, which it argued were contrary to the sixth VAT directive issued by the Council of the European Communities in May 1977. Power to abrogate is conferred by the Decree of 28 November 1983 whenever a regulation 'is unlawful, either because it has been so since the beginning or because it has become unlawful in view of subsequent legal or factual developments'. The Conseil, however, preferred to rest its quashing of the Prime Minister's decision upon a general principle of law: the principle of *légalité* imposed upon him, as the competent authority, the duty to abrogate any measure which was unlawful *ab initio* or because of subsequent legal or factual circumstances: the VAT directive was such a legal circumstance.

In NICOLO (CE 20 October 1989) the Conseil d'Etat was invited to hold that the French statute of 1977, providing for direct elections to the European Parliament, violated Article 227 of the EEC Treaty, and that, consequently, the 1989 elections should be declared void. Nicolo argued that the treaty provision for direct elections extended only to metropolitan France, whereas the statute applied also to the French overseas *départements* and territories. The Conseil d'Etat disagreed as to the scope of the treaty. But it abandoned its previous case-law in being willing to consider an argument that a statute was incompatible with a treaty. The decision itself offers little by way of argument, but in his conclusions the Commissaire du gouvernement, M. Frydman, considered that the times of the absolute supremacy of a statute were past and that Article 55 of the Constitution authorized the courts to review the conformity of a statute with a treaty. After NICOLO, therefore, the Conseil d'Etat and the lower administrative courts must review the conformity of statutes in relation to treaties, even if the plaintiff does not request this: conformity with international law becomes a *moyen d'ordre public* (see p. 224).

The principle in NICOLO was applied in BOISDET (CE 24 September 1990), where the Conseil d'Etat accepted that an EEC regulation (of 1972) must prevail over a French statute (of 1980) which conflicted with it. Finally, the lottery of litigation having provided the occasion, in SOCIETE ANONYME ROTHMANS ET SOCIETE ANONYME PHILIP MORRIS (CE 28 February 1992), the Conseil d'Etat held that an EEC directive (of 1972) must prevail over a statute of 1976 (concerning the state's tobacco monopoly). Accordingly, certain price-fixing measures under the statute were annulled as contrary to the directive, and to the prejudice of the applicants, both importers into France of tobacco products. This decision (of the Assemblée du Contentieux) has laid to rest the retrograde decision in COHN-BENDIT.[32]

---

[32] See further A. Tatham, 'Effect of European Community Directives in France: The Development of the *Cohn-Bendit* Jurisprudence' (1991) 40 *ICLQ* 907.

Belatedly, therefore, the Conseil d'Etat has had to come to terms with the supremacy of Community law, just as the House of Lords has now done in the *Factortame* case,[33] or as did (many years earlier) the Belgian Court of Cassation, in *Belgian State* v. *Fromagerie Franco-Suisse 'Le Ski'* [1972] CMLR 330, and the Italian Constitutional Court, in *Frontini* v. *Minister of Finance* [1974] CMLR 386.

9 CONCLUSION

As an early commentator has said[34]

the presence of French trained jurists and counsellors in the Court of the Common Market has served to bring the municipal practices to the regional-supranational level; in addition, the influence of the legal training of the drafters cannot be overlooked, since the legal background of the jurists, who organised the Communities, will continue to exert a strong influence—indirectly.

It is natural, moreover, by the nature of the topics dealt with by the Community treaties and of the jurisdiction of the European Court, that the strongest influence should come from the public law of the Six,[35] and as that public law is itself so strongly based on the French *droit administratif*, the importance of the latter becomes obvious.

Napoleon may perhaps be thought of as the principal, though involuntary, inspiration of the European Court; unlike the short-lived empire he founded by the sword, the Communities have a more firm basis in the solemn agreements of their members, buttressed by the principles of administrative law that his own institution, the Conseil d'Etat, has evolved during the last 190 years.

---

[33] See n. 22 above.
[34] W. Paul Gormly, 'The Significant Role of French Administrative Jurisprudence as Presently Applied by the Court of the European Communities, With Emphasis on the Administrative Law Remedies Avoidable to Private Litigants' (1963) 8 *South Dakota Law Review* 32.
[35] According to L. J. Brinkhorst and H. G. Schermers, *Judicial Remedies in the European Communities* (2nd edn., Deventer, 1977) 308, the 'astonishing fact remains that principles and rules derived from the private laws of the Six play an important role as well' (as that played by the administrative laws of the Six).

11

Conclusions

1 THE CHARACTER AND SUCCESS OF THE FRENCH SYSTEM

The task of administrative judges has been described by Rivero as that of 'submitting the totality of French public life to an ethic whose content it defines without any written text'.[1] In other words, judges set standards of good government. As Odent suggested,[2] their ability to perform such a role depends on a close connection between the administration's judges and the administration itself:

> the interpenetration of administrative and judicial functions is a good thing. If administrative judges were isolated from the active administration, if they ceased to be in constant contact with the needs and constraints of administrative life, they would lose their specific character. Instead of building a law adapted to the necessities of the public service, they would be inspired by a fossilized law bearing no relationship to the realities of the active administration. Administrative judges must have an administrative training, and they have to sustain it to retain an understanding of administrative life.

This interpenetration between the administration and its judges is seen in the 'mixed' composition of the Conseil d'Etat and in the combination of judicial and administrative functions both in the institution and in the careers of its personnel.

The principal question is whether such standards of good government are developed appropriately. Loschak agrees with Rivero that the function of the judge is not to follow the momentary opinions of politicians or administrators, but to give effect to traditional liberal principles of government of which the

---

[1] J. Rivero, 'Le juge administratif français: un juge qui gouverne', *D.* 1951 Chr. 21 at p. 24.
[2] R. Odent, *Contentieux administratif* (Paris, 1981), 746–7.

administrative courts are the guardians.[3] At the same time, she rightly notes that protecting citizens against the administration is only one aspect of the work of the administrative courts: 'for the judge is both too aware of the requirements of administrative action, and too imbued with "a sense of the State" to accept that the protection of individual interests should act to the detriment of the public interest'.[4] This is well illustrated by the recognition of continuity of the public service as a general principle of law (see p. 210). To that extent, the administrative judges do not really impose their own code of ethics on the administration, but 'like any body of people "situated" in a given social and political milieu, the judge merely translates into his case-law the dominant ideological options of that milieu'.[5] In other words, the principles of good administration which administrative law articulates are not idiosyncratic, but reflect the best elements of good practice recognized within the French governmental system. The closeness between administration and judge of which Odent speaks is seen as the foundation of the ability of the administrative courts to set standards which will be seen as appropriate and be implemented by the administration. Indeed, commentators remark on the curious way that the administrative judges, even where they do not formally enjoy the independence of their counterparts in the ordinary courts, actually have demonstrated more independence in the protection of the individual.[6]

Confidence in the administrative courts is shared by the administration and public alike. Members of the Conseil d'Etat are not only judges, they are also fully trained in the expertise of administration, and there is considerable movement of personnel between posts inside and outside the Palais-Royal. In its relations with the public, the Conseil d'Etat commands surprisingly full press-reporting of its activities[7] and enjoys considerable respect, at least comparable to that shown to the senior judiciary in the United Kingdom.

[3] D. Loschak, *Le Rôle politique du juge administratif français* (Paris, 1972), 296.
[4] Ibid. 237. [5] Ibid. 322.
[6] F. Dreyfus and J.-M. Galabert, in CERAP, *Le contrôle juridictionnel de l'administration, Bilan critique* (Paris, 1991), 41–2 and 144–5.
[7] But even in *Le Monde* one does not find the verbatim citation of complete decisions such as one can read in *The Times* or the *Independent* when important English cases are reported.

This confidence in the administrative expertise of the administrative courts has prevented the proliferation of British-style administrative tribunals: as new justiciable issues arise in the field of the administration, they pass quite naturally within the jurisdiction of the general administrative courts. But in Britain, we tend to create yet another specialized tribunal to which the citizen may appeal if he is not to be left largely at the mercy of the *ipse dixit* of the official (see, for example, the General and Special Commissioners of Income Tax, and, more recently, Immigration Adjudicators and Immigration Appeal Tribunals). Although the French have established the office of Médiateur as a redress for the citizen against the administration, this has not significantly affected the work of the administrative courts.

In 1979, President Giscard d'Estaing complimented the Conseil d'Etat on its achievement:

The essential contribution of the Conseil d'Etat is to have succeeded, without any intervention of the legislature, simply by virtue of a slow and judicious development of case-law to impose a spirit of legality on the French administration, that is to say, respect for the law, and then to imbue it progressively with this spirit.[8]

This ambition of setting standards of good administration is much the same as that of the English courts. As Lord Donaldson MR put it,

A specialist administrative or public law court is a post-war development. This development has created a new relationship between the courts and those who derive their authority from public law, one of partnership based on a common aim, namely the maintenance of the highest standards of public administration.[9]

He also accepted that public law judges would have and use an appreciation of the realities and difficulties of public administration in making their decisions in judicial review. But the approach remains different from that of France. The English approach is that review is exercised by the informed outsider, be it the judge or an ombudsman. The French approach is that

---

[8] Speech, 24 Nov. 1979, *Etudes et Documents du Conseil d'Etat*, No. 31 (1979–80) 277.
[9] *R. v. Lancashire CC, ex parte Huddleston* [1986] 2 All ER 941 at 945; also J. Laws, 'The Ghost in the Machine' [1989] PL 27.

review is conducted by the institutionally detached insider. Both are independent, but their backgrounds and starting points may differ.

To a significant extent, the close connection between the administration and its judges is not reflected as much in the lower administrative courts. Especially after the reforms of 1987, recruits have been drawn from outside the administration, and there is, in any case, less movement between the administrative courts and the active administration. The corps of members of the Tribunaux Administratifs and the Cours Administratives d'Appel do not enjoy the same high standing within the administration when compared to the élite corps of the Conseil d'Etat. In important respects, the lower administrative courts are more distinct from the rest of the administration and, in setting standards of good administration, act more like judges in Germany or England, who are very much outsiders in relation to the administration.

2 MERITS OF THE FRENCH SYSTEM

The French administrative courts, with the Conseil d'Etat at their head, would agree with President Giscard d'Estaing that they have been successful in subjecting the administration to the Rule of Law. Apart from the composition and functions of the Conseil d'Etat itself, this success may be attributed to a combination of the following factors:

1. The specially adapted character of the substantive law created and applied by the administrative courts
2. The flexible and principled character of their case-law and remedies
3. The special procedure they have evolved
4. The willingness to introduce administrative-law principles into the administrative process itself

1. As the quotation from Odent suggests, the French find a justification for the distinct character of their *droit administratif* in its capacity to adapt the principles of administrative legality and administrative liability to the differing needs of the various public services, a capacity which they claim could only be found in judges who are also trained administrators.

Yet the gap between *droit administratif* and *droit civil* must not be exaggerated. Even though there may be a formal distinction between the rules applied and the courts which apply them, difference is not sought for its own sake, and care is taken to ensure that valuable lessons are learned from the other side of the public–private divide. For example, in tort law, the principles of liability achieve very similar results, albeit sometimes through different concepts. There are, however, differences, such as on the scope of vicarious liability, where *droit administratif* is willing to consider more activites of the employee as lying within the scope of the employer's liability than *droit civil*. All the same, the distinct position of the administration in contract and in the exercise of its special powers does set it apart from the private citizen or company, a feature which is reinforced by the requirements of Community law. By contrast, the late Professor Mitchell underlined the backwardness of English law in its failure to recognize 'the special position of governmental agencies in their contracts'.[10]

In an era in which commercial concepts of efficiency, accountability, and consumerism dominate the vocabulary used to evaluate the performance of the administration, the need for a distinct set of administrative-law principles might seem under threat.[11] At the same time, the undoubted power of the administration to affect people's lives without their consent grows apace. A distinct administrative law, with its more rigorous scrutiny of the legality of administrative decisions and its more generous system of liability in situations where individuals suffer for the benefit of the public, provides a counterweight to this power. In developing principles in this way, French administrative law fully justifies its distinct existence.

The general right of appeal from the Tribunaux Administratifs either to a Cour Administrative d'Appel or to the Conseil d'Etat, together with the *recours en cassation*,[12] is seen by the French as a guarantee that the law will be uniformly applied throughout the entire system of administrative courts, whether of general or

---

[10] *The Contracts of Public Authorities* (London, 1954), 242.
[11] See J. Caillosse, 'La Modernisation de l'Etat', *AJDA* 1991, 755.
[12] Cassation may be sought against decisions of a Cour Administrative d'Appel as well as those of the special administrative jurisdictions listed in App. D: see further Ch. 3, p. 55.

special jurisdiction. In the British system of administrative tribunals, each with their specific jurisdiction (*une compétence d'attribution*, as the French would say), this guarantee of uniform justice is imperfectly met by such institutional devices as the Council on Tribunals and the presidential system for some important groups of tribunals: applications for judicial review or (where available) statutory appeals to the ordinary courts cannot ensure that tribunal decisions are uniform, only that they are lawful.

Although in France the man in the street looks rather to the civil courts to safeguard his traditional rights of property and personal liberty, as well as his more modern rights to social security, he also needs the administrative courts to police intelligently the complex administration of the modern state. It is true that (tax-payers apart) the biggest single category of citizens who resort to the administrative courts are civil servants, and this is itself a testimony, from those best competent to judge, to the efficacy of administrative justice. But, as we have seen, the Conseil d'Etat is also the court of last resort in disciplinary matters for such important public callings as those of the doctor, the dentist, the pharmacist, the architect, the accountant, and the teacher (at all levels). Moreover, members of the general public come into personal contact with the administrative courts in such everyday matters as income tax, planning (*urbanisme*), or negligence in the public services; and indirectly it is they who ultimately benefit from the protection given by the administrative judge to such of the *principes généraux du droit* as equality before the law, or the fundamental rights referred to in the Preamble to the Constitution, as they do also from his enforcement of proper standards of public administration. In one specialized sphere, that of local elections, the control exercised by the administrative courts has given them a significant influence on the political life of French democracy, and their influence here is all the greater because their case-law in this sphere has been adopted practically in its entirely by the Constitutional Council when supervising the conduct of parliamentary elections.

2. A major characteristic which distinguishes *droit administratif* from *droit civil* is that its general principles and framework have developed from case-law. Legislation has a significant role in specific areas, but, unlike in *droit civil* and *droit pénal*, there is no code of general principles and so it has been for the courts

CONCLUSIONS 277

to integrate the various elements into a coherent system. Unrestricted by a strict doctrine of *stare decisis* or the strait-jacket of a code, the Conseil d'Etat has proved itself agile in holding the scales evenly balanced between, on the one hand, the shifting needs of public administration in an era of rapid economic and social change and, on the other, the rights of the individual in a free society. Indeed, through the general principles of law, the Conseil d'Etat was in the vanguard of developing protections for fundamental freedoms well before the Constitutional Council took on that role in 1971.

The merit of the case-law of the Conseil d'Etat is that it has been principled and flexible. It has been principled in that the procedure and remedies have been relatively coherent and simple compared with the confusion of choice which used to bedevil the common law before the reforms of the Rules of the Supreme Court, Order 53, in 1977. The use of just three *recours*: the *recours pour excès de pouvoir* and the *recours en cassation* for issues of legality, and the *recours de pleine juridiction* where damages or fuller relief is sought, have proved to be very flexible over time. The general *recours en cassation* has enabled the Conseil d'Etat to exercise control over all inferior tribunals in a more coherent and systematic way than the High Court currently exercises over tribunals in England. The *recours pour excès de pouvoir* has shown itself remarkably adaptable to the changing organization and scope of administrative powers, enabling the administrative judge to demarcate the boundary between what is legality and what is policy, but without trespassing upon the latter. Ideas such as proportionality and *erreur manifeste* have been incorporated gradually and effectively into the armoury of the courts to cope with the need for greater scrutiny of administrative powers.

The French creation has an *elegantia iuris*, the study of which can yield much intellectual satisfaction. But it is also a practical product: the late Professor Hamson aptly contrasted the fine precision-tool that the Conseil d'Etat had at its command with the blunt instrument with which the English judge has to make do—although he also observed that blunt instruments, suitably applied, can be wondrously effective on occasion (as, for instance, the criminal prosecution of the offending official). Even recent developments in English law have not effaced this contrast.

Untrammelled by binding precedent, the Conseil has been able to reconsider its case-law in order to adapt to changing circumstances or to abandon old solutions which have revealed unforeseen problems. The restrictive approach to the economic activities of public authorities earlier this century has given way to a greater willingness to permit the administration to engage in economic ventures (see p. 128). More recently, a dualist attitude to the impact of treaties, especially in Community law, has given way to a more monist position (see p. 267). Such changes are typically foreshadowed by conclusions of Commissaires du gouvernement and are then made by higher judicial formations of the Conseil d'Etat. In this way the changes can be made coherently and decisively. Unlike the *cours d'appel* in civil and criminal law, the Tribunaux Administratifs have not resisted decisions of their supreme court, the Conseil d'Etat, very frequently and have been less agents for change in the case-law. But the Cours Administratives d'Appel, by the quality of their decisions, have already shown an ability to innovate.[13]

3. The administrative courts have evolved their own procedure, the special characteristics of which were examined in Chapter 5. Like the remedies just mentioned, the procedure is simple and highly effective. The process of *instruction* and the conclusions of the Commissaire du gouvernement provide the plaintiff with both an investigation into the facts and a survey of the law which have no counterpart in English proceedings for judicial review. The procedure is also remarkably inexpensive for the plaintiff, especially where, as in the *recours pour excès de pouvoir*, legal representation is not obligatory.

4. Judicial control is, of course, only one method of controlling administrative action. It has to be set alongside a vigilant public opinion (served by a free press and independent radio and television services), a watchful parliament, a self-disciplined civil service, and what we have termed 'judicialization' (see p. 55), that is the introduction within the administrative process of statutory tribunals and inquiries. Perhaps because in England these other controls function reasonably well, we have not evolved so effective a judicial control as the French. By contrast,

[13] See, for example, a number of decisions on the liability of the administration reached by the Lyons Cour Administrative d'Appel: *AJDA* 1991, 164–8, esp. CAA Lyon, 21 December 1990, GOMEZ.

in France the administrative courts are expected to bear a heavier share of the load, which is why the French are more conscious of what is an inevitable shortcoming of judicial control in any legal system, namely, that it can only operate after the event. As Professor Weil observed:

French administrative law often appears more aesthetically satisfying to the lawyer than to the ordinary citizen. This is because it comes into operation afterwards, and does not always make its presence felt in the *tête-à-tête* between the Administration and the 'assujettis'.[14]

As if to compensate for the *ex post facto* character of judicial control, the French legislature has always tended, over a wide area of administrative activity, to impose upon the administration the duty to consult the interested parties before the decision is taken, and the case-law of the administrative courts has shown a similar tendency. Sometimes, as in the legislation governing the grant of public assistance (*aide sociale*), formal tribunals are created in the British manner; sometimes 'commissions' are set up which are regarded as acting administratively, rather than judicially, as in the legislation on the regrouping ('remembrement') of scattered farmland; sometimes, the process of consultation remains wholly informal, in response to the widely held belief that too much judicialization might both unduly complicate public administration and curtail judicial control proper—the power of the Conseil d'Etat *en cassation* being less, we have seen, than when acting *pour excès de pouvoir*.

In so far as this legislative trend reflects the influence of the administrative sections of the Conseil d'Etat in the *travaux préparatoires* of legislation, it is a reminder that the guiding hand of these sections is as important a factor in French public life as the corrective jurisdiction of the Section du Contentieux.

Nevertheless, although the courts can effectively control the administration only after the event in any particular case, it must always be remembered that each successful *recours* has a preventive influence generally upon future administrative behaviour.

---

[14] See P. Weil, 'The Strength and Weakness of French Administrative Law' [1965] CLJ 242 at p. 257.

3 DEFECTS OF THE SYSTEM

In drawing up a balance sheet for French administrative law, it is customary to set prominently against its advantages the jurisdictional conflicts to which two separate court systems inevitably give rise. Undoubtedly, a price has to be paid for maintaining a specialized system of administrative courts: English legal history teaches the same lesson with the struggles between the courts of common law and of equity. But without Chancery there would have been no equity, and without the Conseil d'Etat, France might have had the kind of complicated machinery of judicial review that has been adapted to meet the needs of modern Britain,[15] helped out by a powerless (although not uninfluential) Council on Tribunals and an Ombudsman tied down by parliamentary shackles.[16] In France, the distinction between public law and private law remains a difficult area, in which many subtle distinctions are drawn. As a result, the Tribunal des Conflits is called upon to make about forty decisions a year. But among the volume of litigation before the administrative courts, this is a very small problem.[17] On the whole, the lines of demarcation are well known and understood. Only in relation to complex contractual or employment situations will a problem typically arise, and the basic principles for resolving them are well established. In any case, though entrenching the division between public-law and private-law courts as a fundamental principle recognized by the laws of the Republic, the decision of the Constitutional Council on the Competition Law (CC decision no. 86–224 DC of 23 January 1987) does offer encouragement for greater flexibility

---

[15] Just how complicated this is may be seen by reference to a basic student textbook such as D. Foulkes, *Administrative Law* (7th edn., London, 1990), where the author's careful exposition of judicial review in ch. 11 now requires 84 pages of closely written text.

[16] But the Council has aspired to a more important role, only to find its Special Report, *The Functions of the Council on Tribunals* (Cmnd. 7805; 1980) fall on largely deaf ears in government, and the Parliamentary Commissioner has been greatly reinforced by the Local Commissioners for Administration in England, Wales, and Scotland. See further, JUSTICE report, *Our Fettered Ombudsman* (1977); *The Local Government Ombudsmen, Annual Report 1990/91*.

[17] A leading practitioner states that the division between public-law and private-law courts arises in barely one per cent of all cases on which he has to advise before the Conseil d'Etat and the Cour de Cassation: Lyon-Caen, in CERAP, *Le contrôle juridictionnel*, 217.

in the allocation of issues between administrative and civil courts. Rational regrouping of similar issues before the same courts has been encouraged by the highest authority (see p. 124 or 131). Under the common influence of the Constitutional Council and the Tribunal des Conflits, the problems of the separation of courts, such as they are, will continue to remain marginal in the operation of the system. It is the technicality of the rules which govern the division of judicial competence, and the differences in the rules which are applied on either side of the divide, rather than the division of courts itself, which are the main source of complaint.[18] All the same, there are important voices arguing that a more common approach by judges on both sides of the divide would make control of the administration more effective and less complex for the litigant. This is notable in the case of labour law, where the protection for private-law and public-law employees can be significantly different, even where the employer is the same.[19]

The most significant defects in the current system of administrative justice in France were highlighted by the Minister of Justice, M. Henri Nallet, in his meeting with heads of the administrative courts on 19 February 1991. In his speech, he pointed to three main issues: (1) delay, (2) the failure of the administration to implement decisions of administrative courts, and (3) the need for alternative methods of dispute settlement.[20]

1. The most important problem facing the administrative court system is its very success in attracting litigation. In 1990, some 82,194 *recours* were lodged with the various administrative courts. Because of recent reforms, the Conseil d'Etat and the Cours Administratives d'Appel were able to give more judgments than the number of cases lodged, but these small inroads into the accumulated backlog of nearly 20,000 cases before those higher courts were dwarfed by the position in the Tribunaux Administratifs, where 69,853 new *recours* were lodged, but only 57,629 judgments given. The backlog before the Tribunaux Administratifs stood at 146,914 cases on 31 December 1990, nearly three years' work. Mass litigation threatens to swamp the administrative courts and to reduce their effectiveness by undue delay.

[18] See Weil, *Strength and Weakness*, 252; and, generally, CERAP, *Le Contrôle juridictionnel*.
[19] See J.-E. Ray and D. Cohen, CERAP, *Le Contrôle juridictionnel*, chs. 6 and 9.
[20] *Etudes et Documents du Conseil d'Etat*, No. 42 (1991), 246–7.

Moreover, as Professor Weil observes of such delay:[21] 'Time is never neutral and here it works almost always against the citizen.' An excessive time-lag is always bad. Yet it may help reduce the personal element in the conflict between judge and administrator: by the time the case is reached, the official responsible may, through promotion, transfer, or retirement, be no longer personally involved in defending his decision. Again, the longer the time-lag, the more bold may be the court of judicial review, since the political implications of the decision in question may no longer be important.

The reforms of 1987 went a considerable way to restoring the role of the Conseil d'Etat as a place for deliberation on major issues of law, rather than on routine matters of fact. Greater use of *ordonnances* of the president of a Sous-section and of the *procédure préalable d'admission* could help to reduce still further the number of cases requiring full deliberation from the Conseil d'Etat. The success of the Cours Administratives d'Appel in establishing a reputation for innovative legal thought and for the efficient dispatch of routine appeals which would previously have gone to the Conseil d'Etat has reinforced the quality of the system. But as was predictable,[22] the reforms have left untouched the serious problems in the Tribunaux Administratifs. The average time between a *recours* being lodged and judgment continues to increase and now stands at 2 years 6 months, and this is despite a 27 per cent increase in the number of cases decided per judge between 1987 and 1990. But there has been great difficulty in recruiting to these courts, and in 1990 they were still 20 per cent understaffed, though 14 per cent more new posts were created on 1 January 1991.

This is not to say that the Tribunaux Administratifs are always slow. Under the Law of 2 August 1989 on the administrative expulsion of foreigners (*la reconduite à la frontière*), the administrative courts must decide on an appeal against expulsion within 48 hours and currently do so within an average of 35 hours of the appeal's being lodged. Nevertheless, the vast majority of litigants must wait a long time for judgment, and, as has been noted on p. 120, the situation has occasionally reached such

---

[21] *Strength and weakness*, 257.
[22] See Brown and Bell, (1989) 8 CJQ, at pp. 81–2.

serious proportions that a breach of the European Convention on Human Rights has occurred simply because of the time taken to process a case.

2. Another serious worry is the frequency with which the administration refuses to implement decisions of the administrative courts. This is important not merely because of the injustice caused to individuals, but also as a reflection on the influence of the administrative courts on the administration. If the courts fail to impose themselves on a recalcitrant administration, then respect for their decisions will diminish.

As has been seen above, the Section du Rapport et des Etudes is charged in the final resort with ensuring that judicial decisions are implemented. In its 1990 report,[23] it notes that it received a mere 745 complaints of failure to implement decisions (of which 549 were admissible) compared with the 71,709 decisions of the administrative courts in that year, and it was able to resolve 414 of these in the year. All the same, it did have a backlog of a further 956 cases under examination at the end of the year. A Decree of 15 May 1990 enables the Section to delegate the task of ensuring the implementation of a decision to the local Tribunal Administratif or Cour Administrative d'Appel, which will speed up the process and reduce the backlog. The Section points to local authorities and public bodies associated with them as major culprits, notably in employment and planning matters. In the case of the state, the chief culprit is the Ministry of Education, against whom there was a backlog of 276 cases by the end of 1990.

Very many decisions of administrative courts result in an obligation on the part of the administration to pay money. The Law of 16 July 1980 introducing the *astreinte* also introduced a procedure whereby a judgment could be turned into an executory administrative order to pay (see p. 113). In consequence, there should be no problem in enforcing such money judgments against the administration. But there is evidence that this Law is very rarely invoked, especially against the state. In the first ten years of its existence, only 6 per cent of finance officers had encountered the Law in relation to monies owed by the state, and only 14 per cent with regard to local authorities.[24]

---

[23] Etudes et Documents, No. 42 (1990), 127–44.
[24] J.-J. François, in CERAP, Le contrôle juridictionnel, 131–2.

Where a judicial decision does not require the payment of money, but the quashing of a decision or some form of specific enforcement, then problems of implementation are more frequent. Most cases are not of refusal to implement a decision, but of delay in doing so. Sometimes this may be explained by practical difficulties. For example, how can someone be reinstated when the job is being performed by someone else, or when there are not enough credits to pay for additional employees? Again, bureaucratic procedures for giving effect to the decision may not be speedy. But these cannot be the whole story. Government circulars encouraging compliance, and proposals to increase the disciplinary sanctions which may be imposed in such cases, only touch the symptoms of the problem. The cure would seem to lie in the administrative culture. It remains to be seen whether the Conseil d'Etat is right that more legal training among administrators would improve the situation.[25]

3. The third problem is that of preventing litigation arising in the first place, or introducing a conciliation procedure before litigation or arbitration is commenced. Pre-litigation procedures have been successful in reducing the case-load in tax matters (see p. 27). Unfortunately, in other areas, Decrees of 4 December 1980 on non-contractual liability and of 18 March 1981 on public contracts, which created the option of conciliation by the courts, have been unsuccessful. In the past, the climate of opinion among administrative judges has not been favorable to conciliation.[26]

Article 22 of the Law of 6 January 1986 made it clear that conciliation is one of the functions of the Tribunal Administratif, and the Conseil d'Etat has endorsed the refusal of such a court to proceed with judging a case while conciliation procedures are in progress (CE 23 June 1989, VERITER). The emphasis in the Law of 31 December 1987 was to a similar effect. The provisions of the latter Law have been implemented by a Decree of 25 February 1991 creating a national committee for conciliation in relation to public contracts made by the state and administrative public bodies, as well as local consultative committees for contracts of other state bodies. Given the will to make them work, such procedures could have an important impact on the case-load.

[25] See J. Bell, 'French Judicial Overload' [1987] PL 175 at p. 177, *Etudes et Documents*, No. 42 (1991), 130.
[26] See M. Lévy, 'La conciliation par le tribunal administratif et le rôle du juge dans l'instruction des litiges', *AJDA* 1987, 499 at pp. 503–5.

Apart from these institutional defects, a number of criticisms may be made of the substantive administrative law, especially in terms of procedure. A current member of the Conseil d'Etat and former judge of the European Court of Justice, Yves Galmot, has drawn attention to three respects in which French administrative law compares unfavourably, in his view, with the approach of the European Court of Justice.[27]

He notes first that, when a decision is annulled in France, it is a total nullity and the nullity operates retrospectively. By contrast, the European Court sometimes uses the technique of limited or prospective nullity, in order to protect both citizens and the administration, who have acted in good faith, relying on its validity. Secondly, he notes the great reluctance of the French courts to issue orders to the administration, a reluctance which is not shared by the European Court, which regularly issues orders to the Commission and to member states. Thirdly, he notes that, when faced with a *recours* relating to the legality of an administrative decision, the French courts will limit themselves to that issue and leave ancillary matters, such as compensation, to further litigation. By contrast, the European Court is more willing to deal there and then with all the issues which arise in relation to the wrongful decision of the administration.

A FINAL ASSESSMENT

Galmot's comparison of French administrative law with the administrative law of the European Community is a timely reminder that legal systems are no longer self-contained. European administrative law systems will come increasingly under the common influence of the European Court of Justice in Luxembourg and the European Court of Human Rights in Strasbourg. Indeed, through the notion of 'general principles of law which underlie the constitutional traditions of member states', principles developed in Luxembourg are coming to include values recognized in the European Convention on Human Rights.[28]

---

[27] CERAP, Le contrôle juridictionnel, 236–8.
[28] See *Johnston* v. *Chief Constable of the Royal Ulster Constabulary* [1986] 3 All ER 135 and, more generally, Y. Galmot, 'Réfléxions sur le recours au droit comparé par la Cour de Justice des Communautés européennes', *RFDA* 1990, 255 at pp. 258–9.

Such a process of influence by osmosis will encourage all legal systems to match the best features in the European legal orders or in the national laws of other member states. Whether this will create a general European public law is a matter for speculation.[29] Among the best features of French law, Galmot lists the wide range of administrative decisions which the courts can review, the simple and coherent system of administrative liability, the concepts which organize the law on contracts, and above all the subtle and nuanced system of principles governing judicial review of administrative discretion.[30] Indeed, in all these respects, the *droit administratif* has appeared stronger than our administrative law, and the Conseil d'Etat has long shown its capacity to police intelligently the complex administration of the modern state. Alas, Galmot sees little to be learned from Anglo-Saxon systems of administrative justice.[31] Here he neglects the remarkable advances in Britain in recent years, relating not only to the substantive law and procedure of judicial review, but also to the structure and composition of the court chiefly concerned with administrative justice—in England, the Queen's Bench Division of the High Court, and the Court of Session in Scotland[32]. In both courts, a specialist list of judges now sits and this has produced the specialist public law court to which Lord Donaldson alluded. The genius of those who shaped the *droit administratif* was to have anticipated this development in France by nearly two centuries. Despite these developments, there is undoubtedly a need for a more principled structure and a greater simplicity in the system of administrative law in the United Kingdom. We may well have a distinct system of administrative law which is effective, but it has yet to match the certainty and clarity which more established systems like the French exhibit.

---

[29] See T. Koopmans, 'European Public Law: Reality and Prospects' [1991] PL 53; F. Jacobs, 'The Principle of Legality—towards a European Standard', in G. Hand and J. McBride, *Droit sans frontières: Essays in Honour of L. Neville Brown* (Birmingham, 1991), 235.

[30] CERAP, *Le Contrôle juridictionnel*, See also J. Schwarze, *European Administrative Law* (London, 1992), an encyclopaedic guide in 1,547 pages, published jointly by Sweet and Maxwell Ltd. and the Office for Official Publications of the EC.

[31] Ibid. 231.

[32] See L. Blom-Cooper, 'The New Face of Judicial Review' [1982] PL 250; A. W. Bradley, 'Applications for Judicial Review: The Scottish Model' [1987] PL 313.

Appendix A

The Division of Jurisdiction Between the Ordinary Administrative Courts

* *Renvoi*: 'referral of a new point of law presenting a serious difficulty and arising in a number of cases' (Art. 12)

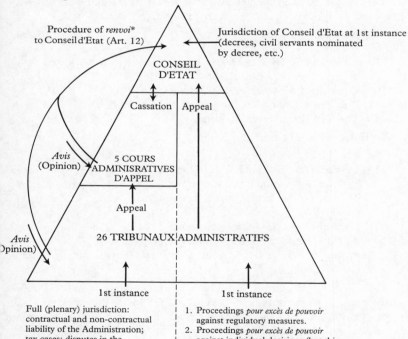

Appendix B

The Membership of the Conseil d'Etat

The official *Annuaire du Conseil d'Etat 1990* lists the active membership (*membres en activité*) on 1 February 1990 as:

  28 Auditeurs
  63 Maîtres des requêtes
  12 Conseillers d'Etat *en service extraordinaire* (who work exclusively in the administrative sections)
100 Conseillers d'Etat *en service ordinaire*
   7 Présidents de Section (one being supernumerary)
     The Vice-Président (M. Long)

To this total of 211 must be added:

52 *détachés* (18 Conseillers et 34 Maîtres des requêtes)
30 members *en disponibilité* (e.g. on compassionate leave or other absence for personal reasons ('pour convenances personnelles'). This group is made up of 4 Conseillers and 26 Maîtres des requêtes)
 1 Conseiller *en délégation* (attached to le Ministre de l'Equipement, du Logement, des Transports et de la Mer)
 3 members *hors cadre* (e.g. as Secretary-General of the Banque Crédit-Lyonnais)

Thus without including 13 senior *fonctionnaires* attached to the Conseil, the overall membership is 297, but of those, 86 are not working within the Conseil, and of the active members a few are only in part-time service with the Conseil.

Appendix C

The Organization of the Conseil d'Etat

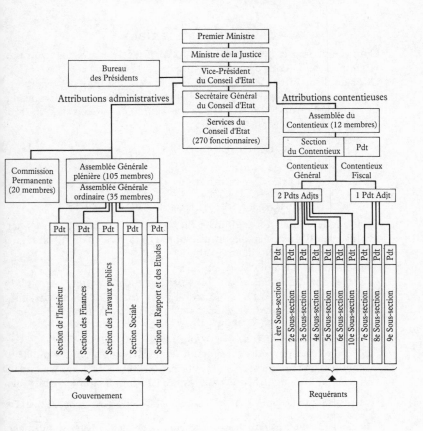

Appendix D

The Principal Specialized Administrative Jurisdictions [1,2]

A. PUBLIC FINANCE TRIBUNALS

* Cour des Comptes (which since 1982 also receives appeals from the Chambres régionales des comptes; see pp. 34–5.
* Cour de discipline budgétaire et financière (which may impose monetary penalties on officials guilty of financial mismanagement causing loss to public funds or of failure to execute court decisions, p. 115 above).

B. DISCIPLINE OVER JUDGES

* Conseil supérieur de la Magistrature (*statuant en matière disciplinaire*, under Art. 65 of the Constitution of 1958).

C. TRIBUNALS RELATING TO EDUCATION

Conseils de l'éducation nationale (at level of *département*), which act as tribunals of first instance in disciplinary matters affecting teachers in state schools, with appeal to the
* Conseil supérieur de l'éducation nationale.

Conseils d'administration des universités (and of equivalent higher educational establishments) exercising in their 'Sections disciplinaires' jurisdiction over academic staff and students in higher education, with appeal (since 1989) to the
* CNESER (Art. 23 of Law of 10 July 1989).
* Juridiction disciplinaire des personnels enseignants et hospitaliers des centres hospitaliers et universitaires (CHU). (The equivalent of 'clinical' staff in British university medical schools come within this

---

[1] Those subject to control of the Conseil d'Etat by way of cassation are marked with an asterisk.

[2] For an authoritative and exhaustive account, see Colette Même, 'Juridictions administratives spécialisées', *Encyclopédie Dalloz; Répertoire de contentieux administratif*, ii (1985).

separate disciplinary system because of their dual responsibility to the medical schools and the hospitals where they work.)

D. TRIBUNALS RELATING TO THE PROFESSIONS

The French distinguish those independent (or 'liberal') professions with a code of conduct and compulsory membership of a self-regulatory body or 'Order' at national level exercising discipline over their members: the disciplinary tribunals of such professions are styled *juridictions ordinales*. Other professions (e.g. nursing, architecture, banking) where members are not in a majority on their disciplinary tribunals are classified separately. Similarly, the social insurance tribunals which discipline doctors, midwives, and paramedicals for breaches of service to the public are separately classified because the profession does not form a majority of their members.

Juridictions Ordinales include:

Conseils régionaux de l'ordre des médecins, acting as disciplinary tribunals of first instance, with appeal to the
* Conseil national de l'ordre des médecins (section disciplinaire).

Conseils régionaux de l'ordre des chirurgiens-dentistes, with appeal to the
* Conseil national de l'ordre des chirurgiens-dentistes (section disciplinaire).

Conseils régionaux et conseils centraux de l'ordre des pharmaciens, with appeal to the
* Conseil national de l'ordre des pharmaciens (chambre de discipline).

Conseils régionaux de l'ordre des vétérinaires, with appeal to the
* Conseil supérieur de l'ordre des vétérinaires (chambre de discipline).

Chambres régionales de l'ordre des experts-comptables, with appeal to the
* Chambre nationale de l'ordre des experts-comptables.

Conseils Régionaux de l'ordre des géometres-experts, with appeal to the
* Conseil supérieur de l'ordre des géometres-experts.

Other Professional Tribunals include:

Architects:
Chambres régionales, with appeal to the
* Chambre nationale de discipline des architectes.

Company auditors:
Chambres régionales, with appeal to the
* Chambre nationale de discipline des commissaires aux comptes de société.

Doctors, midwives, and paramedicals treating patients under the social insurance system: complaints are heard at first instance by the

Sections des assurances sociales des conseils régionaux de discipline de l'ordre des médecins (compare in Britain the Service Committees of Family Health Service Authorities), with appeal to
* Section des assurances sociales du Conseil national de l'ordre des médecins.

Similarly, appeals on social insurance complaints respecting dentists or pharmacists lie from regional disciplinary tribunals to the
* Section des assurances sociales du Conseil national de l'ordre des chirurgiens-dentistes, or the
* Section des assurances sociales du Conseil national des pharmaciens.

Nurses:
Commissions régionales, with appeal to the
* Commission nationale de discipline des infirmiers et infirmières.

Bankers:
* Commission bancaire (a statutory body replacing, in 1984, the Commission de contrôle des banques set up in 1941 but with similar powers to supervise banking law and practice and to impose penalties upon banks in breach).

Journalists:
* Commission supérieure de la carte d'identité des journalistes (when acting as a disciplinary body to withdraw a journalist's professional identity card).

E. SOCIAL WELFARE

* Commission centrale d'Aide Sociale (the final appellate jurisdiction from Commissions départementales in complaints concerning what in Britain has been successively, and euphemistically, called poor relief, public assistance, national assistance, supplementary benefit, and, now, income support).
* Section permanente du Conseil supérieur d'Aide Sociale (the final appellate jurisdiction from Commissions régionales in complaints concerning hospital and home-help charges; but, where the amount in dispute is below a certain level, the Commission régionale is the tribunal of both first and last instance, subject only to a *recours en cassation devant le Conseil d'Etat*)
* Commissions départementales de contrôle de l'emploi obligatoire des mutilés de guerre (the seats once reserved for these victims of war will be remembered by older travellers on the Paris Metro).
* Commissions départementales des handicapés.

F. MILITARY PENSIONS

Commission spéciale de cassation des pensions militaires: this receives *recours en cassation* against decisions of the

* Tribunaux départementaux and Cours régionales (tribunals concerned with military invalidity pensions). The Commission spéciale is a jurisdiction closely linked to the Conseil d'Etat by its composition—it includes several senior members of the Conseil—as well as by its power to refer the most difficult and important pension cases to the Conseil d'Etat for decision.

G. REFUGEES

* Commission de recours des réfugiés, for aliens or stateless persons seeking to obtain under French law the status of refugee on account of political views, racial or religious persecution, etc.; this tribunal hears at first and last instance appeals against adverse decisions of the director of the French Office for Protection of Refugees. It has an ever-increasing case-load: on 1st January 1991, 16,589 cases were pending before it.

H. PRIZE COURT

* Conseil des prises. This jurisdiction concerning ships seized as prizes of war dates back to legislation of l'an VIII but last sat over 25 years ago: appeal lies to the Conseil d'Etat but the doctrine of *la justice retenue* survives in this instance, so that the case, after *instruction* by an administration section (not the Section du Contentieux), comes before the Assemblée Générale, which then proposes a draft decree to the President of the Republic (for *la justice retenue*, see p. 45).

I. COMPENSATION FOR FRENCH CITIZENS

dispossessed of their property in a territory previously under French sovereignty: twelve Commissions du contentieux de l'indemnisation decide compensation claims arising in different territories, with a right of appeal before 1988 to the Conseil d'Etat, but now to the Cours Administratives d'Appel under Art. 1 of the Law of 31 December 1987 (with review on cassation by the Conseil d'Etat in the normal way).

J. CONSCIENTIOUS OBJECTORS

* Commission juridictionnelle des objecteurs de conscience. This special tribunal, set up in 1963, was abolished in 1983 when jurisdiction to determine claims for the status of conscientious objector was transferred to the Tribunaux Administratifs. But to determine in what way those young persons serving prison sentences are to fulfil their national-service obligations, Art. L.51 of the Code du service national has established a special tribunal, the
* Commission statuant sur le cas de certains jeunes devant effectuer leur service national.

APPENDIX D

Statistics of Selected Administrative Jurisdictions Listed Above

Name of administrative jurisdiction	Cases decided			Appeals on cassation to Conseil d'Etat 1985
	1973	1981	1990	

DISCIPLINE OF PROFESSIONS

Juridictions ordinales

Conseil national de l'ordre des médecins (section disciplinaire)	43	148	206	18
Conseil national de l'ordre des chirurgiens-dentistes (section disciplinaire)	7	52	51	12
Conseil national de l'ordre des pharmaciens (chambre de discipline)	27	27	51	3
Chambre nationale de l'ordre des experts-comptables	–	11	1	5
Conseil supérieur de l'ordre des géometres-experts	8	13	14	2

Other disciplinary bodies

Chambre nationale de discipline des architectes	4	3	5	1
Chambre nationale de discipline des commissaires aux comptes	1	3	19	0
Section des assurances sociales du Conseil national de l'ordre des médecins	9	60	138	2
Section des assurances sociales du Conseil national des pharmaciens	27	27	6	0
Section des assurances sociales du Conseil national de l'ordre des chirurgiens-dentistes	3	25	38	1

SOCIAL MATTERS

Commission centrale d'Aide Sociale	3030	2955	3203	4
Section permanente du Conseil supérieur d'Aide Sociale	70	65	–	0

APPENDIX D

Statistics of Selected Administrative Jurisdictions Listed Above

Name of administrative jurisdiction	Cases decided			Appeals on cassation to Conseil d'Etat 1985
	1973	1981	1990	
OTHERS				
Commission de recours des réfugiés	224	2633	51,585	308
Commission statuant sur le cas de certains jeunes devant effectuer leur service national	57	807	1,552	2

Appendix E

Statistics of Cases Decided by the Conseil d'Etat Statuant au Contentieux

	1979–1980[a]	1990[a]
GENERAL ANALYSIS (numbers of cases)		
Contentieux général	3933	5833
Pensions[b]	231	162
Elections[c]	161	490
Tax cases[d]	875	1149
TOTAL	5200	7634
ANALYSIS BY FORMATION OF THE CONSEIL D'ETAT (numbers of cases)		
Assemblée Plénière	52	25
Section du Contentieux	97	57
Sous-sections		
Single sous-section	150	2618
Sous-sections réunies[e]	4345	3044
Ordonnances of president of the Section du Contentieux	556	1316
Ordonnances of presidents of Sous-Sections	–	1249
Commission d'admission des pouvoirs en cassation	–	592
ANALYSIS BY JURISDICTION EXERCISED		
Cases judged at first (and last) instance (*recours directs*)	1128	1317
Appeals (from Tribunaux Administratifs)	2965	3765
Cassation (of Cours Administratives d'Appel)	–	614
Other Cassation	192	932
Interprétation, révision et rectification[f]	46	25

	1979–1980[a]	1990[a]
ANALYSIS BY FORM OF RECOURS (%)		
Recours directs	26.04	17
Appels	68.45	45
Cassation	4.43	20
Others (*recours en interprétation* etc.)	0.29	17
TOTAL	100.00	100
RESULTS OF CASES (OTHER THAN TAX CASES) (%)		
Dismissed	41.44	55
Complete annulment	21.56	17
Partial annulment	10.36	6
Non-suits and discontinued appeals	26.64	22
TOTAL	100.00	100
ANALYSIS OF CASES BY SUBJECT-MATTER[g]		
Fonction publique (Civil Service) Disputes concerning the status, career, and conditions of service generally of *fonctionnaires*	703	923
Travaux publics (public works) Especially compensation for such temporary occupation of land as is allowed for the purpose of public works and for any permanent damage to property resulting from public works	266	162
Expropriation et urbanisme (compulsory purchase and town planning). These heads do not refer to compulsory purchase and planning inquiries of the English kind but to the review of irregularities in the administrative phase of procedures such as *expropriation pour cause d'utilité publique*[h]	378	585
Victimes de guerre right to title and privileges of 'Member of the Resistance', 'déporté', etc., under special legislation	15	28
Marchés et contrats (contracts for public works and supplies)	70	164
Agriculture: mostly, but not exclusively, the compulsory exchange (*remembrement*) of agricultural holdings	243	353
Professions[i]	97	192
Police Cases brought agains the police, including claims for damages in tort.[j]	294	71

	1979–1980[a]	1990[a]
Contributions:		
tax cases, especially income tax and corporation tax	800	1,321
Responsabilité	–	163
Health:		
including cases involving public hospitals	246	29
Étrangers	–	954

[a] The statistics for these two periods have been included for purposes of comparison, where available.

[b] i.e. pensions payable to *fonctionnaires* upon retirement etc., but not retirement pensions arising under Social Security.

[c] i.e. *recours électoraux* in respect of elections to all public office other than parliamentary elections; the incidence of local elections causes much variation in number from year to year.

[d] i.e. *recours fiscaux* in respect of *contributions directes*.

[e] Including *plénière fiscale*.

[f] i.e. limited appeals against previous judgments of the Conseil d'État or applications for their interpretation: see p. 116.

[g] Only the more important categories, comparatively or numerically, are listed.

[h] See Rendel, 'How the Conseil d'Etat Supervises Local Authorities, [1966] PL 213.

[i] For the governing bodies of professions subject to the Conseil d'Etat by way of cassation See App. D.

[j] See Ch. 8.

General Notes

1 The Special Cassation Commission for Military Pensions (see App. D for its link with the Conseil d'Etat) decided a further 1169 cases.

2 During the judicial year 1990 a total of 8070 cases was entered in the Conseil d'Etat Registry; for further details see accompanying chart.

3 During the same judicial year the local Tribunaux Administratifs decided 58,302 cases; the number of cases pending on 31 December 1990 was 146,914.

4 Discrepancies in the totals of some of the respective tables are due to deliberate omissions, in the interest of simplicity, of a few minor categories.

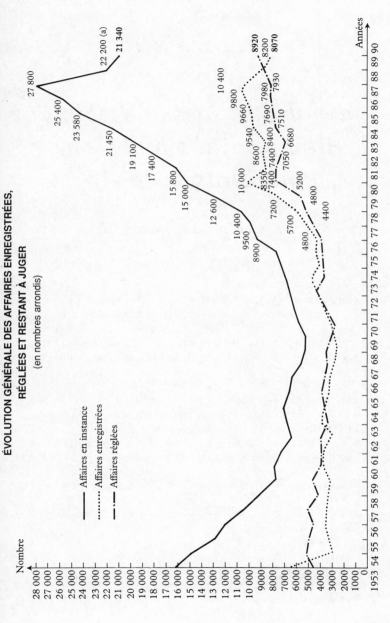

(a) La baisse qui apparaît en 1989 est la conséquence de la réforme du contentieux.

Appendix F

Judicial Statistics of Tribunaux Administratifs and Cours Administratives d'Appel

I. JUDICIAL STATISTICS: TRIBUNAUX ADMINISTRATIFS, YEAR 1990

Summary

New cases registered: 69,853
Cases disposed of: 58,302 cases were either decided or otherwise concluded by discontinuance, non-suit or referral
Cases pending at 31 December 1990: 146,914 (it has been estimated that in theory it would take 30 months to clear this backlog)

Analysis of Cases Disposed of by Categories

The number of cases disposed of during the year 1990 was 58,302, i.e. 83 per cent of the number of cases registered during the same period. The 58,302 disposals fell into the following categories:

- 11,478 désistements et non-lieu (cases discontinued and non-suits)
- 3,950 sursis à exécution (stays of execution): see p. 117 above.
- 3,329 référés (interlocutory orders): see p. 116 above.
- 1,400 renvois (referrals to other courts, e.g. for preliminary rul-ings): see p. 119 above.
- 1,287 constats d'urgence (decisions on matters of urgency)
- 36,859 other categories of decision

Classified by subject-matter (%):

Tax cases	22.6
Fonctionnaires et agents publics	14.4
Urbanisme et aménagement du territoire	10.6
Polices autres que municipales	9.2

Distribution of Cases Disposed of Between the Tribunaux Administratifs
in Metropolitan France and the Overseas Departments and Territories
(indicating cases in series[*] and comparisons with the year 1989)

JURIDICTIONS	BRUT	SERIES Nombre de Séries	SERIES Nombre d'affaires composant les séries	NET	Affaires traitées en 1989	VARIATIONS En nombre	VARIATIONS En pourcentage
AMIENS	1405	0	0	1405	1162	243	20.91%
BASTIA	683	non	précisé	683	512	171	33.40%
BESANCON	967	non	précisé	967	925	42	4.54%
BORDEAUX	1546	0	0	1546	1979	−433	−21.88%
CAEN	1451	0		1451	1447	4	0.28%
CHALONS/MARNE	1131	1	468	664	996	135	13.55%
CLERMONT-FD	1459	1	21	1439	1488	−29	−1.95%
DIJON	825	1	19	807	1088	−263	−24.17%
GRENOBLE	3190	8	1042	2156	2302	888	38.58%
LILLE	2668	3	310	2361	2377	291	12.24%
LIMOGES	770	3	30	743	905	−135	−14.92%
LYON	2627	7	182	2452	2751	−124	−4.51%
MARSEILLE	4376	non	précisé	4376	4301	75	1.74%
MONTPELLIER	2388	3	149	2242	2371	17	0.72%
NANCY	1354	non	précisé	1354	1322	32	2.42%
NANTES	2707	1	35	2673	2629	78	2.97%
NICE	2858	2	240	2620	2423	435	17.95%
ORLEANS	2004	4	159	1849	1878	126	6.71%
PARIS	8373	non	précisé	8373	9100	−727	−7.99%
PAU	707	1	36	672	1108	−401	−36.19%
POITIERS	1716	8	101	1623	1602	114	7.12%
RENNES	1727	non	précisé	1727	1940	−213	−10.98%
ROUEN	1268	non	précisé	1268	1464	−196	−13.39%
STRASBOURG	2065	non	précisé	2065	1890	175	9.26%
TOULOUSE	1807	2	44	1765	1635	172	10.52%
VERSAILLES	3854	non	précisé	3854	3895	−41	−1.05%
TOTAL METROPOLE	55926	45	2836	53135	55490	436	0.79%
BASSE TERRE	281	0		281	278	3	1.08%
CAYENNE	111	0		111	129	−18	−13.95%
FORT DE FRANCE	338	0		338	391	−53	−13.55%
ST PIERRE ET MIQUELON				0 0	17	−17	−100.00%
REUNION	764	1	64	701	740	24	3.24%
NOUMEA	369	1	69	301	235	134	57.02%
PAPEETE	513	2	201	314	620	−107	−17.26%
TOTAL DOM-TOM	2376	4	334	2046	2410	−34	−1.41%
TOTAL GENERAL	58302	49	3170	55181	57900	402	0.69%

[*] 'Series': this term indicates series or blocks of cases, all of which raise the same point of law or fact and can therefore be disposed of by one decision.

II. JUDICIAL STATISTICS: COURS ADMINISTRATIVES D'APPEL, YEAR 1990

Summary

	Paris	Lyon	Bordeaux	Nantes	Nancy	Total
New cases registered	1129	977	774	687	704	4271
Orders for 'renvoi' to the Conseil d'Etat	196	126	80	47	141	590
Cases disposed of	1560	970	792	597	746	4665
Cases pending at 31 December 1990 in all five CAA:						7229

Appendix G

Judicial Statistics: Tribunal des Conflits 1980*, 1990, and 1991

	1980	1990	1991
Total of cases registered	35	35	45
Total of cases decided	31	44	41
Analysis of cases by category			
Positive conflicts (p.146 above)	3	10	17
Negative conflicts (p.147 above)	2	0	1
Conflict of decisions (p.148 above)	1	0	1
Preliminary rulings:			
To forestall negative conflict (see p. 149 above)	20	30	18
Issue of 'serious difficulty' (see p. 149 above):-			
requested by Conseil d'Etat	2	1	4
requested by Cour de Cassation	3	3	0

* Year 1980 included for purpose of comparison.

Appendix H

An Arrêt of the Conseil d'Etat (with conclusions)

CONCLUSIONS OF THE COMMISSAIRE DU GOUVERNEMENT IN THE CASE OF DEMOISELLE MOSCONI

64,732.—*Demoiselle Mosconi*

I. La commune de Conca, dans l'arrondissement de Sartène (Corse), compte un peu plus d'un millier d'habitants et sans doute n'est-elle pas très riche. Elle s'est pourtant offerte le luxe, il y a une dizaine d'années, de recruter deux secrétaires de mairie, le sieur Mannarini et la demoiselle Mosconi; il est vrai que ces deux agents n'étaient employés qu'à temps partiel et qu'en outre ils étaient mal payés, puisqu'ils percevaient une rémunération annuelle de 150.000 francs (anciens) qui, même compte tenu d'une prime de 15 francs par habitant, ne pouvait satisfaire que des ambitions limitées.

Il semble, sans qu'on puisse l'affirmer à coup sûr, que la composition donnée en 1956 au secrétariat de la mairie de Conca répondit à de savants dosages familiaux et politiques. La demoiselle Mosconi, en tous cas, est la nièce d'une personnalité locale, qui appartenait alors au conseil municipal et comptait parmi les amis du maire; mais les événements de mai 1958 divisèrent profondément la population de Conca et, tandis que le maire, un sieur Leccia, restait fidèle aux institutions, son ami, le conseiller Mosconi, embrassait la cause des émeutiers.

Soit qu'il ne lui pardonnât pas cette attitude, soit qu'il le soupçonnât d'avoir voulu profiter des circonstances pour s'emparer lui-même de la mairie, le sieur Leccia rompit avec le sieur Mosconi à l'approche du renouvellement de 1959; mais il fit plus: il donna l'ordre à son adjoint de tenir la demoiselle Mosconi à l'écart jusqu'à la date des élections. L'adjoint s'exécuta et, par une décision du 15 décembre 1958, il plaça la demoiselle Mosconi dans la position de congé jusqu'au 15 mars 1959; il spécifiait toutefois, dans ladite décision, qu'il ignorait la raison pour laquelle le maire avait pris ce parti, et il ajoutait: 'personnellement, je n'ai pas de reproche à faire à la secrétaire dans son travail'.

Les élections eurent lieu le 8 mars 1959. Le maire fut réélu, mais il avait chassé de sa liste l'oncle et protecteur de la demoiselle Mosconi. Quant à la demoiselle Mosconi elle-même, elle réintégra la mairie à l'expiration de son congé de trois mois, le 16 mars, mais elle ne devait pas garder ses fonctions très longtemps. Le jour même, en effet, le maire lui joua un tour de sa façon; il la convoqua dans son cabinet et lui dicta une lettre au sous-préfet de Sartène: cette lettre est au dossier, et le fait est qu'on y trouve quelques fautes d'orthographe qui, s'il faut en croire la municipalité de Conca, eurent pour effet d'indigner le sieur Leccia. En tous cas, le sieur Leccia mit fin, séance tenante, aux fonctions de la demoiselle Mosconi.

Plus précisément, la demoiselle Mosconi fut démise de ses fonctions par un arrêté du 16 mars 1959, avec un préavis d'un mois, et—par égard sans doute à d'anciennes traditions gênoises—le maire crut devoir motiver explicitement sa décision: 'Considérant', dit l'arrêté du 16 mars 1959:

> que la demoiselle Mosconi (Rose Marie), secrétaire de la mairie de Conca, en fonctions depuis le 1er juin 1956, ne remplit plus, en ce moment, les conditions requises pour exercer lesdites fonctions d'une façon convenable; considérant en effet que si elle a pu, jusqu'ici, accomplir son travail avec le concours de l'adjoint au maire, il n'est plus possible aujourd'hui de lui adjoindre une aide quelconque pour remplir des fonctions dont elle doit assurer l'entière responsabilité; considérant qu'à cette inaptitude s'adjoint une mauvaise volonté dans son travail, de nature à rendre intolérables les rapports entre elle et le maire...

Autrement dit, c'est l'adjoint au maire qui, pendant deux ou trois ans, a fait ou corrigé le travail de sa secrétaire et, le nouvel adjoint n'étant pas disposé à se montrer aussi bienveillant, le sieur Leccia est tout naturellement conduit à se passer des services de la demoiselle Mosconi.

Telle est, du moins, la manière dont il convient d'interpréter les motifs de l'arrêté du 16 mars 1959, si l'on tient à leur donner un sens.

Quoi qu'il en soit, la demoiselle Mosconi saisit le tribunal administratif de Nice d'une demande tendant à l'annulation de l'arrêté du 16 mars. Cette demande, introduite le 2 mai, était accompagnée de conclusions tendant à ce que la commune de Conca fût condamnée à payer à la demoiselle Mosconi une indemnité de 2.500.000 francs, l'équivalent d'environ 15 ans de traitement. La commune répondit au fond sur la demande d'indemnité, et le contentieux se trouva lié sur l'un et sur l'autre points.

Par un jugement du 10 juin 1964, le tribunal administratif annula l'arrêté du 16 mars 1959 pour violation des droits de la défense et condamna la commune à payer à la demoiselle Mosconi une indemnité de 2.000 francs (nouveaux). La demoiselle Mosconi a fait appel de ce jugement en tant que, par ledit jugement, le tribunal lui a alloué une indemnité qu'elle estime insuffisante. La commune de Conca, de son côté, a présenté un recours incident tendant à la suppression et, subsidiairement, à la réduction de l'indemnité mise à sa charge par les premiers juges.

II. La requérante expose d'une part que la faute commise par le maire oblige la commune à réparer l'intégralité du dommage qu'elle a subi du fait de son éviction. Elle soutient, d'autre part, que ce dommage est considérable, compte tenu notamment de l'atteinte portée à sa réputation par les circonstances dans lesquelles le maire l'a privée de son emploi.

a. Sur le premier point, c'est-à-dire en ce qui concerne le principe et l'étendue de la responsabilité, il est exact qu'une décision illégale présente en elle-même un caractère fautif et que, par suite, elle engage la responsabilité de la puissance publique.

Toutefois, ainsi qu'on l'observe pour la commune, votre jurisprudence n'a jamais admis qu'un vice quelconque d'une décision oblige automatiquement la collectivité responsable à réparer l'intégralité du dommage: c'est là, sans doute, la solution qui prévaut lorsque la décision ne pouvait être légalement prise; mais, en règle générale et, notamment, dans le cas d'un vice de forme ou de procédure, l'intéressé ne peut prétendre qu'à une réparation partielle et il arrive même parfois que toute réparation lui soit refusée.

Nous ne discuterons pas ce système, qui dans l'ensemble est équitable et qui, d'ailleurs, a pour lui de très nombreux arrêts; mais, si la jurisprudence est excellente dans son principe, elle n'est pas toujours heureuse dans sa mise en œuvre. Ainsi, c'est une erreur, à notre sens, de mesurer l'étendue de la réparation à la gravité de la faute, d'abord parce qu'il est difficile d'affirmer que tel vice est plus ou moins grave que tel autre et, surtout, parce qu'il est de règle, dans la théorie de la responsabilité, que la faute commise par l'auteur du dommage l'oblige à réparer intégralement les conséquences de son acte. En ce sens, toute faute est de nature à engager la responsabilité de son auteur, bien que beaucoup d'arrêts semblent dire le contraire (voyez, par exemple, 7 juin 1940, *Dame veuve Hoareau*, p. 194;—14 juin 1946, *Ville de Marseille*, p. 164;—28 juillet 1951, *Dame Le Saux*, p. 458, etc.).

La solution correcte consiste à rechercher s'il existe un lien de causalité entre la faute et le dommage, c'est-à- dire si la décision et, par suite, le préjudice n'auraient pas été les mêmes dans le cas où la procédure aurait été régulière: c'est ce que font quelques-uns de vos arrêts et, notamment, une décision de section du 19 mai 1960, *Dame veuve Pitiot*, p. 305. C'est ce qu'il convient de faire en l'espèce.

Il convient, par conséquent, de reprendre l'ensemble des circonstances de l'affaire et de rechercher si le licenciement, intervenu à la suite d'une procédure irrégulière, était néanmoins justifié par le comportement de la requérante.

Nous ne le croyons pas. A supposer même, en effet, que l'arrêté du 16 mars 1959 ne soit pas entaché de détournement de pouvoir, la responsabilité de la commune ne pourrait être atténuée ou supprimée que si l'insuffisance professionnelle de la demoiselle Mosconi était établie. Or, cette insuffisance professionnelle ne ressort pas des pièces du dossier et, quant aux fautes d'orthographe dont la requérante s'est rendue coupable le 16 mars 1959, à Dieu ne plaise que nous en sous-estimions la gravité: mais peut-être prouvent-elles tout simplement que le sieur Leccia, comme le grand Cornelius Sylla, ne savait pas dicter et ce serait, croyons-nous, manquer au respect que nous devons à la Haute Assemblée que d'y voir, ou feindre d'y voir, la vraie raison du licenciement de la requérante.

Il faut donc admettre que, dans les circonstances de l'espèce, la commune de Conca est entièrement responsable des conséquences dommageables de l'arrêté du 16 mars 1959.

b. En ce qui concerne en revanche l'évaluation du préjudice, il nous semble que les premiers juges ont été bien généreux. Certes, l'éviction de la demoiselle Mosconi est définitive: l'emploi qu'elle occupait a été supprimé, et l'intéressée elle-même ne songe pas à reprendre du service à la mairie de Conca; mais, ainsi que l'observe le ministre de l'intérieur, l'usage est de verser aux agents recrutés à titre précaire, en cas de licenciement, un mois de traitement par année de service. En outre, la demoiselle Mosconi, qui n'avait pas 30 ans à la date de son licenciement, n'était pas à ce point spécialisée dans les affaires administratives qu'il lui fût impossible, si elle le désirait, de trouver un autre emploi en rapport avec ses capacités: il nous semble, dans ces conditions, que l'indemnité allouée par le tribunal administratif excède de beaucoup le montant du préjudice subi par la requérante, y compris l'atteinte portée à sa réputation.

Pour être tout à fait sincère, nous vous proposerions néanmoins, compte tenu de la modicité intrinsèque de l'indemnité, de confirmer le jugement attaqué et de mettre les dépens à la charge de la commune, si le recours principal émanait de celle-ci; mais, puisque c'est la demoiselle Mosconi elle-même qui a choisi de remettre en cause la décision des premiers juges, nous n'avons pas les mêmes scrupules et nous vous demanderons de faire droit, au moins en partie, au recours incident de la commune de Conca.

Nous concluons à ce que l'indemnité due par la commune de Conca à la demoiselle Mosconi soit ramenée à 1.000 francs, à la réformation en ce sens du jugement attaqué, au rejet de la requête et du surplus des conclusions du recours incident et à ce que les dépens soient supportés par la requérante.

THE JUDGMENT (*ARRÊT*)

REPUBLIQUE FRANÇAISE: AU NOM DU PEUPLE FRANÇAIS

Conseil d'Etat: Statuant au Contentieux—No. 64732.

M. Kéréver, Rapporteur; M. Kahn, Commissaire du Gouvernement.
Adopté le 4 Jan. 1967.
Lu le 18 Jan. 1967.

Demoiselle Mosconi

Le Conseil d'Etat Statuant au Contentieux (Section du Contentieux, 4e et IIe Sous-Sections réunies). Sur le rapport de la IIe Sous-Section de la Section du Contentieux.

[*The judgment then gave a brief summary taken from the report, of the* mémoires *presented to the Conseil on behalf of Dlle Mosconi and the Maire of Conca respectively—these are not here reproduced.*]

Ouï M. Kéréver, Maître des Requêtes, en son rapport;
Ouï Me de Ségogne, avocat de la demoiselle Mosconi et Me Sourdillat, avocat de la commune de Conca, en leurs observations;
Ouï M. Kahn, Maître des Requêtes, Commissaire du Gouvernement, en ses conclusions;

Considérant que le licenciement de la demoiselle Mosconi de ses fonctions de secrétaire de mairie a été annulé par le Tribunal administratif en raison du vice de forme dont il était entaché; que la commune doit réparation à la requérante du préjudice qui a été causé par cette mesure irrégulière; que toutefois, compte tenu, d'une part, de ce qu'il résulte de l'instruction que le licenciement était justifié par l'insuffisance professionnelle de l'intéressée et, d'autre part, de ce que la rémunération dont elle a été privée s'élevait seulement à 13.821 anciens francs par mois, le montant de l'indemnité due par la commune devait être fixé à 1.000 francs y compris tous intérêts; que, dès lors, il y a lieu de ramener de 2.000 à 1.000 francs, y compris tous intérêts, le montant de ladite indemnité et, par voie de conséquence, de rejeter tant les conclusions de la requête de la demoiselle Mosconi tendant à une augmentation de l'indemnité qui lui a été allouée par les premiers juges que le surplus des conclusions du recours incident de la commune tendant, à titre principal, à la suppression de toute indemnité et subsidiairement à une réduction plus élevée que celle qui résulte de la présente décision;

DECIDE:

Article 1er—Le montant de l'indemnité que, par l'article 2 du jugement susvisé, la commune de Conca a été condamnée à payer à la demoiselle Mosconi à défaut de réintégration de celle-ci est ramené de 2.000 à 1.000 francs.

Article 2—L'article 2 du jugement susvisé du Tribunal administratif de Nice en date du to juin 1964 est réformé en ce qu'il a de contraire à la présente décision.

Article 3—La requête susvisée de la demoiselle Mosconi et le surplus des conclusions du recours incident de la commune de Conca sont rejetés.

Article 4—La demoiselle Mosconi supportera les dépens exposés devant le Conseil d'Etat.

Article 5—Expédition de la présente décision sera transmise au Ministre de l'Intérieur.

Délibéré dans la séance du 4 janvier 1967 où siégeaient: M. Lefas, Président-adjoint de la Section du Contentieux, Président; Barbet, Ordonneau, Présidents de sous-section; Barjot, Leroy-Jay, Agid, Merveilleux du Vignaux, Conseillers d'Etat; Kéréver, Maître des Requêtes-rapporteur et Aberkane, Maître des Requêtes.

Lu en séance publique le 18 janvier 1967.

Le Président:

Le Maître des Requêtes-rapporteur:

Le Secrétaire du Contentieux du Conseil d'Etat, Secrétaire des 4ème et 11ème Sous-Sections réunies:

Appendix I

Examples of Typical Requêtes Before the Administrative Courts*

Annexe I is a *recours pour excès de pouvoir* brought by the *requérant* before his local Tribunal Administratif against the decision of the local mayor granting an adjoining owner a building permit for the erection of a shed on land adjacent to that of the *requérant*.

Annexe II is a *demande de sursis à exécution* by the same *requérant* to stay the execution of the mayor's decision, pending the outcome of the principal proceedings.

Annexe III is a *requête fiscale* to quash a decision of the Director of Taxes rejecting the appeal of the *requérant* against assessment for income tax.

Annexe IV is a *recours en indemnité* brought by the *requérant* to his local Tribunal Administratif, claiming damages against the local municipality for his personal injuries etc., when he fell at night into an unlit and unfenced trench which had been dug across the pavement.

Annexe V is a *requête en appel* to the Conseil d'Etat to quash the decision of the Tribunal Administratif annulling the mayor's grant of a building permit (see Annexe I above).

Annexe VI is a *recours en appel* to a Cour Administrative d'Appel against the decision of the Tribunal Administratif, challenging the amount of damages awarded (see Annexe IV above).

ANNEXE I

Exemple de requête devant le Tribunal administratif en annulation d'une décision (ou Requête en excès de pouvoir)

* These *requêtes* are reproduced, by permission of *La Documentation Française*, from *La Justice administrative en pratique: comment faire valoir vos droits en cas de conflit avec l'administration* (1989), published as a guide for bringing proceedings before the Conseil d'Etat, the Cours Administratives d'Appel, or the Tribunaux Administratifs.

APPENDIX I

Jean Dupont
(adresse)
   Le (date)
   à
   Monsieur le Président
   et Messieurs les Membres
   du Tribunal administratif de...

Il faut en joindre copie.

J'ai l'honneur de vous demander d'annuler pour excès de pouvoir l'arrêté du maire de X... en date du... qui a accordé à M. Y... un permis de construire pour édifier un hangar sur son terrain contigu au mien. Cet arrêté est en effet illégal tant au fond qu'en la forme.

En la forme:

Cet arrêté aurait nécessité une consultation de l'architecte des bâtiments de France. Or il n'a pas été précédé d'une telle consultation.

D'autre part, cet arrêté a été signé par un adjoint au maire qui n'avait pas reçu délégation.

Au fond:

Précisez autant que possible ces violations

Cet arrêté viole le plan d'occupation des sols de la commune de X... approuvé le... par...

Il autorise en effet une construction dont le volume (...) est supérieur à celui autorisé par le coefficient d'occupation des sols...

Le permis est accordé en violation des règles générales d'aménagement et d'urbanisme prévues au Code de l'urbanisme et qui sont reprises exactement par le plan de la commune: la construction est en effet trop haute compte tenu de sa proximité de la limite parcellaire la plus rapprochée: ...

La construction ne respecte pas les règles d'affectation et d'aspect imposées par le plan: il s'agit en effet d'un hangar d'aspect disgracieux dans une zone réservée par le plan à des habitations individuelles.

Par ces motifs, je vous demande d'annuler l'arrêté du maire de X...

   Signature:

ANNEXE II

Exemple de demande de sursis à exécution

Jean Dupont
(adresse)
Le (date)
à
Monsieur le Président
et Messieurs les Membres
du Tribunal administratif de...

Je demande qu'il soit sursis à l'exécution de la décision attaquée par ma requête ci-jointe jusqu'à l'intervention de votre jugement. Les deux conditions nécessaires à ce sursis sont en effet réunies.

C'est un des cas où il est possible d'obtenir le sursis à exécution, c'est-à-dire la suspension de l'application de la décision.

Vous pouvez donc ordonner qu'elle ne sera pas exécutée jusqu'à ce que vous ayez jugé.

Son exécution aurait d'autre part des conséquences difficilement réparables. La réalisation de cette construction, qui est placée devant la mienne par rapport à la mer, me priverait en effet de toute vue.

Mais vous devez le faire par requête distincte.

Enfin, les arguments avancés contre cette décision sont extrêmement sérieux, incontestables et confirmés par toute la jurisprudence administrative.

Signature:

ANNEXE III

Exemple de requête fiscale devant le Tribunal administratif

Jean Dupont
12, rue...
Le (date)
à
Monsieur le Président
et Messieurs les Membres
du Tribunal administratif de...

J'ai honneur de vous demander l'annulation de la *décision* de M. le Directeur des Impôts reçue le 15 septembre 1986 (pièce jointe) qui a rejeté ma réclamation concernant l'imposition de 1 850 F mise à ma charge pour l'année 1984 au titre de l'impôt sur le revenu.

Donnez bien les références exactes de l'impôt contesté.

Je suis représentant de commerce, et j'ai demandé que soient déduits de mon revenu brut les frais professionnels réels que j'ai exposés dans mon activité. L'administration estime que ces frais sont trop élevés, et ne veut pas m'appliquer une autre déduction que celle qui résulte de l'annexe IV article 5 du Code général des impôts.

La réclamation préalable au Service, qui est indispensable, a été faite, et la décision doit être jointe. L'argumentation repose sur des motifs de droit.

Or, l'article 83 du Code général des impôts me donne la possibilité de demander la déduction de mes frais réels. Il faut qu'ils soient justifiés et qu'ils correspondent à des dépenses professionnelles. Tel est mon cas, et vous trouverez ci-joint, outre *deux copies* de ma requête:

Ne pas oublier.

1) Des *pièces justificatives* attestant de l'ensemble des dépenses que j'ai supportées, soit 11 500 F. Ces pièces vont de la note de restaurant à la facture d'essence.

Il faut prouve tout ce qu'on avance. Le Tribunal ne peut vous croire sur parole.

2) Une *attestation* de mon employeur établissant d'une part la grande étendue du secteur dont je suis responsable, d'autre part la nécessité d'avoir de fréquents contacts avec les clients.

3) Les *attestations* des clients que j'ai visités, prouvant l'exactitude des déplacements effectués et le fait que tous ces déplacements étaient nécessités par mon activité professionnelle.

Il faut signer.

Signature:

ANNEXE IV

Exemple d'une demande d'indemnité devant le Tribunal administratif

Jean Dupont
12, rue...

Le (date)
à
Monsieur le Président
et Messieurs les Membres
du Tribunal administratif de...

Il faut relever les faits importants et les prouver.

J'ai l'honneur de porter à votre connaissance les faits suivants: le 9 septembre 1986, à 20 h 45, je marchais sur le trottoir de la rue Durand, côté numéros pairs, dans la ville de X... Il faisait nuit et la rue était *mal éclairée*. J'ai soudain glissé dans une tranchée qui coupait le trottoir sur toute sa largeur. Vous trouverez ci-joint les *témoignages* de M. Bertin, qui a assité à l'accident, et de M. Albin qui m'a sorti de ma fâcheuse position.

Il faut que l'obstacle ait une certaine épaisseur ou une certaine profondeur.

Il est incontestable que la ville de X... est responsable de cet accident. Dans cette rue mal éclairée, cette tranchée n'était absolument pas signalée, ni protégée. Par le fait même qu'elle se trouvait sur la voie publique, sa présence entraîne la responsabilité de la ville de X... quelle que puisse être par ailleurs la part qu'a prise dans cet état de choses l'entreprise qui a pu creuser cette tranchée, *profonde d'un mètre*. C'est pourquoi je vous demande de déclarer la ville de X... entièrement responsable de cet accident.

Il faut individualiser chaque élément du préjudice et le justifier. Votre déclaration de revenus vous est opposable lorsque vous demandez réparation de votre perte de revenus.

J'ai subi un préjudice important sur le plan vestimentaire. J'ai déchiré dans ma chute mon pantalon et mon manteau (voir témoignages), soit un préjudice de 800 F.
Par ailleurs, j'ai subi des frais médicaux (j'eu entre autre une jambe fracturée) s'élevant à 2 000 F (voir pièces justificatives jointes) et les dix jours pendant lesquels j'ai été immobilisé ont représenté pour moi un préjudice de 900 F (voir *copie de ma déclaration de revenus*).

C'est le « pretium doloris ».

Enfin j'estime à 800 F la réparation qui m'est due pour les *souffrances* que j'ai supportées. Je précise que je suis *assuré social*, affilié à la Caisse de..., sous le n°...

APPENDIX I 315

Vous devez indiquer si vous êtes assuré social ou non lorsque vous invoquez un dommage corporel. Il faut les demander pour les obtenir.

C'est pourquoi je vous demande:

1) De déclarer la ville de X... responsable de l'accident dont j'ai été victime.
2) De la condamner à me verser la somme de 4500 F représentant le montant de mon préjudice.
3) De la condamner à verser cette somme et les *intérêts de droit* à compter du jour de ma demande.

Signature:

ANNEXE V

Exemple d'une requête devant le Conseil d'État en appel d'un jugement d'annulation rendu par un Tribunal administratif

Votre nom
Votre adresse

Le (date)
à
Monsieur le Président
de la Section du Contentieux
Conseil d'État
Place du Palais-Royal
75100 PARIS RP

Objet: appel d'un jugement du Tribunal administratif de...

PJ: Le jugement attaqué—1 copie de la présente requête.

J'ai l'honneur, par le présent pourvoi, de demander au Conseil d'Etat l'annulation du jugement, rendu le 1er février 1987 par le Tribunal administratif de..., qui m'a été notifié le 1er mars.

Par ce jugement, le Tribunal administratif a annulé, sur la demande de M. X..., l'arrêté du maire de ... qui m'a accordé, le 5 janvier 1985, le permis de construire un hangar sur la parcelle, dont je suis propriétaire, voisine de celle de M. X...

(Dans un mémoire complémentaire que je produirai ultérieurement, je démontrerai que).

C'est à tort que le Tribunal administratif a annulé le permis de construire dont j'étais bénéficiare.

En effet, M. X... a saisi le Tribunal après l'expiration du délai de recours, l'affichage ayant été fait à compter du...

De toute façon, j'estime que c'est à tort que, pour annuler le permis, le Tribunal s'est fondé sur les dispositions de l'article R.111–2 du

Code de l'urbanisme, pour affirmer que la construction envisagée porterait atteinte au caractère des lieux avoisinants et mettrait en cause, par son implantation, l'harmonie et l'esthétique du paysage naturel alors que les pièces versées au dossier montrent que toutes les précautions avaient été prises pour qu'aucune atteinte ne fût portée par la construction litigieuse au caractère pittoresque des lieux avoisinants.

C'est par suite à tort que le Tribunal administratif a déclaré que le maire avait commis une erreur manifeste d'appréciation en m'accordant le permis de construire contesté par M. X...

Je demande à recevoir communication des observations qui seront produites par l'administration et par M. X... afin d'y répliquer. Je demande également a être informé du jour de l'audience.

(Sous réserve de la production du mémoire complémentaire), Je conclus à ce qu'il plaise au Conseil d'Etat:

—annuler le jugement du Tribunal administratif de...
—rejeter la demande de M. X...

                Signature:

ANNEXE VI

Exemple d'appel devant une Cour administrative d'appel en cas de rejet par un Tribunal administratif d'une demande d'indemnité

Votre nom
Votre adresse   Le (date)
                             à
                             Monsieur le Président
                             de la Cour administrative
                             d'appel de...

Objet: appel d'un jugement du Tribunal administratif de...

PJ: Copie du jugement du 30 juin 1987
Copie de la présente requête

Le 9 septembre 1986 à 20 heures 45, je marchais sur le trottoir de la rue Durand, côté numéros pairs, dans la ville de X... J'ai soudain glissé dans une tranchée qui coupait le trottoir sur toute sa largeur, et qui n'était pas signalée.

J'ai saisi, dès le 10 octobre, le Tribunal administratif de..., pour être indemnisé du préjudice subi à l'occasion de cet accident, dont la responsabilité incombe à la ville de X...

Par un jugement du 30 juin 1987, le Tribunal administratif de... a déclaré la ville de X... responsable de l'accident dont j'ai été victime et l'a condamnée à me verser la somme de 1000 F en réparation du préjudice subi, avec les intérêts de droit.

J'estime que cette somme est insuffisante. En effet j'avais demandé (vous rappelez ici vos demandes formées devant le Tribunal administratif, puis vous indiquez les raisons pour les quelles vous estimez que c'est à tort que le juge de première instance ne vous a pas donné raison. Vous pouvez vous référer à l'argumentation développée devant celui-ci).

Je demande en conclusion à la Cour administrative d'appel de...

1–de réformer le jugement par lequel le Tribunal administratif de... a fixé à 1000 F le montant du préjudice subi lors de l'accident dont j'ai été victime;

2–de condamner la ville de X... à me verser la somme de 4 500 F;

3–de la condamner à verser les intérêts de droit sur cette somme à compter du jour de ma demande.

Signature:

Appendix J

An Avis of the Conseil d'Etat

N° 345.400–M. LAMBRON, rapporteur Séance du 21 mars 1989

Le Conseil d'État (section sociale) saisi par le ministre de la Solidarité, de la Santé et de la Protection sociale de la question de savoir si, au regard des dispositions de la loi n° 86–33 du 9 janvier 1986 modifiée, portant dispositions statutaires relatives à la fonction publique hospitalière et du décret n° 88–976 du 13 octobre 1988 relatif à certaines positions des fonctionnaires hospitaliers, un fonctionnaire hospitalier peut ou non être détaché dans un emploi contractuel de sa propre administration:

Vu la loi n° 83–634 du 13 juillet 1983 modifiée portant droits et obligations des fonctionnaires;

Vu la loi n° 86–33 du 9 janvier 1986 modifiée portant dispositions statutaires relatives à la fonction publique hospitalière;

Vu le décret n° 88–976 du 13 octobre 1988 relatif à certaines positions des fonctionnaires hospitaliers;

Est d'avis de répondre à cette question dans le sens des observations qui suivent:

Il résulte d'une jurisprudence constante du Conseil d'État statuant au contentieux que, sauf disposition expresse contraire, un fonctionnaire titulaire ne peut être recruté comme agent contractuel par sa propre administration, fût-ce après détachement.

Par suite, un fonctionnaire hospitalier, en l'absence d'une dérogation expresse, que ne prévoit aucune disposition législative ou réglementaire ne peut être détaché dans un emploi contractuel de sa propre administration, c'est-à-dire de l'établissement dont il relève.

Appendix K

Select Bibliography

Books and Materials in English

FRENCH PUBLIC LAW AND GOVERNMENT

Bell, J., *French Constitutional Law* (Oxford, 1992).
Hamson, C. J., *Executive Discretion and Judicial Control* (Hamlyn Lectures, London, 1954).
Hayward, J. E. S., *Governing France: The One and Indivisible Republic* (2nd edn., London, 1983).
Rendel, M., *The Administrative Functions of the Conseil d'Etat* (London, 1970).
Ridley, F. F., & Blondel, J., *Public Administration in France* (2nd edn., London, 1970).
Schwartz, B., *French Administrative Law and the Common Law World* (New York, 1954).
Wright, V., *The Government and Politics of France* (3rd edn., London 1989)

OTHER WORKS ON FRENCH OR COMPARATIVE LAW

Amos and Walton's Introduction to French Law, eds. F. H. Lawson, A. E. Anton, and L. Neville Brown, (3rd edn., Oxford, 1967).
Bell, J., and Bradley, A. W., *Governmental Liability: A Comparative Study* (UKNCCL, London, 1991)
David, R., and Brierly, J. E. C., *Major Legal Systems of the World Today* (3rd edn., London, 1985).
Hand, G., and McBride, J., *Droit sans frontières: Essays in Honour of L. Neville Brown* (Birmingham, 1991).
Harris, D., and Tallon D., *Contract Law Today* (Oxford, 1988).
Kahn-Freund, O., Lévy, C., and Rudden, B., *A Source-book of French Law* (3rd edn., Oxford, 1991)
Lawson, H., and Markesinis, B., *Tortious Liability for Unintentional Harm in the Common Law and the Civil Law* (2 vols., Cambridge, 1982).
Merryman, J., *The Civil Law Tradition* (2nd edn., Stanford, 1985).

Nicholas, B., *French Law of Contract* (London, 1982; 2nd edn., Oxford, 1992).
Schwarze, J., *European Administrative Law* (London, 1992).
Von Mehren, A. T., and Gordley, J., *The Civil Law System* (2nd edn., Boston, Mass., 1977), esp. chs. 5 and 6.

CASE-REPORTS

Some cases are reproduced in Kahn-Freund, Lévy, and Rudden, and in Von Mehren and Gordley, above. Recent cases are often summarized in a section 'Recent Decisions of the Conseil d'Etat', published regularly in *Public Law* since 1985 by R. Errera, Conseiller d'Etat.

Books and Materials in French

BOOKS

Auby, J.-M., and Drago, R., *Traité de contentieux administratif* (2 vols., 4th edn., Paris, 1991).
—— and Fromont, M., *Le Recours contre les actes administratifs dans les pays de la Communauté Economique Européenne* (Paris, 1971).
Braibant, G., *Le droit administratif français* (1st edn., Paris, 1984; 2nd edn., Paris, 1988).
—— Letowski, J., and Wiener, C., *Le contrôle de l'administration en Europe de l'est et de l'ouest* (Paris, 1985).
—— Questiaux, N., and Wiener, C., *Le Contrôle de l'administration et la protection des citoyens* (Paris, 1973).
CERAP, *Le Contrôle juridictionnel de l'administration* (Paris, 1991).
Chapus, R., *Droit administratif général* (2 vols.: i, 5th edn., Paris, 1990; ii 5th edn., Paris, 1992).
—— *Droit du contentieux administratif* (3rd edn., Paris, 1991).
Colliard, C.-A., and Timsit, G., *Autorités administratives indépendantes* (Paris, 1988).
Le Conseil d'Etat: Livre Jubilaire (published (in Paris) in 1952 to commemorate the 150th anniversary of the Conseil d'Etat).
Le Conseil d'Etat (1799–1974): Son histoire à travers les documents d'époque (published in 1978 to celebrate the 175th anniversary of the Conseil d'Etat).
Delvolvé, P., *L'Acte administratif* (2nd edn., Paris, 1991).
Encyclopédie Dalloz: Répertoire de contentieux administratif (updated regularly, with articles on specific issues).
Encyclopédie Dalloz: Répertoire de la responsabilité de la puissance publique (updated and organized in the same way).
Flogaïtis, S., *Administrative Law et droit administratif* (Paris, 1986).

SELECT BIBLIOGRAPHY

Gabolde, Ch., *La Procédure des tribunaux administratifs* (5th edn., Paris, 1990).

Gentot, M., and Oberdorff, B., *Les Cours administratives d'appel* (Que sais-je?, Paris, 1991).

Lachaume, J.-F., *Les grandes decisions de la jurisprudence. Droit administratif* (6th edn., Paris, 1991).

Laubadére, A. de, *Traité de droit administratif* (i, 11th edn., Paris, 1990; ii, 9th edn., 1992; iii, 4th edn., 1989).

Letourneur, M., Bauchet, J., and Méric, J., *Le Conseil d'Etat et les tribunaux administratifs* (Paris, 1970).

Long, M., Weil, P., Braibant, G., Delvolvé, P., and Genevois, B., *Les Grands Arrêts de la jurisprudence administrative* (9th edn., Paris, 1990).

Mestre, J., *Introduction historique au droit administratif* (Paris, 1985).

Moreau, J., *Droit administratif* (Paris, 1988).

Odent, R., *Cours de contentieux administratif* (Les Cours de Droit, 1977–81, Paris).

Pacteau, B., *Contentieux administratif* (2nd edn., Paris, 1989).

Philippe, X., *Le Contrôle de proportionnalité dans les jurisprudences constitutionnelle et administrative françaises* (Paris, 1990).

Rivero, J., *Droit administratif* (13th edn., Paris, 1990).

Rougevin-Baville, M., and Labetoulle, D., *Leçons de droit administratif* (Paris, 1989).

Stirn, B., *Le Conseil d'Etat: Son rôle, sa jurisprudence* (Paris, 1991).

Université de Paris 2, *Le Conseil constitutionnel et le Conseil d'Etat* (Paris, 1988).

Vedel, G., and Delvolvé, P., *Droit administratif* (2 vols., 11th edn., Paris, 1990)

Waline, M., *Droit administratif* (7th edn., Paris, 1957; 9th edn., Paris, 1963)

ARTICLES AND CASE REPORTS IN:

Actualité Juridique: Droit Administratif (AJDA): a specialized monthly journal to which members of the Conseil d'Etat regularly contribute notes and articles. The law reports published here give full and rapid coverage of decisions of the administrative courts.

Etudes et Documents du Conseil d'Etat: published annually since 1947, this is an official publication of the Conseil d'Etat, giving reports on its activities, and commentaries on current development. As such, it has special authority.

Recueil Dalloz (D.): published since 1845 and combined with the *Recueil Sirey (S.)* since 1965, it is divided into a number of sections providing articles (Chronique), full case-reports (Jurisprudence), and texts of legislation (Législation), as well as summaries of cases (Sommaires and

Informations rapides). Its reporting of administrative-court decisions is very selective, but often has a signed commentary.

Recueil des décisions du Conseil d'Etat ('Recueil Lebon': *Leb.*): published since 1821, this is a semi-official series of law reports for all decisions of the Conseil d'Etat and the Tribunal des Conflits and selected decisions of the other administrative courts, including some conclusions of Commissaires du gouvernment, but without any commentaries.

Revue du Droit Public (*RDP*): published since 1834, this contains mainly articles and reviews of legal developments in specific areas (Chroniques), but also contains some reports of cases, often including the conclusions of the Commissaire du gouvernement.

Revue française de droit administratif (*RFDA*): published since 1985, this bi-monthly journal contains articles on current legal developments typically grouped around a theme, together with reports of cases, often with conclusions of the Commissaire du gouvernment or a commentary. It also contains short summaries of very recent cases of the Conseil d'Etat.

Index

act of state:
  doctrine of 134–5
actes de gouvernement:
  acte détachable, theory of 156
  judicial review, exclusion of 155–6
  war, state of 156–7
administration:
  judicial control of 28, 41
  Médiateur, role of 29–30
administration of justice:
  judiciary, organization of 133
  public authority, exercise of 132
administrative act:
  annulment, seeking 168
  bilateral 153
  content, controlling 230
  English courts, quashed in 232–3
  exercise of discretion, object behind 231–2
  house-rules distinguished 154
  judicial review, *see* judicial review
  legality, challenge to 142–4
  motivation, questions of 233–4
  notion, extension of 153
  unilateral 153
administrative agencies:
  staffing from outside civil service 21–2
  tribunals, being 55–6
administrative courts:
  administration, acting at request of 170–1
  career structure within 80–5
  confidence in 272–3
  Conseil d'Etat, *see* Conseil d'Etat
  Constitutional Council, relationship with 21
  control by 6
  costs 121
  Cours Administratives d'Appel, *see* Cours Administratives d'Appel
  criminal court, acting as 170–1
  decisions, administration failing to implement 283–4
  delays in 281–2
  general jurisdiction 26

Germany, in 259–60
Greece, in 261–2
hierarchy of 41
history of 42–7
judicial control by 278–9
jurisdiction, *see* jurisdiction of administrative courts
legal representation in 89–90
litigation before: *contentieux de l'annulation* 168; *contentieux de l'interprétation* 169–70; *contentieux de la répression* 170–1; *contentieux de pleine juridiction* 168–9; traditional classification, criticism of 171–2
Netherlands, in 255–6
ordinary courts, separation of functions from 131
procedure, *see* procedure of courts
reform of 27
requêtes, examples of 309–16
role of 25
standardization of decisions 120–1
supervision of 41
Tribunaux Administratifs, *see* Tribunaux Administratifs
uniform application of law by 275–6
administrative decision:
  acte juridictionnel, as 153–4
  administrative jurisdiction, from 153–4
  conflicts 148–9
  incompétence, void for 226–7
  inexistant 136, 224–6
  interpretation of 169–70
  le bilan, doctrine of 249
  motivé, being 239
  reasons, duty to give 218
  review, range subject to 286
administrative law:
  comparison of 1–2
  remedies 167–8
  unequal parties, regulating relations between 167
administrative legality:
  case law, extension by 205

administrative legality (*cont.*):
  cassation 235–7
  minimum and normal review 250–1
  principes généraux du droit, see *principes généraux du droit*
  principle of 172–3, 202; content of 203–5; previous decisions, rights created by 204
  recours en indemnité, remedy of 202–3
  review, grounds for: *détournement de pouvoir* 229–35; fair hearing, failure to give 228; *incompétence* 226–7; *inexistence* 224–6; *moyen d'ordre public* 224; person raising 224; *vice de forme* 227–8; *violation de la loi* 228–9;
  sources of 203–4
  ultra vires doctrine, similarity to 202, 204
  violation of principle 223–4
administrative liability:
  administrative, within jurisdiction of 176
  English law, in 174
  exclusion of 192
  fault, without: risk, theory of 183–4, *see also* risks
  legislation, arising from 189–91
  non-contractual 190
  principle of 173–4
  public authorities, of 175–6
  public interest, abnormal burdens suffered in 188–9
  rules of 176
  system of 180
  torts, for: contribution, right of 179–81; *cumul*, notion of 177–9; damages 191; examples of 175; expansion of 175; *faute de service* 176–8, 181–2; *faute personnelle* 177; indemnity 179–81; late feasance 181; legal principles governing 176; personal fault in public service 178–9; vicarious 176, 275
administrative tribunals
  agencies being 55–6
  Britain, in 273
  Commission de recours des réfugiés 57, 293
  conscientious objectors, for 293
  Conseil d'État, supervision of 57
  creation of 55
  education, relating to 290
  French and British compared 26, 56
  French citizens, compensation claims 293
  judges, discipline over 290
  military pensions 292
  professions, relating to 290–2, 294
  public finance 290
  refugees 293
  social welfare 292
Ancien Régime 8
association:
  liberty of 16
audi alteram partem:
  principle of 217–18

Belgium:
  Conseil d'Etat 252–4
Bonapartist tradition 8
broadcasting:
  case-law on 17

cassation:
  Conseil d'Etat, jurisdiction of 235–7
  jurisdiction subject to 235
  mistake of fact, review of 236
  proceedings 52–3, 57–8
  review, grounds for 235
Chambre Régionale des Comptes:
  functions of 34–5
civil courts:
  issues allocated to 131
  matters traditionally reserved to 139–40
civil liberties:
  constitutional provisions, development of 17
civil servants:
  administrative courts, resort to 276
  career structure 24
  contracts of employment 200–1
  discretions given to 24–5
  local government positions 24–5
  powers, growth of 24
  schoolmasters as 37
  status 38
civil service:
  career opportunities in 37
  influence of 39–40

INDEX

local and central government
  distinguished 37
post-entry training schools 39
technical and non-technical
  branches 39
tradition of 38
Code rural:
  criminal sanctions, introduction of
    15–16
Commissaire du gouvernement:
  appointment 101
  conclusions of 102–3, 108
  function of 46
  independence of 101
  office of 101
  role of 87, 102
  workload 101–2
Commission d'accès aux documents administratifs:
  rights of access to documents,
    advice on 88
Commission de recours des réfugiés:
  work of 57, 293
common law:
  codification of 2
communes:
  financial control 33–6
  fusions 32
  population 32
Community law:
  supremacy of 270
compétence liée:
  doctrine of 237–8
competition:
  jurisdiction 141
conciliation:
  administrative courts, by 27
conscientious objectors:
  tribunals 293
Conseil d'État:
  achievement of 273
  administration, as part of 77
  administrative courts, supervision
    of 41
  administrative role: advice,
    government seeking 62, 68; bills
    submitted to 61–2; decrees
    submitted to 62; delegated
    legislation, inspection of 62;
    drafting supervision 70–1;
    general legal adviser, as 62–3;
    President, annual report to 63;
    Standing Committee 69–70

administrative sections 64–7
administrative tribunals,
  supervision of 57
annual reports 63, 73
appeals, filter system for 51
arrêt of 303–8
astreinte, remedy of 113–14
audience publique 104
avis of 317
backlog of cases 48, 50
Belgium, in 252–4
CANAL case 54
case-law: merits of 277;
  reconsideration of 278; reversal
  of 267–70; statistics 296–8
cassation proceedings 52–3, 57–8
Commissaire du gouvernement 46
Commission du Contentieux,
  creation of 45–6
Commission permanente 69–70
compliance with general principles,
  imposing 12
Conseil du Roi, resemblance to
  42–3
Conseil Supérieur 84–5
counsel in 90
court, jurisdiction of 45
court of appeal, as 49
current role of 54
decisions of 120; reconsidering
  116
decrees challenged before 16
double role of 59
drafting supervision 70–1
early powers of 44
establishment of 43
exclusive club, atmosphere of 59–60
French and community law,
  reaction to conflict of 267–70
General Assembly 67–8, 77
general jurisdiction 46
Greece, in 261–2
grievances, redressing 47
implementation of decisions 111
international relations, lack of
  competence in 134
internal division 60
Italy, in 257–8
judicial role: Section du
  Contentieux, *see* Section du
  Contentieux, *below*; Section du
  Rapport et des Etudes 73–6
litigants in person 105

Conseil d'État (*cont.*):
  members 103, 288; active administration, recruitment by way of 79; administrative expertise 77–8; administrative sections, of 60; career structure 80–3; civil servants, as 81; élite, as 60; grades 80–1; National School of Administration, from 78–9; number of 60; physical division of 76; promotion 82; recruitment 78–80; salaries 82–3
  method of working, improvement in 50
  ministre-juge, doctrine of 44–5
  nature of 5–6
  Netherlands, in 254–6
  oral argument, absence of 86
  parliament, conflict with 133–4
  parties, presence of 107
  present jurisdiction 51–3
  proposed legislation, examination of 65–7
  questions of law submitted to 51
  recours de cassation 115–16
  recours, use of 277
  Report Commission 73–4
  Report Section 61, 73–6
  rigid precedent, no doctrine of 166–7
  séance de jugement 105–6
  secretary-general 85
  Section du Contentieux 46–7, 55; administrative sections, Conseillers attached to 64; administrative sections, detachment from 76–7; judgments 72–3; l'Assemblée du Contentieux 71–3, 106; members of 71; responsibilities within 60; *séance de jugement*, present at 106; Sous-sections 71–2
  Section du Rapport et des Etudes 73–6
  sections, work of 44
  Standing Committee 69–70
  strength of 6
  structure of 61–76, 289
  success of 59
  travaux préparatoires, influence in 279
  treaties, not adjudicating on 134

Conseil du Roi:
  abolition 43
  role of 42
Conseils de Préfecture:
  court of first instance, as 48
  establishment of 47
  interdepartmental Conseils, replacement with 48
  Tribunaux Administratifs, becoming 49
constitution:
  division of powers 8–13
  Fifth Republic, of 8–9, 12
Constitutional Council:
  administrative courts, relationship with 21
  adverse ruling, decrees made after 16
  case-law, development of 17
  compatibility of international conventions, adjudication on 14
  constitution, interpretation of 12–13
  creation of 13
  decisions of 10, 15–16
  declassification, process of 11
  formal objections made to 11
  functions of 13–14
  increased activity of 18
  influence, growth in 20–1
  investigation, procedure as 19–20
  judges of administrative courts, status of 18
  jurisdiction 14
  members 15, 20
  nature of 19
  references to 20
  rights of citizen, defending 21
  rights of parliament, safeguarding 17–18
  role of 6
  sixty deputies or senators, seised by 18–19
  statutes, interpretation of 12–13
contentieux administratif:
  meaning 5
contract:
  administrative: concessions 194; contractor, rights of 200; criterion 192; employment, of 200–1; English law, in 195, 201; equilibrium 196; *fait du prince*, doctrine of 197–8; *force*

INDEX

majeure, effect of 199–200;
formation 193–5; *imprévision*,
doctrine of 198–9; jurisdiction
193; modification 196–7; other
agency, approval of 200; public
procurement 194; regulation by
193; rules on 193; terms of
195; types of 192; unequal
parties, between 192–3
clauses exorbitantes 137–9
public authority, of 137–9
costs:
principles for 121
counsel:
Conseil d'Etat, before 90
Cour des Comptes:
public money, control over
expenditure of 57
Cours Administratives d'Appel:
audience publique 104
Conseil d'Etat, questions
submitted to 51
creation of 50
inspection, commission of 53
judges of 53
judicial functions 59
judicial statistics 299–301
legal representation in 90
members 84–5, 103
parties, presence of 107
work of 50
criminal courts:
exception d'illégalité 142–4

damages:
loss of a chance, for 191
prior decision, rule of 157
tort, in 191
dangerous operations:
risks arising from 186–8
décision exécutoire:
privilege of 173
decrees:
government, made by 12
specific delegation, under 12
décrets, see *lois et décrets* cited
deportation:
prefect, limited power of 241–2
Dicey, A. V. 4–5
droit administratif:
case-law, development of
principles from 276–7
common lawyers, study by 1

droit civil, separation from 275–7
European Court, unfavourable
comparison with 285
fully developed system as 3
influence outside France 252
judge-made 2
legality, principle of 172–3, 202,
see also administrative legality
public law in Community, basis of
270
reform of 22–3
responsabilité, principle of 173
rule of law, comparison of 4
source materials 6
state, liberties of individual against
23
strength of 23
substantive: case-law 166–7;
evolution of 166; legality, see
administrative legality; liability,
see administrative liability
system of: character of 271–4;
defects of 280–5; merits of
274–9; success of 271–4
translation of 5
vicarious liability, scope of 275

education:
central government function, as 37
tribunals relating to 290
employers' liability:
jurisdiction 141
environment:
protection of 214–15
equality:
law, before 215–17
racial 217
sexes, of 216–17
erreur manifeste d'appréciation des faits:
case-law, adaption of 249–50
doctrine of 240, 245–50;
application of 246
proportionality, link with 247
European Court of Justice:
administrative law systems,
influencing 285
appearances before 252
case-law of 253
composition 266–7
Court of First Instance 266
droit administratif, influence of 262
jurisdiction 264–6
national law, application of 263

European Court of Justice (*cont.*):
  preliminary rulings 265
  procedure 267
  substantive law of 263–4
evidence:
  expertise 94
executive:
  powers of 9
extradition:
  review of 134

flagrant irregularity:
  doctrine of 135–6
France:
  civilization of 2
French law:
  comparison of 1

Germany:
  administrative law in 258–8
government:
  bill outside parliamentary competence, objection to 10
  decrees of 11–12
  parliament's power to overthrow 9
  pouvoir réglementaire 9, 62
  previous autonomous powers 9–10
  stability, record of 9
Greece:
  administrative courts in 261–2
  Council of State 261–2
  judicial review, system of 261

highways:
  on land, contraventions relating to 171
human body:
  inalienability of 222–3
human rights:
  economic and social 213–14
  individual, liberties of 211–13
  principes généraux 208–9

impartiality:
  principle of 217
independent agencies:
  policy areas, regulation or supervision of 25
interim order procedure:
  use of 116–17
interlocutory orders 96
international conventions:
  Constitution, compatibility with 14

international relations:
  administrative courts, outside competence of 134–5
Italy:
  Consiglio di Stato 257–8

judges:
  administration, close connection with 274
  administrative courts, of: conciliation by 27; status of 18
  Cours Administratives d'Appel, of 53, 84–5
  discipline over 290
  judicial control by 41
  task of 271–2
  Tribunaux Administratifs, of 53, 83–5
judgment:
  audience publique 104–8
  délibéré 108–9
  execution 110–15
  séance de jugement 103
judicial functions:
  administrative functions, separation from 123–4
judicial review:
  commencement of proceedings, time-limit for 161–2
  conditions precedent: act under review, nature of 152–7; list of 152; parallel relief, absence of 160–1; plaintiff, locus standi of 158–60; prior decision, rule of 157
  délai 161–2
  disciplinary decisions, of 154
  discretion of administration: absolute 238–40; law, within limits of 240; limited 240–5; none 237–8
  English approach to 236
  English law, in 164–5
  erreur manifeste d'appréciation des faits 242, 245–50
  exclusion of 162–5
  extent of 237
  governmental acts, exclusion of 155
  Greece, in 261
  introduction of bills, not extending to 134
  judicial review, *see* judicial review

meaning 5
minimum and normal 250–1
mistakes of fact, of 236, 244–5
parliamentary proceedings not
  subject to 133–4
principes généraux, of 222
jurisdiction of administrative courts
  conflicts: case-law 122; decisions,
    of 148–9, 151; defect of system,
    as 280; disadvantage of 150;
    negative 147–8; positive 146–7;
    procedure 144–6; Tribunal des
    Conflits, *see* Tribunal des
    Conflits
  criterion, search for 124–31
  division of 287
  English 122
  exceptions: administration of
    justice 132–3; civil courts,
    matters traditionally reserved to
    139–40; competition 141;
    employers' liability 141;
    exception d'illégalité 142–4;
    flagrant irregularity, doctrine of
    135–6; generally 131; illegality,
    defence of in ordinary courts
    142–4; indirect taxation 140–1;
    international relations 134–5;
    nuclear accidents 141;
    parliamentary proceedings
    133–4; postal service 141–2;
    private character, administrative
    acts of 136–9; running-down
    cases 141; statutory 140–2;
    summary of 144
  existentialist approach 130
  guiding principle 130–1
  judiciary, organization of 133
  private firms subject to 129–30
  public service, *see* public service
  specialized 290–5
jurists and authors cited:
  Atkin, Lord 110
  Auby 253
  Bernard, A. 178–9
  Blondel, Professor J. 22, 37, 39
  Blum, Léon 103, 110, 178, 196
  Braibant, M. Guy. 71, 75, 76,
    115, 126, 220, 238, 245, 248
  Brinkhorst, L.J. 270
  Bouffandeau, President 206
  Chapus, R. 55, 56, 92, 95, 96,
    195, 117, 122, 224
  Chenot, B. 130
  Dale, Sir William 71
  David, René 3, 103, 110
  Delvolvé, P. 59, 122, 123, 124,
    129, 130, 131, 135, 205, 209,
    226, 228, 230, 234, 238, 242,
    250
  Denning, Lord 3, 110, 164, 165,
    233
  Dicey, A.V. 4, 5, 47, 173, 203, 216
  Diplock, Lord 4, 110, 227
  Donaldson, Lord 273, 286
  Duguit, Léon 172, 184
  Errera, R. 28, 154, 175, 213, 222,
    242, 250
  Favoreu, L. 11, 13, 17, 21
  Foulkes, D. 280
  Fournier 208
  Fromont 253
  Franks 240
  Galabert, J.-M. 272
  Galmot, Y. 165, 285–6
  Garner, Professor J.F. 216
  Gaudemet 10
  Hamson, C.J. 166, 187, 277
  Hauriou, M. 77
  Holdsworth 57
  Jennings, Sir Ivor 4, 5
  Koopmans 264, 267, 286
  Laferrière, J. 125, 168, 177
  Lawson, F.H. 5
  Letourneur, M. 74, 205, 235
  Loschak, D. 272
  Maleville, M. Georges 81
  Marimbert 21, 74, 75, 96, 97, 103
  Massot 21, 74, 75, 96, 97, 193
  Même, C. 56, 290
  Mitchell, J.D.B. 3, 196, 198, 275
  Odent, President R. 122, 126,
    236, 271, 272, 274
  Purdue, Professor M. 236–7
  Reid, Lord 3
  Ridley, F. 22, 37, 39
  Rivero, Professor J. 16, 155, 271
  Romieu 102, 110
  Schermers, H.G. 270
  Schwartz, Professor B. 177, 181,
    224
  Thompson, Professor 184
  Vedel, Professor G. 59, 122, 123,
    124, 129, 130, 131, 135, 160,
    173, 205, 209, 226, 228, 230,
    234, 238, 242, 250

jurists and authors cited (*cont.*):
  Wade, Professor H.W.R. 2, 238
  Waline, Professor M. 126, 138,
    169, 176, 206, 230, 236, 238
  Weil, Professor P. 23, 156, 279,
    281, 282
  Weston, M. 43
  Wright, V. 24, 39

legislation:
  retrospective effect, with 220–1
  state liability arising from 189–91
  travaux préparatoires, influence of
    Conseil d'État 279
  unlawful regulations, duty to
    abrogate 221–2
liability:
  administrative, *see* administrative
    liability
  delictual or quasi-delictual 174–5
  fault, without 183–4
  risk, for, *see* risks
litigants in person 105
local government:
  autonomy 34
  budgetary control 34–6
  central government, dominated by
    32
  civil servants in positions of 24–5
  democratization of 33
  local initiative, allowing 36
  officials, status of 34
  regional councils 33
  tiers of 32
  traditional 31
loi:
  Constitutional Council, powers of 11
  parliament, passed by 10
  règlement, dividing line with 11
lois organiques:
  meaning 14
lois et décrets cited:
  16–24 August 1790 43, 44, 123,
    124
  10 *vendémiaire an* IV (1795) 183
  28 *pluviôse an* VIII (1799) 183
  2 November 1864 46, 228
  24 May 1872 44, 45
  5 April 1884 34
  16 April 1914 151
  20 April 1932 148
  31 December 1957 141
  17 November 1958 134

  25 July 1960 149
  13 April 1962 54
  15 January 1963 54
  10 July 1963 75
  20 October 1968 141
  16 September 1972 90
  3 January 1973 29, 30
  10 July 1976 215
  24 December 1976 30
  17 July 1978 28, 88
  11 July 1979 28, 88, 218, 239
  10 January 1980 72
  16 July 1980 75, 113
  4 December 1980 284
  18 March 1981 284
  2 March 1982 32, 37, 193
  28 November 1983 228
  6 January 1986 84, 102, 284
  1 December 1986 141
  31 December 1987 50, 51, 84, 284
  2 September 1988 121
  23 September 1988 116
  2 August 1989 241
  25 June 1990 104
  2 July 1990 141
  3 January 1991 195
  25 February 1991 284
  10 July 1991 90
  22 January 1992 97
  17 March 1992 50, 51, 52, 287

mayor:
  agent of *commune*, as 34
  agent of state, as 34
  discipline 33
  police powers 33
  role of 33
Médiateur:
  complaint to, effect of 30
  decision, no judicial review of 153
  establishment of 29
  functions of 30
  impact of 30–1
  number of complaints to 31
  reports of 30
military pensions:
  tribunals 292

natural justice:
  principes généraux du droit,
    comparison of 205
Netherlands:
  administrative courts 255–6

INDEX

administrative justice, system of 256
  Council of State 254–6
  droit administratif, influence of 254–5
  Rule of Law 255
nuclear accidents:
  jurisdiction 141

parliament:
  executive, surrender of powers to 24
  government, restriction on power to overthrow 9
  legislation by 10
  pouvoir législatif 9
  rights, safeguarding 17–18
parliamentary officials:
  accidents to 134
parliamentary proceedings:
  judicial review, not subject to 133–4
parliamentary sovereignty:
  principle of 23–4
parliamentary tradition 8
personal liberty:
  protection of 139
personal status:
  questions of 139
plaintiff:
  locus standi 158–60
police:
  activities, challenge to 132
  administrative and *judiciaire* 132
postal service:
  jurisdiction 141–2
  loss of mail, exclusion of liability for 192
precedent:
  doctrine of 166–7
prefects:
  budget, imposing 35
  Corsica, in 32
  mayor, disciplining 33
  responsibility of 32
President:
  regulations, issuing 9
  role of 9
press:
  case-law on 17
Prime Minister:
  parliament, responsible to 9
principes généraux du droit:
  administration, prerogatives of 210–11

  audi alteram partem 217–18
  categories of 209
  descriptions of 206
  economic and social rights 213–14
  environment, protection of 214–15
  equality before the law 215–17
  human body, inalienability of 222–3
  human rights 208–9
  impartiality 217
  import of 205
  individuals, liberties of 211–13
  judicial review, right to 222
  natural justice, comparison to 205
  new emphasis on 208
  non-retroactivity 220–1
  proportionality, notion of 218–20
  reasons, duty to give 218
  sources of 207
  statutory interpretation, as 206
  unlawful regulations, duty to abrogate 221–2
private law:
  public law, separation of 4
privilège de juridiction:
  meaning 173
prize court:
  jurisdiction 293
procedure of courts:
  administrative documents, right of access to 88
  advocate, role of 107–8
  appeal 115–16
  arrêt 110
  audience publique 104–8
  case, calling 107
  commencement of proceedings: example of 87–8; *recours* 89–91; time for 91
  Commissaire du gouvernement, role of 87, 101–3
  contradictoire 86, 97
  costs 121
  decision 109–10
  délibéré 108–9
  English procedure, differing from 86
  evolution of 278
  execution: administrative courts, criticism of 110–11; *astreinte*, remedy of 113–15, 283; complaints 112; illegal decision, injunction against implementation of 114–15;

procedure of courts (*cont.*):
  judicial methods 113;
  mandamus, issue of 111; reform of law 112; Section du Rapport, intervention of 112–13; stay of 117–19
  fresh examination of matters 120
  inquisitorial approach 86
  instruction: case undergoing 91; expert evidence 94; facts, inquiry into 95, 98; further questions, *rapporteur* asking 95; information, requests for 92–3; inquisitorial procedure 96–8; interlocutory orders 96; meaning 91; new head of claim, court not introducing 96; purpose of 92; *rapporteur* 98; site visits 94
  interim orders 116–17
  judgment, *see* judgment
  legal problems, discussion of 98
  length of 120
  oral observations 86–7
  preliminary ruling, adjournment for 119
  proof of case 97
  reasons, duty to give 88
  référé 116–17
  report 99–100
  séance d'instruction 100–1
  standardization of decisions 120–1
  stay of execution 117–19
  triangular characteristic 87
  written 86
professions:
  tribunals relating to 290–2, 294
property:
  civil courts, jurisdiction of 140
  expropriation of 140
  public authority, owned by 137
  sanctity of 17
proportionality:
  'balance-sheet' doctrine of 220
  erreur manifeste linked with 247
  importance, increasing 218
  requirement of 218–19
provident societies:
  social-insurance system, administration of 127
public authority:
  administrative jurisdiction, statutory exceptions 140–2
  contracts 137–9
  liability of 175–6
  private character, administrative acts of 136–9
  private property, ownership of 137
public inquiries:
  use of 28–9
public interest:
  abnormal burdens suffered in 188–9
public law:
  private law, separation of 4
public need:
  decision of public authority, defined by 126
public service:
  administration of justice as 132–3
  administrative law regime, exclusion of 128
  assisting in, risks of 185
  body carrying out 127
  commercial and industrial character, with 129
  continuity of 272
  definition 125
  doctrine, application of 130
  elements of 126
  fault, liability for 182–3
  forest fire-fighting as 137
  guiding principle 130–1
  injury arising out of 125
  principle of 125
  social 129
  special methods and prerogatives 128

rapporteur:
  assignment of case for 91
  cases, pace of 99
  facts, responsibility to establish 95
  legal problems, discussion of 98
  report 99–100
référé:
  use of 116–17
refugees:
  tribunals 57, 293
regions:
  division of 32
  regulation by decree 10
réglement:
  constitutionality, view on 11
  loi, dividing line with 11

risks:
  dangerous operations, arising from 186–8
  prisoners, escape of 187
  public service, assisting in 185
  theory, emergence of 183–4
Rule of Law:
  administration, success of 274
  equality before the law, requiring 216–17
  Netherlands, in 255
running-down cases:
  jurisdiction 141

separation of powers:
  administrative courts, matters removed from 132
  French conception of 123
  principle of 123
  theory of 43
site visits:
  conduct of 94
social security:
  French system of 27
  tribunals 56
social welfare:
  tribunals 292
social-insurance system:
  administration of 127
sporting bodies:
  decisions, review of 127–8
state:
  legal representation of 90
statuant au contentieux:
  meaning 5
statutes:
  Constitutional Council, interpretation by 12–13
  constitutionality, no judicial review of 12
stay of execution:
  application for 117–18
  order for 118
  sursis, availability of 118–19
surrogacy:
  principes généraux 222–3
sursis, availability of 118–19

taxation:
  fiscal measures, interest of tax-payer in attacking 159–60
  indirect, jurisdiction as to 140–1
  local, challenge to 160
torts:
  administrative 174–83; *see also* administrative liability
  damages in 191
treaty:
  review of acts under 134–5
Tribunal des Conflits:
  cases before 150
  conflicts before: decisions, of 148–9, 151; negative 147–8; positive 146–7; types of 146
  establishment of 145
  judicial statistics 302
  members of 145
  paired nature of 145
  preliminary rulings 149
  proceedings before 145–6
  substantial law, role in development of 150
tribunals, *see* administrative tribunals
Tribunaux Administratifs:
  appeals from 50–2
  back-log of cases 281–2
  conciliation procedure 284
  Conseil d'Etat, questions submitted to 51
  Conseils de Préfecture becoming 49
  double role of 59
  inspection, commission of 53
  judicial statistics 299–301
  judiciary, status of 53
  legal representation in 90
  members 83–5, 103
  oral hearing in 104
  parties, presence of 107
  right of appeal from 275

ultra vires doctrine
  légalité, comparison of 202, 204
  procedural 227–8
  substantive 226

voies de fait:
  doctrine of 135–6, 139–40, 225
voluntary associations:
  restrictions on formation of 16

welfare state:
  administrative tribunals 26